The Literature of Waste

The Literature of Waste

Material Ecopoetics and Ethical Matter

Susan Signe Morrison

palgrave
macmillan

THE LITERATURE OF WASTE

First published in 2015 by
PALGRAVE MACMILLAN®
in the United States—a division of St. Martin's Press LLC,
175 Fifth Avenue, New York, NY 10010.

Where this book is distributed in the UK, Europe and the rest of the world, this is by Palgrave Macmillan, a division of Macmillan Publishers Limited, registered in England, company number 785998, of Houndmills, Basingstoke, Hampshire RG21 6XS.

Palgrave Macmillan is the global academic imprint of the above companies and has companies and representatives throughout the world.

Palgrave® and Macmillan® are registered trademarks in the United States, the United Kingdom, Europe and other countries.

ISBN: 978–1–137–40566–1

Library of Congress Cataloging-in-Publication Data

Morrison, Susan Signe, 1959–
The literature of waste : material ecopoetics and ethical matter / by Susan Signe Morrison.
pages cm
Includes bibliographical references and index.
ISBN 978–1–137–40566–1 (hardback)
1. Literature and society. 2. Refuse and refuse disposal in literature. 3. Waste (Economics) in literature. 4. Consumption (Economics) in literature. 5. Recycling (Waste, etc.) 6. Ecocriticism. 7. Excess (Philosophy) I. Title.

PN51.M676 2015
809′.93358—dc23 2014047263

A catalogue record of the book is available from the British Library.

Design by Newgen Knowledge Works (P) Ltd., Chennai, India.

First edition: June 2015

10 9 8 7 6 5 4 3 2 1

To Mr. Alfred Short, who took me to the dump in South Dennis, Massachusetts, in his old truck piled high with sandy seaweed and garbage cans

Contents

Part III Recycling and Composting: Form as Restitution

Part IV Source Reduction and Reuse: Compassion through Generous Metaphor

Acknowledgments

The insights and support of many people helped bring this book to light. At Palgrave Macmillan, I was fortunate to work with the enthusiastic and supportive Brigitte Shull and Ryan Jenkins, who made the production process flow smoothly, as well as Deepa John and the fine copyediting staff at Newgen Knowledge Works. I am indebted to the anonymous reader for excellent suggestions to improve the manuscript.

At Texas State University, I would like to thank colleagues who endured my "wasteful" puns and insisted on producing them, including Miles Wilson; Vicki Smith; my beloved mentor Edgar Laird; Charles Britt Bousman; Molly Dunaway; Cathy Jaffe; Steve Wilson; Dan Lochman; Susan Hanson; Mike Hennessy; and Rebecca Raphael with her guest, John J. Collins. I found Eric Leake's work on difficult empathy and Rob Tally's on world literature especially helpful. Diann McCabe and Heather Galloway facilitated my Waste Studies Honors seminar.

At Alkek Library, Margaret Vavarek, librarian extraordinaire, undertook research on Chaucer and censorship, as did my sister-in-law and fellow swimmer, Barbara Hunt. Kudos to Leo Landrey for sharing his Latin scatological wisdom and Catullus translation. Don and Steven Singer informed me about various aspects of Jewish culture, for which I am grateful. The Poetry Library, London, was forthcoming with pertinent information.

Many students deserve thanks for not only bravely entering my classroom, knowing that waste would be their subject for a semester, but for taking the material seriously and respectfully, all the while lending good humor and insight to the class. They include Winifred Hunton-Chan, Madeline Barnes, Keri Fitzgerald, Rachel Gibbs, Emily Rothbauer, Teresa Mack, Amanda Sabo, Camrie Pipper, Heather Robinson, Ray Stockstead, Andres Lopez, Victoria Le Grice, Zachary O'Connell, Shelby Armstrong, Courtney Glenn, Matt Hudson, Shannon Shaw, Jennifer Jones, Richard Ramon, Christina Caro, Allison Estrada-Carpenter, Deanna Rodriguez, Alex Addington, Devin Garza, Rachel Snow, Rachel Bondurant, Richard Guzman, Holly Ratcliff, and Maria Calcaben. Many thanks to Jeff Wilson, aka Professor Dumpster, who engaged my students with the Dumpster

Project and "Green is the New Black" initiative at Huston-Tillotson University.

At various conferences I have been stimulated and buoyed by the indomitable Eileen Joy, Myra Seaman, Michael Johnson, Dan Kline, and Ken Tompkins, along with all the BABEL folks. Ieva Reklaityte's work in Kalamzoo on sewers unclogged my mind, Martha Bayless's latrinal insights were pleasantly pungent, and Kevin J. Wanner guided me to *Eyrbyggja Saga*. Through TEMA, John Friedman shared alchemical sources. At the MLA in Boston, thanks to Kathy Lavezzo, Ruth Evans, Jody Enders, James Simpson, and Jonathan Hsy. I am grateful for the generous invitation to speak at Texas Christian University and remain indebted to Jill Havens, Scott Williams, Brad Lucas, Teresa Ann Ellis, Gabby Raymond, David Colón, and Sarah McNeeley. At the virtual Compass Interdisciplinary Conference, Kivmars Bowling, Dennis Mazur, John Scanlan, Valerie Allen, Scott Noegel, and Wendy Turner kindled lively dialogue. My waste work met Sustainable Development Discourse through Emilia Ferraro, whose panel, "Rethinking Humanism," at St. Andrews proved vital for my thinking process. Francesca Knox and her conference on "The Power of the Word: Poetry, Theology and Life" at Heythrop College, London, helped spark my exploration of the ethics of waste.

As Visiting Research Fellow, Institute of English Studies, School of Advanced Study, University of London (2010–11), I was assisted by Conor Wyer, Wim Van Mierlo, Sandra Clarke, Pete Smith, Jay Thomson, Nicholas Sagovsky, Gillian Whiteley, Rosie Cox, Ruth Mateer, and David and Pam Neville-Sington. Special acknowledgment to Carey Robinson who produced "Human Waste: A Filthy Encounter" (http://soundcloud.com/landspeed/filthy-encounter1/s-NxKME). At the UCL Dean's Seminar, people enthusiastically asked excellent questions that focused my thinking: Katia Pizzi, Dean Roger Cain, Peter Niven, Natalie Joelle, Isabel Davis, and John Snelling.

Various folks have lent insight and sensitivity to the subject of waste. Scott Slovic unreservedly supported the Waste Cluster for *ISLE: Interdisciplinary Studies in Literature and Environment*, a cluster that would never have existed but for Véronique Bragard's reaching out to me in 2011. Without the encouragement of Gary Waller, I never would have plumbed the depths of *Hamlet*. The work of Michael Uebel, generous and wise, continues to inspire. Others who encouraged me include Herman and Uta Romer, Nick Allred, Judith Plaskow, and Amélie Junqua. Scot Erin Briggs, computer technician extraordinaire, not only installed software, but also had a deep insight into why I have embarked on my journey to understand waste. I told her how I spent a summer in 1985 in the Black Forest to improve my German speaking skills. As a "Mädchen für Alles" (a girl for everything) at a hotel and restaurant, I cut chives, washed dishes, changed sheets, skinned game, and cleaned toilets. Scot asked me, “Do you think that’s why you’ve been so fascinated with excrement?” I’d never thought about it before, but that experience may well have been a fundamental incentive for my field of research.

Finally, deep appreciation to my family. My dad, Bob Morrison, for jokes about “garbage juice” when taking out the trash. My mom, Joan Morrison, for giving voice to “the wretched refuse of your teeming shore” in her book *American Mosaic*. My husband, Jim Kilfoyle, master at recycling, salvaging, and consolidating material and written detritus. My daughter, Sarah Kilfoyle, for suggesting Walt Whitman passages for this project. My son, John Kilfoyle, who still has not come to terms with eating leftovers. Guess what’s for supper!

Permissions

I gratefully acknowledge the following permissions:

Two articles appear in various chapters of this book.

"Waste Aesthetics: Form as Restitution." *Interdisciplinary Studies in Literature and Environment* 20.3 (2013): 464–478, courtesy of Oxford University Press.

"Postmedieval Fecopoet[h]ics." *postmedieval: a journal of medieval cultural studies* 1 (2010): 150–156, courtesy of Palgrave Macmillan.

A section of chapter 2 was originally published as "Waste Space: Pilgrim Badges, Ophelia, and Walsingham Remembered," in *Walsingham in Literature and Culture from the Middle Ages to Modernity*, edited by Gary Waller and Dominic Janes (Farnham, England: Ashgate, 2010), 49–66, and appears courtesy of Ashgate Publishing.

INTRODUCTION

The Waste-ern Literary Canon in the Waste-ern Tradition

Waste may be described as simultaneously a most harrowing problem and a most closely guarded secret of our times.

Zygmunt Bauman, *Wasted Lives*[1]

In a world in which material prosperity and life itself are inevitably linked to pollution and the production of waste, how can we humans—ourselves sources of waste in terms of all that we discard—understand and cope with waste? A special exhibit at the Wellcome Collection (2011), titled "Dirt: The Filthy Reality of Everyday Life," presented six urban locations designed to test the dictum the anthropologist Mary Douglas made famous: "Dirt is essentially disorder... matter out of place."[2] An eclectic exhibit, items ranged from seventeenth-century paintings of Delft—famous for its cleanliness—to slabs made of human excrement from India; from videos of dust mites to a jewel and pearl–encrusted broom; from vials of Victorian cholera-ridden diarrhea to anti-Semitic posters accusing Jews of being carriers of typhus during the Second World War. The exhibit included a "Scratch and Sniff" card redolent with odors from an eighteenth-century tannery and medieval sewage sludge from the BBC2 Filthy Cities series. One could put one's own dust into an envelope to be transformed into a brick as part of the "Laid to Rest" time capsule project. The Dirt Exhibit at the Wellcome was a crash course in waste (rubble, rubbish, trash, garbage, litter, filth, and excrement). The repercussions from and interpretations and ethics of waste were the focus of this study.

From the garbage-filled moats of the Middle Ages to the overflowing landfills of today, waste has been and continues to be an enduring issue. *The Literature of Waste* builds upon my interdisciplinary monograph, *Excrement in the Late Middle Ages: Sacred Filth and Chaucer's Fecopoetics*, which contextualized Chaucer's use of scatology within medieval cultural, scientific, and societal understandings and representations of excrement. Coining the word "fecopoetics," I drew attention to the ecological and poetic dimensions of feces in Chaucer's poetry, examining how they resonated within medieval culture in terms of the law, history, theology, gender, concepts of nature, and foodways. The book concluded by arguing for the critical exploration of waste and the broader project of "waste studies"—both conceptually and in terms of the texts it explores.[3] Fecopoetics exposes a wide spectrum of attitudes to excrement in the Middle Ages. Parallel to "foodways"—the anthropological exploration into food as representative of a culture—I employed the term "wasteways," referencing how cultures treat waste, to see "how the world orders filth."[4] In this book, we will see that the wasteways of poets and writers reveal a vast array of attitudes and perceptions of waste. *The Literature of Waste* serves as a diptych with *Excrement in the Late Middle Ages*. While excrement in Chaucer allowed me to adhere to one author's corpus and take it in many directions, this work sees excrement as just one of a number of wasted elements under the umbrella term of "Waste Studies." This study attempts not a *mastery* of, but a *foray* into, waste studies as a literary and cultural critical approach.

Core works on the ethics of waste influencing my project are Zygmunt Bauman's *Wasted Lives*, Gay Hawkins's *The Ethics of Waste*, and John Scanlan's *On Garbage*, which varyingly interrogate the history and philosophy behind waste in Western culture. All concur on one key point. They characterize the human intention toward the world as different things: for Bauman and writer Wendell Berry, limitlessness; for Hawkins, mastery; and for Scanlan, perfectibility. Despite the different nomenclature, they concur: human belief in and acting on limitlessness, mastery, and perfectibility causes and exacerbates the specter—material and ineffable—of waste in modern life.

The special cluster focused on Dirt Theory and Material Ecocriticism in *Interdisciplinary Studies in Literature and Environment* asks us to think about "our storied bodies" and "the agentic capacities of matter."[5] Using Bruno Latour's concept of the "actant" that she applies to matter, Jane Bennett asks, "How, for example, would patterns of consumption change if we face not litter, rubbish, trash, or 'the recycling,' but an accumulating pile of lively and potentially dangerous matter?"[6] Along the lines of vibrant materialists, I see waste matter not as "passive stuff... raw, brute, or inert,"[7] but vital. Waste, pulsating with threat, attracts and repels us. Why? My critical approach, situated within the literary and cultural theoretical realms of ecocriticism and biopolitics, adheres to the "ethical turn,"[8] hence my subtitle: *Material Ecopoetics and Ethical Matter*. Material ecocriticism needs to examine, as Pippa Marland and John Parham have suggested, more than the "effects of humanity's impact on the earth." By exploring the "emotional affect," we can see how "cultural forms might dramatise those effects or posit alternative outcomes."[9] Focusing on literature in the Western tradition from a "waste-oriented"[10] material ecocritical perspective can help us to understand how we theorize, manage, and are implicated in waste. Literature reflects the ways in which humans commonly perceive waste, yet can also offer complexly textured models for individual and communal behavior and relationships with the world around us. Not always negatively charged, waste contains the potential to *charge*, catalyzing ethical behavior and profound insights, even compassion.

Can and should literary criticism be politically and/or ethically driven, and, if so, how much? Using presentist theory can be threatening. In the April 24, 2009, edition of *The Chronicle of Higher Education*, Mark Edmundson rails against the use of "readings," by which he means "the application of an analytical vocabulary ... to describe and (usually) judge a work of literary art." Edmundson goes on to perform readings explicitly allied with Longinus, Tennyson, Matthew Arnold, and Sontag—who presumably are not as tainted as are the enemies he cites—"[Marx, Freud, Foucault, Derrida, or whoever]." He clearly feels under siege, threatened, and disappointed in the turn away from what he calls, for example, an ideally Blakean reading of Blake's verse or an "Eliotic" reading of T. S. Eliot's verse.[11] How we can achieve such readings is not specified. We can, after all, only interpret in our own time. The reclamation of aestheticism and formalism in the early twenty-first century, after a perceived theory overload in the late twentieth century, responds to the real and abiding angst that theory as a process is suspected of having decayed; hence special issues of *Critical Inquiry* and *PMLA* that responded to anxieties about the future of theory. There seems to be a sense that presentist theory obfuscates the past and enables us to even disremember the past. Yet, living in the "Garb-age"[12] as we do, it is incumbent upon us to utilize not just presentist theory, but also *futurist* theory, for our own and our society's well-being.

The Waste-ern Canon

> What was their civilization? Vast, I allow: but vile. Cloacae: sewers. The Jews in the wilderness and on the mountaintop said: *It is meet to be here. Let us build an altar to Jehovah.* The Roman, like the Englishman who follows in his footsteps, brought to every new shore on which he set his foot (on our shore he never set it) only his cloacal obsession. He gazed about him in his toga and he said: *It is meet to be here. Let us construct a watercloset.*[13]

James Joyce's comment reflects what seems to be true: the Western canon is sedimented in waste. Fundamentally problematic, waste attracts deeply emotional reactions from those reading about it, including comments such as "I guess this is what happens when you no longer take Western

Civilization seriously."[14] My endeavor is to show how Western civilization *is*, in fact, *filled* with waste; indeed, has *always* been filled with waste. That this clearly threatens only suggests to me that there is something there worth exploring further.

But "where should literature establish itself?"[15] Heinrich Böll asks. In his 1952 essay, "In Defense of 'Rubble Literature [*Trümmerliteratur*],'" Böll identifies and extols the first writer of the Western tradition as having been the "progenitor" not only of the European epic, but also of rubble literature itself. Homer "tells of . . . the destruction of Troy . . . a literature of war, rubble, and homecoming."[16] That Böll's own papers have now been destroyed in the wreckage of the Cologne Library 2009 collapse due to fire is a tragic irony Böll himself would best appreciate. Since Homer, waste has haunted the European canon, from Grendel's arm to Yorick's skull to the "dust" in Dickens's *Our Mutual Friend*. Waste, present in cultural artifacts, has been a concern of the Western canon since its inception.

Much of the work in vibrant materialism, waste and discard studies, and other thing-oriented approaches (to be discussed in chapter 9) deals with literature from recent centuries, the capitalist system, and our current environmental plight. Waste speaks to "rubbish ecology," Patricia Yaeger's term for the obsession with trash in postmodern art, wherein the "binary trash/culture has become more ethically charged and aesthetically interesting than the binary nature/culture."[17] Within the "trash aesthetic" as qualified by Yaeger, "rubbish becomes a strange vale of soul making and creativity,"[18] such as the work of a chewing gum artist.[19] The "detritus aesthetic," Yaeger argues, goes back to Dickens, though I argue that the sacred sublimity inherently possible in trash was present in the Middle Ages as well.[20] I push these interests back in time, thickening the focus to include medieval studies.[21] Technological advances are starting to impact our biological selves in terms of our brain function. In commenting on new software, Steven Johnson writes how innovative search engine tools make associations, which change the way we think.

> The other day I ran a search that included the word "sewage" several times. Because the software knows the word "waste" is often used alongside "sewage" it directed me to a quote that explained the way bones evolved in vertebrate bodies: by repurposing the calcium waste products created by the metabolism of cells. That might seem like an errant result, but it sent me off on a long and fruitful tangent into the way complex systems—whether cities or bodies—find productive uses for the waste they create.[22]

Johnson contends that this software may affect what we write, not for linear arguments and narratives, but for associative organizations; in other words, for rhizomatic thinking and understanding of the world, a kind of "medium theory."[23] The very word for designating what is medieval comes from the Latin *medium aevum*. Medieval literature, replete with waste, can help us to understand how we are increasingly coming to perceive the world itself today.

Waste critics scavenge familiar textual landfills. The core books, the "Biggies," cited repeatedly, include Dickens's *Our Mutual Friend*, James Joyce's *Ulysses*, especially the outhouse scene; Don DeLillo's *White Noise* (with its chemical spill) and *Underworld*, whose central character is a waste management executive; Upton Sinclair's *The Jungle*, the outrage- and disgust-inducing exposé of the meat-packing industry in Chicago; Jonathan Swift's *Gulliver's Travels*, whose protagonist voids urine to save the Lilliputian palace; Samuel Beckett's plays, including *Happy Days*, whose protagonist Winnie is buried up to her neck; the Czech novelist Ivan Klíma's *Love and Garbage*, with its street-sweeping protagonist; Sartre's *Nausea*; Patrick Chamoiseau's *Texaco* about a shanty town in Martinique; Ayi Kwei Armah's *The Beautyful Ones Are Not Yet Born*; and Cormac McCarthy's entire oeuvre. The Anglophone canon since the medieval and early modern periods focuses on the disruptive body, capable of political, social, and cultural discomfort, and a society immersed in filth. Melville's *Moby Dick* shows how the whale's vomit can be treasured as ambergris; Emma Lazarus's iconic "The New Colossus" invokes rhetoric analogizing immigrants and rubbish. Native American writers likewise respond to the subject of trash, as in Sherman Alexie's "What You Pawn I Will Redeem." Nobel Laureates in Literature tackle waste. From Portugal, José Saramago's *Blindness* imagines the filth created when all go blind; from Austria, the feminist Elfriede Jelinek's plays and other writings create protest and shame. These lie at the tip of the rubbish heap.

Elizabeth V. Spelman playfully wonders if "a life without trash is not worth examining?"[24] A life with trash certainly merits examination, though I do not attempt to exhaust the exploration into the intersection of waste and literature. Indeed, I hope to provoke other scholars from a diversity of fields to apply and further develop the model I propose in this book, one that allows us to focus on scraps of text that may have been ignored, and add depth and texture to our understanding of literary works and the cultures that spawned them. Though I initially planned to write a comprehensive volume on Waste Literature, it rapidly became clear to me that virtually every piece of literature contains waste, depending upon one's definition. Waste is so pervasive a theme, topic, metaphor, and element, that I have chosen to analyze a limited number of works to illustrate how literary scholars can make both manifest and latent waste visible. Here I forage a stable of literary works as case studies, selected from various time periods (biblical, medieval, modern, and postmodern). A family of literary works inhabits each chapter: *Leviticus*, *The Book of Ruth*, *Beowulf*, the

Icelandic *Eyrbyggja Saga*, Geoffrey Chaucer's *Canterbury Tales*, the Pearl-Poet's *Cleanness*, William Langland's *Piers Plowman*, *Till Eulenspiegel*, François Rabelais's *Gargantua and Pantagruel*, William Shakespeare's *Hamlet*, Jonathan Swift's *Gulliver's Travels*, Laurence Sterne's *Tristram Shandy*, T. S. Eliot's *The Waste Land*, F. Scott Fitzgerald's *The Great Gatsby*, Samuel Beckett's *Endgame*, H. G. Adler's *The Journey*, Italo Calvino's "La Poubelle Agréée," Dorothy Allison's *Trash*, Gerald Vizenor's *Landfill Meditation*, Miles Wilson's "Everything," A. R. Ammons's *Garbage*, poems by Walt Whitman, and Paul Fleischman's *Seedfolks*. I will attempt a criticism by accretion in hopes of modelling as case studies some of the ways waste has been approached in the humanities and how one could approach waste in literature. Do not look for a sustained reading of one literary text here, though *Beowulf* and *Hamlet* recur, echoing through the dumpsters, landfills, and sewers I frequent.

The heart of comparative literature is quite simple: comparison. While the "ethics of comparison" is under pressure,[25] there remains the "imperative to compare."[26] Using Levinas, Thomas Claviez proposes a different kind of comparative literature, a "dis-parative" literature, one rooted in metonymy and contingency. "Only Levinas's ethics precludes comparison because Levinas conceives of the other as unequal, nonequivalent, and thus by definition incomparable—at least to the self, which thus is always more responsible to the other than it can presume the other to be to it."[27] Claviez's metonymic model for dis-parative literature matches the metonymic "string of bunnies, trees, stars, space, toothbrushes, skyscrapers" Timothy Morton points out make up the mesh, the "potentially infinite series of alterities" we encounter when we look for the self.[28] Waste Studies exemplifies the metonymic charge by looking at texts as a series of non-hierarchized literary actants. We perceive waste in a chain of composted linguistic nonhuman actors—not *against* ourselves, but alongside.

While the many texts I cite function as mere examples from the rubbish tip that contributes to the waste canon, they coexist in what Djelal Kadir calls "'contrapuntal juxtaposition.'" I follow Kenneth Reinhard's suggestion that we undertake a comparative "reading in which texts are not so much grouped into 'families' defined by similarity and difference, as into 'neighborhoods' determined by accidental contiguity, genealogical isolation, and ethical encounter."[29] Friedman argues for the metaphor of *collage* in comparing, a suggestion that echoes sympathetically with my approach. Borrowing from "paratactic modernist poetics," collage "maintains the particularity of each, refuses hierarchy and instrumentalism, and fosters identification of new generalities based on what texts share." This "'cultural parataxis' or 'cultural collage,'" this "radical juxtaposition of texts from different geohistorical and cultural locations,"[30] permits me to explore texts from wildly varying time periods that cohere in their inability to escape waste. As Lois Parkinson Zamora has argued against traditional periodization, we can bypass chronology and seek out "unexpected" affinities among works.[31]

The ethics of seeking out more than a single story to articulate a concept, people, nation, or entity has been eloquently argued by Chimamanda Ngozi Adichie. A "presumption of sameness" threatens the stereotyping of vastly diverse peoples. As Adichie says concerning the problem of single stories, "They make one story become the only story." Hence my desire with waste literature to draw on material from various eras, genres, and cultures to nuance the complicated relationship we have with waste. The responsibility ultimately lies with the reader, who "must commit to accumulating multiple accounts of a place and its people"[32]—or, in this case, the troubled responses humans have had to waste.

A thirteenth-century Icelandic saga may cuddle up with a bleak twentieth-century drama, an effusive Chaucerian tale may hold hands with *Hamlet*, and Walt Whitman may sing the body medieval. In my literary neighborhood, all these texts meet at the local dump, to deposit their waste in a communal potlatch to share in a potluck of, I ultimately argue, affinity, a kind of "*network*."[33] Rummaging through these artifacts from different cultures, time periods, and genres facilitates a broad view of Waste Studies to reflect the pervasive presence of waste in our cultural junkyard. Like Victor Frankenstein, I have frequented the charnel house of the literary canon, gleaning bits and pieces to cobble together my creature—this book.

Why "Waste Studies"?

Waste theorists concur on the central role humans play in generating, ignoring, coping with, and analyzing waste in its various manifestations. As Kate Soper has elegantly articulated, "Waste understood as the unused or inutilizable remains of human productive activity and consumption . . . must today count as one of the major markers of the distinction between humanity and the rest of nature. All animals excrete, but only humans create waste."[34] The website, http://discardstudies.com/, maintained by Max Liboiron and founded by Robin Nagle, anthropologist in residence for New York City's Department of Sanitation, explores the field dubbed "Discard Studies." Like "Critical Discard Studies," my work regards waste as

> not automatically disgusting, harmful, or morally offensive, but that both the materials of discards and their meanings are part of wider sociocultural-economic systems. Our task is to interrogate these systems for how waste comes to be, and our work is often to offer critical alternatives to popular and normative notions of waste.[35]

While Liboiron urges the use of *Discard* Studies over *Waste* Studies "to ensure that the categories of what is systematically left out, devalued, left behind, and externalized are left open,"[36] I prefer "waste." The focus on what has been thrown away or is considered worthy of "discard" suggests a sterility that I deliberately do not wish to convey. Always provocative, waste as concept defies sterility and objectivity, since it prompts responses ranging from disgust to pity. Both words stemming from Latin, *waste*—via the Old French *wast(e)* from Latin *vāstus*—dates further back as a term, to about 1200.[37] *Discard* emerges only in the sixteenth century. Rather than closing down categorization, Waste Studies provokes new understandings of categories that exist and can help us see, associate, and be in the world according to a new, more ethical, paradigm.

What constitutes waste? In the earliest usages in English, waste was whatever is not or no longer utilitarian, something squandered, empty or barren, or lacking purpose. Waste has meant desolation, pointlessness, and uselessness, but also excess and surplus; both extremes have been viewed as problematic, void of meaning, and immoral.[38] Valerie Allen's excellent question—"What is the history of waste—a constant battle with the unusable or a shifting index of our sense of value?"—implies that both suggestions are valid. The meaning of waste in earlier English periods has a subtle nuance not necessarily entirely portable to contemporary English.[39] Waste necessarily implicates history,[40] and implies materiality, metaphor,[41] and emotional affect. Though waste has been understood differently in various places over time, certain aspects remain constant: waste is always material (first) and figurative and metaphoric (second). Without the material that is discarded, we cannot enter the realm of the metaphoric, of literature, and of the imagination. Waste is literal *and* literary.[42] Indeed, "matter itself becomes a text" or a "*site of narrativity*."[43] The meaning of waste "exceeds interpretation," and is "radically litteral [*sic*] in that it always derives from the residue of waste which resists quick degradation."[44] Waste, "the sole type of matter that is independent from our intentionality,"[45] is not a passive object, but a subject, an "actant" with "the capacity to modify a state of affairs by making a difference." As Levi Bryant points out, "*all* entities are actors."[46] Even, as we shall see, waste itself.

The history of waste records our mutable relationship to whatever is seen as constituting waste (which can alter over time). Clearly one must be wary of making equivalencies about waste between the medieval period, say, and the twenty-first century. Waste is contextually, historically, and culturally specific—a manure heap by a farm could be useful for fertilizing crops, yet in the city it is a source of disease. In her study on waste in Hungary, Zsuzsa Gille points out that "materials are not 'born' to be waste: they are transformed into waste by identifiable material and social processes"[47] through the course of history. Despite the inevitable necessity of eliminating waste for organisms as small as a protozoan to conglomerations as large as a city, Western culture has long denied waste or marked it as "other." Waste in all periods constitutes "simultaneously civilization's *other*... as well as the *trace or remainder* of civilization,"[48] and is liminal, hybrid, and socially specific.[49] Inherent to waste's identity (indeed, waste could be said to be that which has no identity) is its ability to slip away from one single, clearly articulated designation.[50] Nevertheless, whether we are on a medieval pilgrimage or at a lavish 1920s party, in the convulsions of the Reformation or suffering from nuclear malaise during the Cold War, various tropes linked to waste appear repeatedly over the centuries. By juxtaposing texts that do not spiral within the same constellation, I hope to show the thematic affinities Western literature contains with regard to waste, such as its link to humor, from Chaucerian scatology to Jonathan Swift's vicious use of excrement with the Yahoos. Waste in its myriad forms has caused societal division, alienation, and disgust.

Waste, as an integral aspect of the human condition, has inevitably come to have a place in another human generated product: that of literature. *Writing* about waste differs from the material waste *object.* Poets and writers utilize materiality in their own objects—literary texts. Waste, as a reviled piece of stuff, has been "overmined," what Graham Harman has cautioned against when objects are "reduced upward to their appearance in human consciousness or their environmental effects."[51] As Slavoj Žižek has observed, "The properly aesthetic attitude of a radical ecologist is not that of admiring or longing for a pristine nature of virgin forests and clear sky, but rather that of accepting waste as such, of discovering the aesthetic potential of waste, of decay, of the inertia of rotten material which serves no purpose."[52] This book explores the many ways waste has been written about; these approaches, in turn, reflect the variety of attitudes toward waste and gesture toward how we might confront and even make friends with waste in the future. After all, as Steve Mentz points out, "If the goal of ecological thinking is, in Timothy Morton's phrase, to 'love the non-human,' squalid brown mixtures represent blended things that are hard to love but also important not to ignore."[53] In making "place for the grimy, the sloppy, and the abject,"[54] I hope to show the purposes waste in literature has served and has the potential to serve, by demonstrating the contributions to cultural harmony and understanding that can be discovered in literary moments focused on waste. Poets in particular, as Gary Snyder has argued, "give what she or he has done as nourishment [...] back to the community."[55]

The philosophy of Emmanuel Levinas, who argues that an openness to the "Other" is a sign of the ethical, can provide guidance for how ethical literary criticism can challenge appropriative readings of the past. Levinas's writings can alter how we read literature, not to create a closed system of analysis, but as a way into seeing the utterance of literature as an ethical event. Ethical criticism is dialogical, is "interpretation as interruption."[56] Waste literature enables culture to acknowledge what it has to deny,[57] such as, I would argue, bodily, cultural, and societal waste—material and metaphorical aspects of our world.

Structure of This Book

This book reflects the four ways cultures have dealt with garbage for millennia as determined by William Rathje and Cullen Murphy in their analysis of garbology: "dumping it, burning it, turning it into something that can be useful (recycling), and minimizing the volume of material goods—future garbage—that come into existence in the first place (this last is known technically in the garbage field as 'source reduction')."[58] The United States Environmental Protection Agency (USEPA) states a hierarchy for solid waste management on its website.[59] In ascending order of desirability, the least preferred option is that of Treatment and Disposal; then Energy Recovery; third, Recycling and Composting; and finally Source Reduction and Reuse. This tweaks the hierarchy suggested through Sustainable Development (SD) Discourse of disposal, recovery, reuse, and reduction.[60] If we apply these hierarchies to the study of literature, we might see disposal as the creation of waste literature itself, the reposition of waste in literature. Recovery lies in our discovery of the waste in so-called "rubbish literature." Recycling, composting, and reuse can be found in intertextuality, allusion, and metaphoric composting, the compression of language into image and symbol.

I structure my book according to the USEPA's hierarchy to remind us of the current ecological crisis as we rummage through literature's response to waste in all its material and figurative forms, multiple definitions, and concepts. Codification and the civilizing process have enabled humans to construct rules by specifying how waste is to be controlled and managed. This first part relates established ways of viewing waste and all its cognates. *Treatment and Disposal: Approaches to Disciplining Waste* occupies part 1, organized around approaches to waste that recur in anthropological codification, psychological repression of our bodily decay, sociological civilizing process, historical garbaging of the past, economic conspicuous consumption, urban disposal of bodily waste, religious sin, and philosophical *angst*. By displacing waste, these disciplinary approaches establish themselves by centering attention on validated categories.

Attempts at codification inevitably fail. Codifying clean as opposed to dirty does not get rid of dirt; rather, it creates a coherent space within a system continually susceptible to what it leaves out. "That which we are no longer using remains, nonetheless, part of our lives."[61] Repression of our body's ability to excrete foul effluvia cannot stop our natural cycle of digestion and excretion. The civilizing process refuses waste through chastising its public presence, yet cannot eliminate it. The linkage of wasted ruins and the lament of historical decay in texts as diverse as the Old English "The Wanderer" or "The Ruin" and T. S. Eliot's *The Waste Land* do not address identical moments in history, yet all these works use images of decay and waste to articulate their considerations of historical "garbage." History is impotent to make past events that haunt coherent linear narratives disappear. Orgies of consuming desired stuff and fetishized objects fail to limit the production of things. Urban environments may move the trash created within its limits, but displacement does not signify reduction.[62]

While Bruno Latour points out that "transparent, unmediated, undisputable facts" are rare, the anomaly of waste is its very materiality, whose potency we are all invested in.[63] The materiality of waste—as in landfills, trash cans, garbage dumps, and compost piles—inevitably becomes metaphoric. Metaphor necessarily dyes our discussion. Codification demands that we "trash" certain items to create order. When categorization fails, we become filled with a nameless dread. Human waste such as excrement or menstrual blood adheres morally to the body emitting those exudings. Sociologically, certain classes become seen as trash because of their status. The historian must dispose of or obliterate facets of the past in favor of a coherent linear narrative. In economics, money—itself is a symbolic concept—can be both wasted and the cause of waste. Urban and rural realities confront waste. The city, seen metaphorically as a body, emits filth that cannot be easily reused as in the traditional rural landscape. Consumption and excess have insinuated themselves into our understanding of the world metaphorically, in terms of psychological, moral, or metaphysical waste, corroding socially, spiritually, and ethically. The intrusion of waste inward renders one's spiritual being into a wasteland of sin. Religious codes interiorize immoral acts, eating away at the spirit. Metaphysical waste seeps into our sense that existence is mere meaningless folly.

This section concludes with an extended analysis of the most virulent metaphoric manifestation of waste: that of wasted humans, the wholesale "wasting" of entire classes of human beings as equivalencies for trash, garbage, and filth. The cultural model of waste is dumped onto humans, whereby those who handle filth become tainted by it morally and socially. They are considered to be cognate to waste, the so-called wasted lives, as Bauman so movingly puts it. The equating of humans with waste allows for the disposal of such members of humanity via ostracism, defamation, exclusion, and even death. As Christopher Schmidt points out, "Capitalism's need to name waste as *other*, against which value, profit, and product are then defined, is homologous to the social logics that have produced and subjugated queer identities"[64]—or immigrant identities, Jewish identities,

Muslim identities, female identities, trans identities. The list, alas, goes on. As we will see, this naming as other lies at the cold heart of the Western literary tradition.

Inspired by Ottmar Ette's call for the humanities to "conceive of themselves as sciences for living [*Lebenswissenschaften*],"[65] the second section of the book builds on the humanities focus of the first. Potentials for the "*sciences humaines*" include reflecting upon obligations to initiate dialogue with non-humanities disciplines. Literary and cultural knowledge enables "a more complete understanding of life and of the humanities as part of the *sciences for living*."[66] I explicitly link my work with SD, a field that has rhetorical, policy, and environmental impact and repercussions.[67] Waste Studies as part of Ette's sciences for living can (1) bring the humanities into dialogue with Sustainability Science and SD and (2) suggest new ways of seeing and understanding waste. Geographer Emilia Ferraro urges us "to situate [SD] debates within the realm of the Arts and Humanities."[68] SD discourse, calling for a radical change in the way the West has conceived of the relationship between humankind and its role in the universe, can speak to the humanities. The "humanistic" dimension of Sustainability Science, to which my project contributes, is becoming increasingly urgent in this period of global climate threats and economic instability. It is vital to bridge the humanities and the sciences, so that the "two cultures" C. P. Snow wrote about over 50 years ago can be in dialogue. The study of waste and of waste in literature contributes to this dialogue.

Part 2, *Energy Recovery and the Dynamic Power of Things*, responds to the USEPA's next level in the hierarchical scale. Recovery lies in the discovery of the vibrancy of objects, including waste, trash, garbage, and rubbish. Applying aspects of SD discourse to literature, I plan to show how objects in their "secret other world"[69] suggest the audacity of thingness and the poignancy of materiality. Acknowledging the dynamic agency of waste can lead to ecological awareness and reduction of filth production. Vibrant materialism and ecomaterialism allow us to recover the worth in stuff, objects, and things. Despite humans exploiting seemingly inert stuff to show off one's status, things themselves have dignity. The anthropomorphism of objects and animals can work *against* anthropocentrism. If we readjust our linear perceptions of time and space, we can create transformative metaphors for living more modestly. Arguing against linear progressive narratives of history, the arts of transience and proximity enable trash meditation.

While Mary Douglas's paradigm distinguishing clean from dirty, orderly from disorderly, seems inviolable, I put it under pressure for the last two sections of the book. Part 3—*Recycling and Composting: Form as Restitution*—shows the integration of waste in literature. Recycled literary waste, as in intertextuality, forms the litter-ature of waste texts. Figurative and recycled literary waste, such as digressions, leftovers, puns, parody, and intertextuality—all are peculiarly—even touchingly—human. Gleaning aesthetics reads poetry as a form of reclamation, leading to the creation of community and affinity.

The book concludes with part 4, *Source Reduction and Reuse: Compassion through Generous Metaphor*. In his book *The Poetics of Waste: Queer Excess in Stein, Ashbery, Schuyler, and Goldsmith*,[70] Christopher Schmidt explores the poetics of four poets critical of the high modernist values wherein waste is identified with "feminine and queer bodies."[71] *Pace* Kristeva, Schmidt argues that waste in the "messy" texts he examines is not the source of abjection—the psychological discomfort that arises from the confusion "between subject and object"—rather it is "a source of creative play, innovation, and textured, often eroticized, pleasure." For Schmidt's writers, waste exudes a "mysterious charisma."[72] While I do not touch on these writers, like Schmidt I have found that "waste comes to function as a concept, a material, and most importantly, an aesthetic category"[73] for many authors.

While the metaphor of waste has often been used for destructive purposes, the articulation of a waste aesthetics can reveal the humanity we share. Metaphor—reduced language thickened and intensified like a savory sauce—bridges cultures and can open us to ethical understanding. Compost aesthetics reads poetry that acknowledges the poignancy of materiality. The poethics of metaphor transforms disgust into compassion. Metaphor as reduced, compact language can be used to ethical ends. The concluding chapter, "Poetry as Homeopathy: Poet as Ragpicker," sees the poet as a participant in the circulation of waste. A producer of dirt, perhaps, nevertheless the waste poet homeopathically cures, despite arguments by those censoring, for example, a "dirty" Chaucer. How can the encounter with fragmented and corruptible bodies through grimy verse result in community? I scavenge A. R. Ammons's *Garbage*, who suggests that our aesthetic treasuring of verse not be limited to the glorification of the natural world. Rather, as Yaeger has suggested, "detritus has unexpectedly taken on the sublimity that was once associated with nature."[74] Poetry, itself inherently metaphorical, functions as a kind of homeopathy or social cure for the alienation and disgust we all too often feel toward our own and others' bodily and material waste and decay. In recognizing waste as possessing the potential for loving agency, therapeutic vibrancy, and healing power, we can acknowledge the humanity of the other and the *affinity* the other has to us in an alchemical generosity.

In her book, *The Design of Everyday Life*, Elizabeth Shove describes the accidental process whereby enamel washing up bowls became outdated in favor of plastic ones. Through the unexpected comparison with plastic (a new quieter and softer material), enamel, formally *unmarked*, became *marked*, as when "something previously hidden comes to our attention."[75] Enamel came to signify in a new way. Problems that had hitherto gone unnoticed acquired new and undesirable characteristics such as chipping, scraping, and cracking.[76] I hope my book will "mark" my readers to attend to waste, to keep it "on our radar,"[77] both theoretically and practically. Literature and poetry, even—or *especially*—those animated by waste, are crucial for learning how to lead an ethical life.

PART I

Treatment and Disposal: Approaches to Disciplining Waste

CHAPTER 1

Codification: The Anxiety of Ambiguity

Before our son was born, I told close friends and family not to bother buying newborn clothing for him. "We already have perfectly good infant clothes, some of which have never even been used." They belonged to my daughter, born several years earlier. Now, the thought that our son would be clad in a pink, female-identified wardrobe was too much for older relatives who sent boy clothes, so that my son, not even cognizant of the battle of wills taking place, would be properly identified and identifiable in the public arena. While it mattered little to my husband and me that the little lad would be in pink or even (god forbid!) flowered clothes, so long as he was happy, healthy, and warm, it was essential to others that he be marked from the moment of his entry into this world as a male. The compulsion existed to codify my son even before his birth. Such orderly divisions percolate through human cultures, including distinctions related to class, wealth, and—as in my as yet-unborn son's case—gender.

The Compulsion to Codify and Order

Codification underlies Mary Douglas's anthropological analysis of perceptions of dirt in her seminal work *Purity and Danger*. Something may be designated as dirt in one location, yet an alternate space makes it acceptable.[1] For example, a dung heap on a farm connotes fertility, safety for future crops, and even material well-being, while a dung heap in the lobby of the Metropolitan Opera House in New York City would not be welcomed by high-society matrons. To fail at the attempt to codify pure versus dirty or to establish them as binary opposites threatens. According to Douglas:

> When we honestly reflect on our busy scrubbings and cleanings in this light we know that we are not mainly trying to avoid disease. We are separating,

> placing boundaries, making visible statements about the home that we are intending to create out of the material house. Both we and the Bushmen justify our pollution avoidance by fear of danger.[2]

This "danger" infects the sense of dread hanging over the inability to divide and differentiate. The nameless, the undifferentiated, the hybrid, the ambiguous—all exist in the realm of metaphor, the result of futile codification.

Codification facilitates the establishment of culture and civilization. The Pentateuch firmly articulates stipulations for Judaic culture with distinctions between clean and filthy. Obsessed with divisions between what constitutes proper and improper sacrifice and consumption, *Leviticus* contains many food prohibitions: "It shall be a perpetual statute for your generations throughout all your dwellings, that ye eat neither fat nor blood" (*Lev.* 3: 17). The seventh-century CE *Visigothic Code (Forum Judicum)*, insidiously aware of the importance of food prohibitions, specifically stipulates that Jews are *not* allowed to divide food into dirty and clean *unless* Christians approve (Book XII, Title II, Law VIII, also XVI).

Separation of clean and dirty appears literally and figuratively in the case of the ceremony of animal sacrifice for a priest's sin:[3]

> And the skin of the bullock, and all his flesh, with his head, and with his legs, and his inwards, and his dung, Even the whole bullock shall he carry forth without the camp unto a clean place, where the ashes are poured out, and burn him on the wood with fire: where the ashes are poured out shall he be burnt. (*Lev.* 4: 11–12)

Proper sacrifices to Yahweh must be conducted with unblemished animals. A ram without a blemish atones for the body's and soul's blemishes:

> And the LORD spake unto Aaron, saying, Do not drink wine nor strong drink, thou, nor thy sons with thee, when ye go into the tabernacle of the congregation, lest ye die: it shall be a statute for ever throughout your generations: And that ye may put difference between holy and unholy, and between unclean and clean; And that ye may teach the children of Israel all the statutes which the LORD hath spoken unto them by the hand of Moses. (*Lev.* 10: 8–11)

Angry with Aaron's sons for not obeying proper stipulations, God prescribes order through codification. The dutiful fulfilment of God's ordained divisions reflects His power and might. If the Jews behave according to the Lord's will, they will be clean and be rewarded by living in a land of milk and honey (*Lev.* 20).

Medieval rabbinical discussions frequently deal with issues of cleanliness. The Babylonian Talmud (completed around 700 CE) comments on the conflict between sacred duty and bodily filth.

> Our Rabbis taught: If a man needs to consult nature he should not say the Tefillah, and if he does, his prayer is an abomination . . . [One Rabbi] said: Guard thy orifices at the time when thou art standing in prayer before Me.
>
> Our Rabbis taught: One who is about to enter a privy should take off his tefillin at a distance of four cubits and then enter . . . The question was asked, What is the rule about a man going in to a regular privy with his tefillin to make water? . . . It is forbidden, since we are afraid that he may ease himself in them, or, as some report, lest he may break wind in them.[4]

The thirteenth-century Rabbi Meir of Rothenburg was asked about issues concerning blessings and hygiene. The washing of hands accompanies both dirty physical activities, like defecation, *and* moments of spiritual reverence, like blessings or the holy act of prayer.[5] How a latrine should be constructed is graphically described:

> Q. On the Sabbath, may a Jew use latrines which are built in the city wall and open into a ditch surrounding the wall, so that the feces falling into the ditch are removed (by his force) from one Sabbath domain into another. A. He should fasten a board beneath the seat (within not more than three tefahim below the latrine walls) so that the feces first fall on the board and then into the ditch. Should the board break on the Sabbath, he would still be permitted to use the latrine that day.[6]

The blessing listed in the Talmud to utter upon entering and leaving the privy or outhouse can vary. Upon entering, the speaker asks the angel to protect him:

> Praised be You, Adonai our God, Ruler of the universe, who fashioned the human body in wisdom and created in him many openings and orifices. It is well known before Your glorious throne that if one of them were to be opened (wrongly) or be stopped up (wrongly), it would be impossible to stand before You.[7]

Intense discussion about proper and improper ways to construct or use privies reflects the deeply embedded code dividing clean from dirty in Judaic culture.[8] Discomfited by lack of order when divisions break down, one enters "waste space," a liminal site of ambiguity that vibrates threateningly.

Levitican food prohibitions regarding cleanliness underlie early Christian precepts, as in the *Penitential of Theodore* (late seventh century). Penitentials of the Anglo-Saxon church codify punishments, with special emphasis on sexual, dietary, behavioral, and religious transgressions.

> If anyone accidentally touches food with unwashed hands, or [if] a dog, a cat, a mouse, or an unclean animal that has eaten blood [touches it] there is no offense . . . If a mouse falls into a liquid it shall be removed and sprinkled with holy water, and if it is alive it may be taken [for food]; but if it is dead, all the liquid shall be poured out and not given to man, and the vessel shall be cleansed . . . If birds drop dung into any liquid, the dung shall be removed from it, and it shall be sanctified with [holy] water, and it shall be clean food.[9]

Fornication that occurs while violating incest taboos and heterosexual precepts is likewise punished. "Sodomites shall do penance for seven years, and the effeminate man as an adulteress."[10] Sexual transgressions are fundamentally contrary to Christian medieval proscriptions. "If one commits fornication with his mother, he shall do penance for fifteen years and never change except on Sundays. But this so impious incest is likewise spoken of by him in another way—that he shall do penance for seven years, with perpetual pilgrimage."[11] Food transgressions condemn the ingestion of blood: "A wife who tastes tastes her husband's blood as a remedy shall fast for forty days, more or less."[12] Non-believers are seen as filth, out of order, emanating the power of desecration within the Anglo-Saxon church.

> In a church in which the bodies of dead unbelievers are buried, an altar may not be sanctified; but if it seems suitable for consecration, when the bodies have been removed and the woodwork of it has been scraped or washed, it shall be reerected. But if it was previously consecrated, masses may be celebrated in it if religious men are buried there; but if there is a pagan [buried there], it is better to cleanse it and cast [the corpse] out.[13]

The tainted pagan must be expelled from the sanctified space that it symbolically sullied, a space requiring purification.

Intention is key in the Christian variations on Levitican precepts, clearly affecting the extent to which one must undergo penance: "If without knowing it, one eats what is polluted by blood or any unclean thing, it is nothing; but if he knows, he shall do penance according to the degree of the pollution."[14] Disorder compels condemnation for the acting out of accordance with established rules. Order can only be reestablished through a counteracting punishment, a kind of spiritual homeopathy, that washes clean the guilt of the transgressor. The stipulations laid out in *Leviticus* and Christian penitentials function to instill a fear of social rejection.[15] The many instances cited of transgression suggest the impossibility of maintaining a perpetual distinction between order and disorder, exposing the artifice such distinctions try to impose. Divisions between clean and dirty, proper and improper, defy permanent maintenance.

The obsession of Lothario Dei Segni with distinction and order in the wake of potential chaos spills over into his later papal activities. The Fourth Lateran Council, summoned in 1213 by Lothario after becoming Pope Innocent III, took place in 1215. It famously established numerous decrees, many of which express a deeply visceral desire for order, cleanliness, codification, and purity. The decrees of the Council[16] instituted annual confession for Christians and ranged from matters we might not be—unfortunately—surprised to hear about, including special dress required of Jews and Muslims, as well as an intensification of discipline against heretics. Pilgrimage, with its bodies going on journeys for literal cleansing and healing or for spiritual purification in a ritual of devotion, stands as a physical practice that enacts the Church's obsession with purifying itself—stamping out heresy and branding any troublemakers as heretics. Concerning the link between heresy and insanity, Canon 2 rejects the teachings of Abbott Joachim who called Peter Lombard

> heretical and insane . . . If therefore anyone presume to defend or approve the teaching of the aforesaid Joachim on this point, let him be repressed by all as a heretic . . . We also reprobate and condemn the perverse teaching of the impious Amaury (Almaricus, Amalricus) de Bene, whose mind the father of lies has so darkened that his teaching is to be regarded not so much heretical as insane.

Anyone who defends heretics is to be excommunicated (Canon 3); such "pestilential people" are to be denied sacraments. Penitential metaphors of cleanliness apply to heresy. Canon 3: "If from sufficient evidence it is apparent that a bishop is negligent or remiss in cleansing his diocese of the ferment of heretical wickedness, let him be deposed from the episcopal office and let another, who will and can confound heretical depravity, be substituted." These canons stipulate behavioral norms, as though heresy can be purged physically and spiritually. One of the Anglo-Saxon penitentials similarly states how those deceived by heresy are to be singled out for punishment: "If one flouts the Council of Nicaea and keeps Easter with the Jews on the fourteenth of the moon, he shall be driven out of every church unless he does penance before his death."[17] Codification—who worships properly and who deviates from commanded belief—lies at the heart of these decrees. Those who stray are sullied by their disobedience and need to be cleansed or suffer outright ejection from the communal body.

The obsession with genealogy in Icelandic culture likewise reflects the importance of codification. Only through the linkage of oneself to a previous generation can one's proper place be settled and ascertained. Genealogical ties can only be achieved through the rejection of other potential connections. Eliminating those false avenues enables accurate relationships to be established, and with that solidity of one's place in the world comes security and contentment. The *Eyrbyggja Saga*'s overt concern with the establishment of order is likewise reflected in its obsession with filth and transgression. Thorolf is a devotee of Thor. The "high-seat pillars from the temple [from Norway] with the figure of Thor carved on one of them" float ashore in Iceland.[18] Thorolf declares this landing spot to be sacred. This headland that he dubs Thor's Ness "is a mountain held so sacred by Thorolf that no one was allowed even to look at it without first having washed himself, and no living creature on this mountain, neither man nor beast, was to be harmed until it left of its own accord."[19] Most importantly, the district assembly court area "was so holy that he would let no one desecrate it, either with bloodshed or with excrement; and for a privy they used a special rock in the sea which they called Dritsker [Dirt Skerry]."[20] The Kjalleklings, a family of pugnacious troublemakers, became arrogant and

> declared publicly that they would no longer tolerate the arrogance of the Thorsnessings and meant to ease themselves there on the grass just as they would at any other meeting, even though the Thorsnessings, so full of their own importance, thought their land more sacred than any other in Breidafjord. The Kjalleklings let it be known that they were planning to waste no more shoe-leather on trips to any off-shore skerry whenever they felt the demands of nature.[21]

The two groups fight and many deaths take place, soaking the sacred spot with blood. During arbitration, it is decided that, since blood had spilt, the land could no longer be held to be any holier than any other land. "Though this assembly place was held to be highly sacred, people were not forbidden to ease themselves there."[22] Kevin J. Wanner has pointed out the importance of this episode in the saga, treating it "as a crucial element of a myth of origins for Icelandic society, one in which attempts to construct and reconstruct the body politic are shown to play out through attempts to police and to resist policing of a bodily orifice and its emissions."[23] A pure body politic, in other words, demands the reflection of pure bodies, not ones haphazardly defecating on sacred ground.

Swedish berserkers in the *Eyrbyggja Saga* "were wholly unlike human beings, storming about like mad dogs and afraid of neither fire nor weapons."[24] These men are without civilization, not even able to fear two artifacts under human control—fire and weapons. The King of Sweden had given them to Earl Hakon, the ruler of Norway, informing him that "the berserks were good men to have as allies as long as their tempers remained under control."[25] Visiting from Iceland, Vermund decides they would terrify his enemies back home and requests to have them as a gift. Despite the Earl's warnings, he grants the berserks to Vermund, who soon regrets having taken them since they fulfill their warning that "'if ever you refuse us anything which we want and you have the power to give, we won't be at all pleased.'"[26] Their threat holds good when one of them, Halli, wants to have a wife. Having been given by Vermund to his brother Styr, Halli has taken a fancy to Styr's daughter, Asdis. Styr sets up challenging tasks for the berserks to fulfill, later inviting them to the bathhouse to relax. There, he seals off the trapdoor to the bath with stones, spreads a wet ox-hide over it, and "then had a lot of water poured in through the skylight above the hearth, which made the room so terribly hot that the berserks found it unbearable and began attacking the door."[27] As they rush out, they are stabbed to death. The daughter, who was desired by an uncivilized brute, is married off to Snorri the Priest, the representative of law and religion. The berserkers are ultimately led to their deaths by the epitome of civilized behavior—the bathhouse—home to a cleansing ritual that only succeeds in washing them out of society.

Distinctions externalize how *we* are not *there* with *them*, as in F. Scott Fitzgerald's *The Great Gatsby*, a work "haunted" by materiality, waste, and garbage, as Rex Ferguson has argued.[28] Geographical locations reflect upon the people who inhabit those locales, including East and West Egg, geographic centralities, and margins. Favored individuals are codified by locale in East Egg, whereas those located in West Egg remain indelibly tainted, since it is the "less fashionable of the two." Gatsby's mansion is a "factual imitation of some Hôtel de Ville in Normandy... spanking new," unlike the homes of old money. Nick Caraway, the narrator, remains on the margins, though he has "the consoling proximity of millionaires" and "the white palaces of fashionable East Egg."[29] Tom Buchanan, the brutish husband of Daisy, insists he'll stay in the East: "I'd be a God damned fool to live anywhere else."[30] Daisy herself "was appalled by West Egg, this unprecedented 'place' that Broadway had begotten upon a Long Island fishing village—appalled by its raw vigor."[31] The East in general represents "superiority," with all main characters who are Westerners remaining "subtly unadaptable to Eastern life."[32] Coming from the merchant class (wholesale hardware business), Nick reflects on how "the Middle West now seemed like the ragged edge of the universe—so I decided to go East and learn the bond business."[33] He goes from the margins of society and continent—exemplified by the mushy indistinctness of the middle—for the edges marked by culture and approbation. Indeed, the famous line about "a fresh, green breast of the new world"[34] suggests how the feminized body of North America is lovely along its inviolate edges as opposed to its central entrails. We cannot help entering the world of metaphor when waste is at play. The novel codifies clean (privileged) and dirty (disdained) through geographical codification.

When Order Fails: Hybridity and the Inability to Name

He Who Must Not Be Named

Harry Potter and the Order of the Phoenix, Chapter 38[35]

Georges Bataille suggests that not only do we attempt to control nature by naming it, but also "we have fashioned this humanized world in our image by obliterating the very traces of nature."[36] Indeed, the naming—imposition of control—on nature strips it of its "undifferentiatedness" and manages it—brings it under the purview of culture. Unidentified and unidentifiable, garbage and trash have the power to confound. That which cannot be codified threatens with filth.[37] In "A Load of Shit," John Berger describes himself clearing out "the year's shit"[38] from his outhouse and putting it into a hole. Even though he peacefully starts this annual ritual, a kind of anger eventually rises up in him. "Naming the stuff I'm shovelling finally provokes an irrational anger. Shit!" Indeed, the utterance into discourse creates codification, enabling us to distinguish between what is clean and what is dirty. As the narrator points out, "The dream of purity and freshness was born from the omnipresence of muck and dust. This polarity is surely one of the deepest in the human imagination, intimately connected with the idea of home as a shelter—shelter against many things, including dirt."[39] This classification, he argues, lies deep in our psyches, impossible to dig up and out. Words allow us to know about dirt, waste, and their cognate mortality, but simultaneously cut us off from our direct experience of it, so that we experience death "by proxy."[40] What if there is no word to describe something? Don DeLillo's *Underworld* comments on ineffability: "Something that eludes naming is automatically relegated... to the status of shit. You can't name it."[41]

In the *Eyrbyggja Saga*, after some—literal—bad blood, enemies seek out Odd Kaltason to kill him. His witch-mother manages to transform him into various objects so the seekers overlook him when the avengers come to visit. At one point, when Katla "saw [Arnkel and Thorarin] coming up, she told Odd to take a walk with her. They went outside over to the rubbish-heap, and she told him to lie down beside it. 'Stay here whatever happens,' she said." Later, "Arnkel and his men search for Odd inside and out and could find no living creature except for Katla's pet hog lying by the rubbish dump, so they went away."[42] Ultimately, the sorceress Katla and her son Odd are killed: she is stoned to death and Odd hanged. Undifferentiated rubbish permits Odd to temporarily disappear by the rubbish as a pig, rendering him indistinguishable from the trash. Visible to the physical eye, he is ontologically invisible.

Those kindred souls, cleansing and codifying, while emotionally satisfying, have ominous repercussions, such as that which is "undertaken for reasons of social distinction (including those of race and class)."[43] Codification suggests that what is pure and what is dirty can be readily perceived. "Hybrids and migrant beings,"[44] then, threaten to undermine structure and must be condemned. *Leviticus* Chapter 11 famously lays out what may and may not be eaten. Anything that does not cohere completely to its class—if it is mixed or tainted—is perceived as dirt (matter out of place). Like that which cannot be named, hybrids do not belong to a single codifiable and recognizable category. For example, the Lord says: "This is the law of the beasts, and of the fowl, and of every living creature that moveth in the waters, and of every creature that creepeth upon the earth: To make a difference between the unclean and the clean, and between the beast that may be eaten and the beast that may not be eaten" (*Lev.* 11: 46–47). *Leviticus* and *Deuteronomy* exclaim against confusion, incest, hybridity, or grafting. "And if a man lie with his daughter in law, both of them shall surely be put to death: they have wrought confusion; their blood shall be upon them" (*Lev.* 20: 12). If they don't obey, God threatens:

> And I will make your cities waste, and bring your sanctuaries unto desolation, and I will not smell the savour of your sweet odours. And I will bring the land into desolation...And I will scatter you among the heathen, and will draw out a sword after you: and your land shall be desolate, and your cities waste. (*Lev.* 26: 31–33)

As Douglas points out, "Hybrids and other confusions are abominated."[45] Holiness means to adhere perfectly to a particular class or order.

Hybridity is less admirable than being "pure," even in a perceived "wasted" class or category. Gerald Vizenor's "The Last Lecture" satirizes how crossbloods are seen as worse than being pure Native American. Marie Gee Hailme confesses to her passing on false tribal values as a pretend Indian: "We should have trashed the schools, not ourselves with the delusion that we were helping Indian students."[46] Homer Yellow Snow is described sneeringly as "author, historian, tribal philosopher, and last but not least, a perfect pretend Indian."[47] J. K. Rowling uses scorn to her advantage in the Harry Potter novels, where "mudblood" is the insult of choice for those of mixed "muggle" ("normal" human) and wizarding blood. One means by which the elite, identifying themselves with a higher order of culture, can control the nonelite is by equating the lower orders with nature. Garbage and nature, both feared owing to their controllable and uncanny powers, need to be put in their place.

A standard prejudice has existed for centuries by pale-skinned Europeans and their descendants toward those of darker hues. In *The Great Gatsby*, Daisy's bigoted husband Tom, tainted by the book "The Rise of the Colored Empires," worries, "If we don't look out the white race will be—will be utterly submerged."[48] The privileged and dominant group he identifies with—his own group of whites—could itself become a "wasted" people like these "leftovers of civilization."[49] "We're Nordics . . . We've produced all the things that go to make civilization—oh, science and art, all that."[50] The unfaithful Tom talks about the degradation of his proper life: "'Next they'll throw everything overboard and have intermarriage between black and white.' Flushed with his impassioned gibberish, he saw himself standing alone on the last barrier of civilization."[51] The threat of hybridity—that would make visible the connection between white people and the perceived lower races—drives his anxiety of miscegenation.

"Who's there?" (*Hamlet* 1.i.1).[52] So begins the most famous play in English—with a question of dramatic and metaphysical import. Shakespeare distinguishes between who is and isn't there. *There* is the *not here*. *Here* is the not *there*. The being who *is* there uttering this query could just well have asked who *isn't* there. Codification lies at the foundation of this self-pondering meditation, as is also shown in the evident confusion between "Rosencrantz, and gentle Guildenstern"—or is it "Guildenstern, and gentle Rosencrantz"? (*Hamlet* II.ii.33–34)—highlighting the need and desire for distinction, codification, and difference.

When things cannot be codified, we respond in varying ways. In Miles Wilson's short story, "Everything," Roger collects all sorts of useless stuff that takes over his house, "unencumbered by arguments about utility. An electric letter opener, a collection of gilded and tattooed grasshoppers, a gallon pail of assorted hex head bolts, the complete works of Schuyler Colfax, a magnum of cod-liver oil, two propane-fired coffee makers, a wizard's cape from Guinea-Bissau, and lacrosse goalie pads accumulated in snug rows across the basement."[53] One response for the reader is to laugh at the absurdity of the discontinuity.

Even when language *can* articulate what is clean and what is dirty, what *is* and what is *not*, the very accumulation of language threatens to become exposed as waste itself. Michel Foucault describes his laughter at reading Jorge Luis Borges's "The Analytical Language of John Wilkins" (1941), in the opening passage of *Les mots et les choses*. Borges's story "[breaks] up . . . our age-old distinction between the Same and the Other." As Foucault delineates, Borges's work describes

> a certain Chinese encyclopaedia [in which] animals are divided into: (a) belonging to the Emperor, (b) embalmed, (c) tame, (d) sucking pigs, (e) sirens, (f) fabulous, (g) stray dogs, (h) included in the present classification,

(i) frenzied, (j) innumerable, (k) drawn with a very fine camelhair brush, (l) *et cetera*, (m) having just broken the water pitcher, (n) that from a long way off look like flies.

Foucault wonders at "the exotic charm of another system of thought, [demonstrating] the limitation of our own, the stark impossibility of thinking *that*." Grouping eccentrically "localizes their powers of contagion," an apt choice of wording. Neutralizing the pollution of miscategorization—Douglas's "matter out of place"—"the possibility of dangerous mixtures has been exorcized."[54] Only alphabetization gives coherence to this sorting, since all "common ground . . . has itself been destroyed."[55] Worse than disorder, heterotopias "secretly undermine language, because they make it impossible to name this *and* that, because they shatter or tangle common names."[56] Ambiguity engenders disturbance.

Determined to account for what is nothing, the narrator of Donald Barthelme's "Nothing: A Preliminary Account" lists all sorts of random things that hint at stories behind them: "It's not the yellow curtains, we have determined that, and it's not what is behind the yellow curtains which we cannot mention out of respect for the King's rage and the Queen's reputation." He can only conclude, "None of these is nothing."[57] They are all something identifiable. We respond with an existential bemusement and perhaps a creepy sense of the inexorable meaningless of codifying meaning when we realize "our list can in principle never be completed . . . And even if we were able . . . to exhaust the possibilities, get it all *inscribed*, name everything nothing is not . . . the list itself would remain."[58] The lists, a form of "antiliterature," threaten owing to their "nonteleological nature."[59] Not adhering to our compulsion to center on a subject within a linear narrative, the list compiles detail after detail. Yet the list is oddly consoling. Always endeavoring to codify, to name, to stave off the threat of the not named—a form of dirt—we attempt to evade the gruesome truth: we are matter out of place, never able to secure our proper (*propre*, meaning both proper and clean) place.

CHAPTER 2

The Fragmented and Corruptible Body: Gendered Waste

We're meat wrapped round / knotchbone spine.

A. R. Ammons, *Garbage*[1]

I once went on a blind date with a young man who graduated from my alma mater. He proudly showed me my father's book—the standard organic chemistry textbook of the time. Through it, my suitor had driven a four-inch nail, evidently because he had failed that class, essential for entry into medical school.

That was not the only reason I dubbed him "Nails."

At dinner, I was invited to see his collection, one that had landed him, he proudly claimed, in the *National Enquirer.* A walk-in closet was filled with glass jars, labeled with people's names. Inside, were nails that had belonged to the named parties. Yes, he collected nails. No, not those iron items you hammer into wood.

Nails. Fingernails. Toenails.

Why is everyone to whom I regale this story grossed out? Generally speaking, body products *are* rubbish.[2] At least he didn't collect boogers or turds!

I've told this story many times to amuse, mocking both myself as a naïve young woman and the inconceivable expectations of my hapless nail-fetishizing would-be lover. I also joke with my daughter that I hope she'll take this as a lesson in whom *not* to date.

Oh, and Reader, just in case you were wondering, I did *not* marry him.

Bodily Borders and the Grotesque: Threats to Identity

Life is merely the epitome of everything about which one must be ashamed.

Theodor Adorno[3]

The topic of the body is a fraught and intellectually challenging issue that is always with us. Alice Sebold, concerning the importance of bodies in contemporary US life to commemorate the fourth anniversary of 9/11 and in the wake of Hurricane Katrina, writes:

> Perhaps it is an image you remember having seen. A handmade grave of sheets and bricks. "Here Lies Vera. God Help Us"...Perhaps it is less specific: Bodies falling from high windows, bodies floating in muddy water. Bodies wrapped in dirty bedding and tucked along the sides of bridges and highways. The faces of the missing, taped and tacked up on a wall.[4]

While national events have focused attention on the body, the site of highly politicized debates, from abortion to the Terry Schiavo case, the body has long been of abiding concern and fascination for humanity. In various ways, the body is a central focus in all disciplines of the humanities. A perceived mind/body split informs much in the Western philosophical tradition. Perceptions of the body affect political structures (from the development of democracies to the abolition of slavery) and religious rites. On a physical level, the body crucially affects how we give birth, engage in mating and courtship rituals, and view aging and death.

The split between the mind and body articulated by the Cartesian slogan "Cogito, ergo sum" reinforces a hierarchy privileging reason over matter. In an amusing anecdote from 1789, Christoph Martin Wieland describes sitting on the toilet and using scrap paper to wipe his bottom, only to discover the following words inscribed on the sheet next destined for his posteriors: "What is enlightenment?"[5] The juxtaposition of enlightenment with the act of defecation produces disbelief, outrage, humor—anything but placid acceptance. "Why does *corporeal waste*, menstrual blood and excrement, or everything that is assimilated to them, from nail-parings to decay, represent—like a metaphor that would have become incarnate—the objective frailty of the symbolic order?"[6] Read as impurity and disorder, the most virulent subset of dirt—feces—coalesces the most emotional and extreme reactions to filth and manifests in the figuring of excrement as shit—low, horrifying, disgusting. The privileging of reason within the mind-body duality causes us to reject excrement as (mere) filth and extol the life of the mind. By turning "the body into an object," we come to "see the body as the Other, a threatening container of one's own denied desires."[7] Emmanuel Levinas has argued that we want to escape *becoming*. Undergoing constant flux—from excretion to the plucking of newly discovered gray hairs—we cower in the "shame...primarily connected to our body...It is therefore our intimacy, that is, our presence to ourselves, that is shameful."[8]

We shudder at the filth our bodies can produce, hence the psychological and emotional joy at activities founded in cleansing activities. One critic praised Heinrich Böll's development in literary matter as now "free of the smell of washtubs." This backhanded compliment Böll subtly attacks in "In Defense of Washtubs," by lyrically writing of washday as "a special feast day" of community and conviviality. As Böll reminisces, the sight of "shirts and sheets, handkerchiefs and unmentionables, fluttering on clotheslines [all over the world], a sight which, far from depressing me, has consoled me, signalling as it does the tireless energy of the human race to rid itself of dirt."[9] Susan Strasser's point that "discarding things is taken to be a kind of freedom,"[10] resonates in Italo Calvino's "La Poubelle Agréée," when the narrator describes how satisfying it is to take out garbage. It is

> a rite of purification, the abandoning of the detritus of myself... what matters is that through this daily gesture I confirm the need to separate myself from a part of what was once mine, the slough or chrysalis or squeezed lemon of living, so that its substance might remain, so that tomorrow I can identify completely (without residues) with what I am and have.[11]

Taking out garbage is a way of staving off death, "to confirm that for one more day I have been a producer of detritus and not detritus myself."[12] Division, cleansing, and codification can result in emotional pleasure. Calvino continues that

> the satisfaction I get out of this, then, is analogous to that of defecation, of feeling one's guts unburdening themselves, the sensation at least for a moment that my body contains nothing but myself, and that there is no possible confusion between what I am and what is unalterably alien.[13]

Taking out the trash can be seen as both a "contract and as a rite."[14]

> If the gesture of throwing away is the first and indispensable condition of being, since one is what one does not throw away, then the most important physiological and mental gesture is that of separating the part of me that remains from the part I must jettison to sink away into a beyond from which there is no return.[15]

If I—my body—is not that which I reject—my pared nails, my golden piss, my tawny turd—I reinforce hermetic security in myself.

We feel the compulsion to separate ourselves from that which we consider filthy in order to reassure ourselves that we *are not* that filth. To create a self, you must discharge and reject that which is not you. That which is literally ejected from the body, then, is to be forbidden as *not you*. Julia Kristeva building on Douglas's work, dramatically intones, "If dung signifies the other side of the border, the place where I am not and which permits me to be, the corpse, the most sickening of wastes, is a border that has encroached upon everything. It is no longer I who expel, 'I' is expelled."[16] Bodily issues (pus, menstruation, snot) are all unclean. Perversely, ironically, these bodily fluids expelled from within calm the self into believing it is whole and inviolable because it has got rid of what is filthy and impure.[17]

As Lennard Davis has taught us, "normal" bodies that naturally excrete and die manage to threaten us. This presumed coherent identity, situated in the body and willingly believed in, is exposed as a sham by the "abnormal" body,[18] whereby the whole is proven to be a "hallucination" or "fiction."[19] This contingent nature of all bodies we repress; thus we respond to bodies that seem fragmented with horror and fear. At the start of the Greek tradition we are confronted with the rejected outcast in Sophocles's *Philoctetes*. Philoctetes, whose[20] "foul smell" and "terrible stench"[21] render him unfit for accompanying the Greeks to Troy, remains abandoned on Lemnos. As the Chorus intones, "No one care for him, / he has no one to look to. Wretched. Alone, always. / he suffers this terrible disease."[22] With "no friend, no companion,"[23] Philoctetes is "an outcast,"[24] because of a savagely inflicted wound. A snakebite had rendered his foot diseased, oozing pus and seeping foul secretions. The visitor Neoptolemos, Achilles's son, observes "some rags drying out, pus-heavy."[25] This infected pariah resists the attempt to lure him back to Troy, a clever plot on the part of the duplicitous Odysseus who knows through a prophecy that Greek victory can only be assured through the very person they abandoned. Rather than listen to the riviled, treacherous Odysseus, Philoctetes would rather "listen to what I hate most, / the serpent that made of my foot a useless thing."[26] As Philoctetes says, resisting the attempt to integrate him back among the Greeks, "I, who am nothing now, am long since dead to you. / Most hateful to the gods, / why don't I seem lame and stinking to you, now? / How will you be able to burn sacrifices to the gods, / if I sail with you? How make libations? / For this was exactly why you threw me out, as you said then."[27] At his moment of deepest despair, Philoctetes acknowledges, "I am nothing at all."[28] His diseased body threatens and disgusts the fit and "normal" bodies around him.

Similarly reviled, the leper was particularly singled out as an unclean and unwhole (unwholesome and unholy) body, actual body parts sometimes having been sloughed off.[29] *Leviticus* illustrates how *he* is not *us*:

> He is a leprous man, he is unclean: the priest shall pronounce him utterly unclean; his plague is in his head. And the leper in whom the plague is, his clothes shall be rent, and his head bare, and he shall put a covering upon his upper lip, and shall cry, Unclean, unclean. All the days wherein the plague shall be in him he shall be defiled; he is unclean: he shall dwell alone; without the camp shall his habitation be . . . To teach when it is unclean, and when it is clean: this is the law of leprosy. (*Lev.* 13: 44–46; 14: 57)

Diseased bodies are not the only ones cited for approbation. Dead and decaying bodies function like dung:

> At that time, says the Lord, men shall bring out from their graves the bones of the kings of Judah, of officers, priests, and prophets, and of all who lived in Jerusalem. They shall expose them to the sun, the moon, and all the host of heaven, whom they loved and served and adored, to whom they resorted and bowed in worship. Those bones shall not be gathered up nor buried but shall become dung on the ground. (*Jeremiah* 8: 1–2)

Decay remains the inevitable condition of all of us.

Bodily Decay, Margins, and Death

> In life [man] produced lice and tapeworms; in death he will produce worms and flies. In life he produced dung and vomit; in death he produces rottenness and stench . . . What then is more foul than a human corpse? What is more horrible than a dead man?
>
> Lothario Dei Segni (later Pope Innocent III)[30]

Lennard J. Davis's Lacanian reading understands the self as constructing fragmented pieces of the body into a whole. The disabled body, Davis argues, represents the "repressed fragmented body," "the true self of the fragmented body."[31] Disabled bodies serve to remind possessors of "normal" bodies that we are all "about to come apart at the seams."[32] For Mikhail Bakhtin, who offered an aesthetic of the Middle Ages that revelled in presenting the body in its nonidealized form, the grotesque was associated with the common people. The established order would periodically be turned upside down through the carnival and the carnivalesque.[33] While Bakhtin sees the grotesque body as liberating, a new regularized and orderly view of the body comes about in the early modern period. Baudrillard assesses the grotesque and the classic bodies as fundamentally divergent: "The body does not stop at this porous skin, full of holes and orifices; only metaphysics institutes it as the borderline of the body. This body is denied in the interests of a second, non-porous skin that neither exudes nor excretes . . . above all without orifices."[34] This shift in how the body is viewed has only restricted our understanding of ourselves.[35]

Not all bodily exudings are considered rubbish. Tears and milk[36] are typically valued, with water as the primal cleanser and milk as the nurturer of infants. Only when useful are urine and excrement accepted as more than filth. Urine has been used in the traditional process to prepare wool for weaving, while dung can be a fertilizer or burnt for heat. Christian thought found hope in dead saints' bones. Bodily parts have been seen as useful, even holy, within the Catholic tradition as relics where healing occurs due to another's bodily part. Divided into relics, each saint's vestige miraculously contains the totality of the saint.[37] God's ultimate promise and plan is wholeness; if ground to dust, one's body would be reassembled by God at the Second Coming. The exceptional relic points to the tragedy of our everyday bodies, subject to quotidian decay and capable of eliciting disgust.

In the Old Icelandic *Eyrbyggja* Saga, the widow of Vigfus tries to get his kinsmen to redress her husband's death. All demur until she finally takes Vermund's advice to dig up the corpse and cut off the head. She goes to Arnkel who resists taking vengeance. When, he "said he had already told her what he thought of the case, she pulled the head from under her cloak. 'Here's a head,' she said, 'that would never have shirked action if you'd been killed and its help had been needed.'"[38] Though he later avenges the savage murder, Arnkel sends her and her bloody trophy away, horrified by the divided body and gory residue.

Bodily fragmentation induces uneasiness, an anxiety displaced to corporal edges. Take the nose, for instance. As Mary Douglas points out, noses contain two orifices whence viscous liquid is emitted, liquid that is seen as neither purifying or cleansing as tears are. This part of the body arouses suspicion, and not only for its noxious emissions. It acts as a kind of guard for our safety—the smell of gas alerts us that we may be in danger. The stench of rotten food warns us not to ingest it. Hamlet puns when Rosencrantz tries to find Polonius's body. Cryptically saying, "The body is with the King, but the King is not / with the body" (IV.iii.25–6), Hamlet assures that Claudius will be able to "nose him" (IV.iii.35) as he rots.

The Nose Knows

This superior knowledge simultaneously reminds us of our inevitable production of filth (bogeys! snot!), causing ambiguity and discomfort. Though central to one's features, the holes suggest facial vulnerability. The original edition of *The Great Gatsby* famously neglects to include a nose on the ghostly Jazz Age flapper depicted on Francis Cugat's cover. Fitzgerald reputedly told his publisher he had "written it into" the novel. In the various manifestations of the cover, the nose was "edited out,"[39] yet the novel repeatedly evokes the nasal spectre. Daisy tells Nick how their butler used to be the "silver polisher" of a service meant for two hundred people. All he seemed to do was polish it "from morning till night, until finally it began to affect his nose."[40] Like the cover's shadowy lady, "The eyes of [Doctor T. J. Eckleburg] look out of no face, but, instead, from a pair of enormous yellow spectacles which pass over a non-existent nose."[41] One memorable fellow was "young Brewer, who had his nose shot off in the war."[42] Nick describes Mr. Wolfsheim as "a small, flat-nosed Jew [who] raised his large head and regarded me with two fine growths of hair which luxuriated in either nostril . . . his expressive nose."[43] Wolfsheim is clearly a gangster; his Jewishness stereotypically, anti-Semitically, identifiable by his nose.[44] When Daisy comes to Nick's for tea, set up secretly to meet Gatsby unbeknownst to her, Nick asks her about her chauffeur. "Does the gasoline affect his nose?"[45] A celebrity at the party is with a producer "with the sort of blue nose," says Daisy.[46] Tom breaks Myrtle's nose in a dramatic skirmish. Fitzgerald's obsession with the nose—or lack thereof—points back to class rivalry. The rich look down their noses on those beneath them. The rich, determined to sniff out their superiority, nonetheless reek like the less fortunate.

In his brilliant modernist stream-of-consciousness Holocaust novel *The Journey*, H. G. Adler varies the narrative voices. One describes people as noses:

> Just think, you noses, it didn't take much to seduce you into a dream, one that you hardly could have imagined would turn out this way! Now you root about in mud and muck. But now off with you, for the leader has to call the tired noses back to the barrels and the wagon.[47]

The statue of the general in Leitenberg has his nose missing.[48] Leitenbergers are not to talk to the "ghosts." Paul can look about, though, soldiers leading "a chain of blind men consisting of nothing more than a set of noses hanging down from hollowed-out eyes."[49] Reduced to impotently phallic and ambulatory noses, they are doomed to be wiped away.

Living bodies need to be sequestered from those of the dead. Certainly there are biological, health, and sanitation reasons to avoid a corpse. Seen as hybrid "waste, transitional matter, mixture, it is above all the opposite of the spiritual, of the symbolic, and of divine law."[50] In death rituals in the United States today, unlike the former

> public pilgrimage through town to the graveyard, [funerals] now tread lightly across the tiny tableau of our psyches . . . For the first time in history, the actual presence of the dead at their own funerals has become optional, even undesirable . . . A corpse is a stark reminder that human beings are inescapably embodied creatures.[51]

In warfare, victors have been known to mutilate and desecrate their stockpiled casualties. Murder victims are treated as trash by their killers, hence the mesmerizingly horrific power of Joyce Carol Oates's short story "Landfill," which links the decay of the human body with the stench and filth of a dump: "Tioga County Landfill is where Hector, Jr., is found. Or his 'remains'—battered and badly decomposed, his mouth filled with trash.

He couldn't have protested if he'd been alive, buried, as he was, in rubble and raw garbage."[52] His very resting place on a dump pyre suggests the inevitable state of us all.

The gravedigger's scene famously expresses a reductive and elemental view of the human body. Hamlet's reflections on the skull, now tongueless and eyeless, is reminiscent of a contemporary commentary on death. Luis de Granada's *Of Prayer and Meditation* (1582) reflects on how the rich man is treated by the gravedigger: "And the fine dappered gentleman who whiles he lived might in no wise abide the wind to blow upon him . . . , here they lay and hurl upon him a dunghill of filthiness and dirt."[53] When Hamlet and gravedigger discuss how long a body will keep in the earth, we learn that a tanner endures longer than most bodies by a year. The gravedigger, a corpse connoisseur, curates his clientele.

The body—in its material fleshliness and the body politic—controls *Hamlet* metaphorically. Claudius assures Polonius that he is sending Hamlet to England to retrieve overdue tribute; additionally, this change of scene might "expel / This something settled matter in his heart" (III.i.171–172). When Guildenstern tells Hamlet how angry Claudius is about the play, Hamlet suggests he pass this on to a doctor, "for for me to put him to his / purgation would perhaps plunge him into more choler" (III.ii.298–299). Jenkins points out the double meaning of purgation in both the physical sense of excess humors and the spiritual sense, to purge by confession. The Ghost must be purged (I.v.13), but so must Claudius and his sins—all foul corruption must be jettisoned from individual bodies, souls, and the body politic. Guildenstern refers to the "many bodies . . . That live and feed upon your Majesty" (III.iii.9–10), an almost cannibalistic image of the king's body. Rosencrantz comments, "Never alone / Did the King sigh, but with a general groan" (III.iii.22–23). The entire body politic breathes along with the ruler. Hamlet goes to kill Claudius, only to find him praying and enacting spiritual purgation: "And am I then reveng'd / To take him in the purging of his soul, / When he is fit and season'd for his passage?" (III.iii.84–86). Hamlet comments how he and others admired a play once presented by the players that was "well digested in the scenes" (II.ii.377). He continues to point out, "I remember one said / that there were no sallets in the lines to make the matter / savoury" (II.ii.378–380). Food imagery conveys digestion through the entire play, only to culminate in the ultimate excretion of the body politic—the death of the king.

Multiple words suggest rubbish, filth, and excess. The body is reduced to its elemental basics, centered in its intestines and filth. Hamlet assures Claudius that the dead Polonius is at supper: "Not where he eats, but where a is eaten . . . That's the end" (IV.iii.19, 24). Nobility of birth is reduced to sewage. "[A] king may go a / progress through the guts of a beggar" (IV.iii.29–30). Even humans, "in apprehension how like a god," are ultimately a "quintessence of dust" (II.ii.272, 274). The unredeemed body controls *Hamlet*, as we hear in Hamlet's lament: "O that this too too sallied flesh would melt." (I.ii.129). Even the poison poured into Hamlet's father's ear did "curd" (I.v.69) his blood and cause his "smooth body" to break out "[m]ost lazar-like with vile and loathsome crust" (I.v.72–73). When Hamlet speaks to the Ghost, invisible to Gertrude, she comments on his "bedded hair like life in excrements / Start up and stand on end" (III.iv.117–118). "Excrement," meaning "that which grows out or forth; an outgrowth; said *esp.* of hair, nails, feathers," was first used by Shakespeare, according to the Oxford English Dictionary (OED), in a literal sense in *Love's Labour's Lost* (V.i.109). "Excrements" meant not only feces, but also nails and hair. One sixteenth-century manuscript uses the newly coined word, "excrement." It translates part of the text, a question and answer dialogue about surgery, neatly copied in early modern English.

> Q: Wth are yr members called Excrementoas.
>
> A: Yr nails & yr hairs.[54]

Nails, then, are "excrements" of the body. Excrement suggests a superfluity and nonessentiality that spills out over the closed and clean "properly" functioning body.

As Levinas writes, we feel the need to "transcend the limits of finite being," feeling horror from the "degrading types of servitude imposed on us by the blind mechanism of our bodies."[55] It is just this kind of "insurmountable" nausea that revolts "from the inside; our depths smother beneath ourselves; our innards 'heave.'"[56] This revulsion drives Hamm in Samuel Beckett's *Endgame* to demand that Clov, "Clear away this muck. Chuck it into the sea!"[57] The discards are none other than Nagg and Nell, his parents literally consigned to waste bins. Later Hamm accuses Nagg: "Scoundrel! Why did you engender me?" Nagg: "I didn't know." Hamm: "What? What didn't you know?" Nagg: "That it'd be you."[58] That would be the ultimate freedom—to destroy the ones who engendered you, and, in effect, yourself and your stinking flesh. Such liberty comes at the cost of total annihilation.

The Quintessence of the Filthy Body: Menstr[u]ous Women[59]

> The discomfort people feel about their smelly, decaying, and all-too-mortal bodies has ubiquitously and monotonously been projected outward onto groups who can serve as, so to speak, the surrogate dirt of a community, enabling the dominant group to feel clean and heavenly.
>
> Martha Nussbaum[60]

This self-loathing for the body—even one's own materiality—drives our disgust at our fleshly selves and those of others. While the gender of the speaker and partner remain unclear, Shakespeare's Sonnet 129 shows the horror of shame at the delight, then cowering indignity, at the memory of intercourse: "The expense of spirit in a waste of shame / Is lust in action." Nussbaum thoughtfully acknowledges how visceral outrage at the shame of the body's ability to exude revolting materials turns on the less powerful, suggesting, "A ubiquitous reaction to this sense of one's own disgustingness is to project the disgust reaction outward, so that it is not really oneself, but some other group of people, who are seen as vile and viscous, sources of contamination that we might possibly keep at bay."[61] One example of this imposition of self-disgust onto others can be located in misogyny. Though all bodies exude filth, women's bodies in particular have been identified with what is fluid and excess. To lessen the threat to "one's *own and clean self*,"[62] filth becomes identified with the female. The division of the body into clean and dirty collapses in misogyny, where women's bodies have no chance to be clean. Historically, women have most frequently been the ones to wipe waste from the bodies of others, in caring for incontinent infants and elders, and in washing the bodies of corpses. If women have not practiced cleansing bodies in their own families, they have hired other women to do it for them. Shit and filth are woman's place.

While men and women excrete, urinate, and vomit, women's emissions have been viewed as problematic at least since Plato and Aristotle.[63] Vilified, the monthly sloughing of menstrual blood identifies menstruation as failure; as "waste, scrap."[64] Women have been separated from the rest of society when particularly filthy or ambiguous, such as when menstruating or after having given birth. After delivering a child, the mother is less soiled after a boy is born than after a girl is born (*Lev.* Chapter 12). Chrysostom viewed women's bodies as "nothing less than phlegm, blood, bile, rheum, and the fluid of digested food."[65] A sixteenth-century witchcraft text suggests menstruating women can stain a mirror through their polluted eye rays.[66]

As Nussbaum argues, those who perceive others with disgust refuse to see the affinity in humankind:

> Disgust imputes to the other a subhuman nature... Only by imagining how the world looks through that person's eyes does one get to the point of seeing the other person as a someone and not a something.[67]

In *Beowulf*, Hrothgar was aware of the "shape" of a woman and deformed male, but never was able to make the imaginative leap into their point of view; had he developed his poetic imagination, he would have been able to see the inevitable revenge that Grendel's Mother would have to enact. And this, despite the many instances of songs in the hall being performed that suggest the "wasted lives" of debased people. Perhaps he had imagined Grendel's Mother to be an inert and paralyzed Hildeburh, trapped between two warring realms. "Empathy is simply an imaginative reconstruction of another person's experience." A person could be empathic and evil, Nussbaum argues. Yet the humans in *Beowulf* are not even empathic, certainly not compassionate;[68] they cannot even imagine someone avenging Grendel's death. Grendel's mother is so subhuman in their eyes as to not even countenance vengeance.

Embodied in the female figure of Grendel's mother, the monstrous body is aligned with nature.[69] The giant's sword Beowulf propitiously finds by chance in the mere lair of the monsters succeeds in killing Grendel's mother. Beowulf subsequently uses it to behead the dead Grendel, whereupon the sword's blade melts (1615b–1617b) because of the blood of the "ellorgāst" or alien spirit (1621), a term used to describe both Grendel and his mother (807, 1349, 1617). When Beowulf later regales Hygelac with his adventures, Beowulf only mentions beheading Grendel's mother, not Grendel, in the mere surging with blood or gore (2138–2140).[70] Clearly, he wishes to emphasize *her* dismemberment, not her son's. The blood that succeeds in melting the hilt could be Grendel's mother's blood, not Grendel's own as is usually assumed, given the association with filth that women's blood long had. Consider Isidore of Seville's *Etymologies*. In his discussion of *menstrua*, he writes, "From contact with this blood, fruits fail to germinate . . . plants die, trees lose their fruit, metal is corroded with rust, and bronze-objects go black. Any dogs which consume it contract rabies."[71] The *Penitential of Theodore* (of Tarsus, archbishop of Canterbury, late seventh century) explicitly forbids menstruating women from entering a church or taking the Eucharist on pain of fasting for three weeks.[72] While Pope Innocent III likewise refers to this myth of menstrual blood's corrosive powers, he adds that "When a child is conceived, he contracts the defect of the seed, so that lepers and monsters are born of this corruption."[73] Even the compassionate Pope Gregory in Bede's *Ecclesiastical History of the English People* acknowledges a woman's menstrual flow to be "an illness."[74] Is it too much to conjecture that the monks transcribing *Beowulf* would have known of the negative and legendary corrosive powers of women's blood? Grendel's mother "matters"[75] in this poem.

Indeed, her matter "matters."

Like menstrual blood, shit has the "capacity to disturb the body's boundaries and sense of stability."[76] That male bodies likewise produce filth made little difference to misogynist rhetoricians. In the Middle Ages, women's bodies came to represent what was frail and weak about bodies in general[77] and, in some virulently misogynistic comments, were even associated with excrement itself. The holy harlot Thaïs in the tenth-century play by Hrotsvit von Gandersheim is cured of her sexual sins only by being contained in a cell with her own excrement in a homeopathic ritual of filthing.[78] The misogyny toward women's bodies spills over from patristic writings into Hildegard von Bingen's hellish vision of the Antichrist and the harlot who pretends to be a virgin. In Book Three of the *Scivias*, Hildegard sees a woman who is blemished and has a monstrous head in place of her lower torso: "And behold! That monstrous head moved from its place with such a great shock that the figure of the woman was shaken through all her limbs. And a great mass of excrement adhered to the head; and it raised itself up upon a mountain and tried to ascend the height of Heaven."[79] She continues, *"And a reeking cloud envelops the whole mountain, which wraps the head in such filth that the people who are standing by are thrown into the greatest terror.* For the impure and hellish stench will fill the whole place of his pride, in which that worst of criminals boiled with such uncleanliness."[80] Scatology and eschatology fuse in this image of an excremental end of the world. Alluring female sexuality must be mired with the poison of filth by proxy. In *Wolf's [Wulfstan's] Sermon to the English under Viking persecution 1014*, Wolf condemns those who act against common decency. "And it is shameful to speak of what has too commonly happened, and it is dreadful to know what many too often do, who practise that wretchedness that [men] club together and buy one woman in common as a joint purchase, and with the one commit filth one after another and each after the other just like dogs who do not care about filth."[81] The female orifice becomes the repository of male emissions, polluting, not the man so much as the woman, herself a fetid stank.[82] As the sixteenth-century trickster Till Eulenspiegel suffers his final illness, he talks to a beguine about his three regrets in life, such as not tripping up someone with a fashionably long coat. His final regret is "daß ich nit mocht allen alten Weibern, die ob ihren Jahren seind, ihr Ärsch zuflicken, das ist mir auch leid, wann die seind niemans nütz meh uff Erden, daß sie das Erdrich bescheißen, da die Frucht uffsteht" ["that I can't stitch up the arses of all the old women who are past their prime. That makes me sorry as well, since they're of no use to anybody on earth and just make the ground shitty where fruit grows"].[83] Till singles out old women for defamation, associating them with excrement. Useless since they can no longer breed children, Till denigrates them using conventional misogynistic rhetoric.

The Reformation brought about attacks on what Gary Waller has called the *gynotheology* of the late medieval period with regard to the Virgin. An obsession even about her "sexual and reproductive apparatus and functions"[84] in doctrinal manifestos and writings was dramatized in Catholic popular culture, such as pageants and plays. Protestant zeal, however, condemned such clinical exploration of the Virgin's privates, though English officers sacked Cadiz, during which a statue of the Virgin was mutilated: "They took it into the marketplace, cut the figure of the infant Jesus from the Madonna, stabbed at the statue's breasts, chopped off the arms, slashed the face, and nearly beheaded it."[85] Reminiscent of the intrigue and sexual violence of Reformation politics, female sexuality in *Hamlet* becomes the privileged space sanctioned for the most virulent verbal garbage—talking dirt about women (Gertrude and Ophelia)—and literal wasting (the deaths of the two women).[86] Filth befouls female sexuality. Rosencrantz communicates to Hamlet that Gertrude "desires to speak with you in her / closet ere you go to bed" (III.ii.322–323). "Closet," used for "sewer' in Scots in the 1530s, noxiously wafts into Gertrude's chamber with lethal cunning. These scatological echoes culminate in Hamlet's prediction that, if Gertrude believes Claudius's "unction" (III.iv.144), "It will but skin and film the ulcerous place / Whiles rank corruption mining all within / Infects unseen. Confess yourself to heaven, / Repent what's past, avoid what is to come, / And do not spread the compost on the weeds / To make them ranker" (III.iv.145–150). The bodily decay inherent to ulcers and stinking excremental compost all taint her sexuality.

Late in Act IV a "Gentleman" tells of Ophelia's strange behavior, saying she "speaks things in doubt... Her speech is nothing, / Yet the unshaped use of it doth move / The hearers to collection... Indeed [the words] would make one think there might be thought, / Though nothing sure, yet much unhappily" (IV.v.6–13).[87] The nothingness and shapelessness of her speech uncannily evoke waste and trash. This detritus is interpreted as madness.[88] Ophelia's snatches of song are like recycled garbage. She spreads the detritus of popular discourse, the rags and litter of culture, to make a fecund compost of meaning. The mad Ophelia sings bits and pieces of various ballads; these include an adaptation of the *Walsingham Ballad* (IV.v.23–26, 29–32, 36, 38–40), a song about a dead pilgrim and the living bereaved lover. In invoking Walsingham, the shrine dedicated to the Virgin Mary that was destroyed in 1538, Shakespeare reminds us of the shrine's wasted past. Memory is trash from the past. Ophelia, wasted in body and mind, exists as a trace of the Walsingham Virgin: profaned, laid waste to, destroyed. Just as reformers trashed Our Lady of Walsingham by sexualizing her as "Our Lady of "Falsingham," so, too, by talking trash, Ophelia becomes it herself; speaking detritus, she becomes detritus.

Not consigned to the remote past, women's filth continues to sully the twentieth century.[89] Cognate with dumps, Tom's mistress meets Nick for the first time by "the solemn dumping ground"[90] in *The Great Gatsby*, tainting her through proximity. Unlike Mrs. Wilson, whose body can be illicitly poked and penetrated, Jordan is admired by Nick for her "clean, hard, limited person."[91] The penetrability of women's bodies renders sexually loose women subject to condemnation.

In her book of short stories, *Trash*, Dorothy Allison reclaims the word "trash," "to confront the term and to claim it honorific... What it comes down to is that I use 'trash' to raise the issue of who the term glorifies as well as who it disdains."[92] Though Allison overtly embraces the term "trash," her stories demonstrate pitiful and horrifying examples of women figured and treated as sexual trash. Women, particularly those with nonnormative (hetero)sexuality, are singled out for degradation. In "River of Names" all women are "wasted": "They died and were not missed... Somehow it was always made to seem they killed themselves: car wrecks, shotguns, dusty ropes, screaming, falling out of windows, things inside them."[93] The biggest sign of waste is not to be even worthy of narrative. In a perverse extension of Douglas's dictum ("dirt is matter out of place"), there are objects out of place in bodies in cases of rape: "I heard all the words, big words, little words, words too terrible to understand. DEAD BY AN ACT OF MAN. With the prick still in them, the broom handle, the tree branch, the grease gun... objects, things not to be believed... whiskey bottles, can openers, grass shears, glass, metal, vegetables... not to be believed, not to be believed." In the aftermath of such physical and social wastage, "survivors do hate themselves."[94]

Women, incapable of production, only reproduce. Grandmother Shirley in "Meanest Woman Ever Left Tennessee" has a lot in common with Pope Innocent III. "She called the children devils and worms and trash, and swore that, like worms, their natural substance was dirt and weeds."[95] She hates sex: "I never wanted no man to touch me. I sure never wanted *you* to touch me. You put death and dirt in me every time. Death, you hear me? All I've got out of you is death and mud and worms."[96] She represses the outskirts of the body: "Her five sons and three daughters dreamed often of their mother, dreamed she came in to wash their faces with lye, to cut off the places where their ears stuck out, to tie down their wagging tongues, and plane down their purplish genitals."[97] Calling her own daughter a "whore"[98] for answering a rhetorical comment, Grandmother Shirley identifies her own offspring as "trash."[99] In Allison's "Mama," the mother has a hysterectomy and mastectomy, her toes become calcified, and her teeth get pulled: "'With what they've taken off me, off Granny, and your Aunt Grace—shit, you could almost make another person.' A woman, a garbage creation, an assembly of parts... every part of us that can be taken has been."[100] Allison's women—fragmented, brutalized, oppressed, and oppressing—are indicted by their sex and for their sex.

Sylvia Plath's "Lady Lazarus" substitutes one "wasted" group (Jews in Nazi Germany) with another "wasted" group (women in patriarchy). Her self is reduced *to* the body, though she does not want to be identified *by* her body. In a sense, Plath accepts the dualist mind/body schism whereby the mind is placed above the body. Her suicide attempts to "trash" herself are described as garbage. "What a trash / To annihilate each decade." She can't escape her worthless body, described as a sideshow or commodity to be enjoyed. "There is a charge / For the eyeing of my scars, there is a charge / For the hearing of my heart— / It really goes. / And there is a charge, a very large charge / For a word or a touch / Or a bit of blood / Or a piece of my hair or my clothes." Hair and clothes are no longer like miraculous secondary relics, but instead are disgusting waste as though put on display at a concentration camp museum. What had been valuable and venerated as sacred transforms into mere waste. Women can only be dehumanized in the patriarchy, in which she vengefully promises to eat men "like air."

The appropriation and desire for ownership of women by men is marked by the "filthing," literally and metaphorically, of women. Nobel Laureate Elfriede Jelinek's waste-replete play *Raststätte oder Sie machens alle* [in English *Services or They All Do It*] parodies Mozart's comic opera, *Così Fan Tutte*. The housewives Claudia and Isolde visit a motorway service station in hopes of cavorting sexually with two animals in the toilet. Isolde's husband, Kurt, is confident of her fidelity: "The cover sheet of my bank account guards her against trespassers."[101] He and the other husband Herbert agree that they have "marked out [a track] for [their wives] on earth and in the universe."[102] This marking, this ownership, of their wives only encourages the women to slip away from their husbands' dominance. Ultimately cavorting with their own husbands disguised as a moose and bear, Claudia concludes mournfully, "We run all this way for an animal and in the end it's always just you."[103] As Michel Serres observes, "Since immemorial times, the male seeks the ownership of [the vulva and vagina] where . . . he deposits a product that is not very different from urine, as least in terms of its origin. By ejaculating sperm, he thinks he is appropriating the place where his desire is acted out."[104] The only way out of this dire plight, argues Serres, is for women to "*re-appropriate* the organs of their own bodies, while the male should finally be content with the eminently modern role of tenant."[105]

Meanwhile, women remain the marked gender, scapegoats for the vulnerability of all human bodies.

CHAPTER 3

The Civilizing Process: Divisive Divisions

At a dinner party, the hostess laid out on the bar counter various tempting treats for people to enjoy for hors d'oeuvres: salted almonds, Spanish and French cheeses, green olives stuffed with garlic, artichoke hearts, roasted red peppers, water crackers with sesame seeds, a sliced French baguette, and mozzarella slathered in oil and herbs. A preadolescent boy eagerly pushed his way to the front of members of the group, who graciously waited their turns. He proceeded to pick up a slice of French bread, dip it into the container filled with the mozzarella and stuff it into his mouth, the oil dribbling from his chin into the communal receptacle from which all would procure their edibles. "He's planning to open an etiquette school," his mother quipped, blushingly.

The shame! The horror!

But it could have been worse. At least it didn't involve that most child-identified of missteps: the lack of bowel control. Roald Dahl's Big Friendly Giant cannot understand why, "If everyone is making whizzpoppers," his word for farts, we cannot talk about it.[1] Becoming civilized necessitates negating nature.

While the Bible commences with the shame of being naked, as Georges Bataille points out,

> One doesn't even speak of the horror of *excreta*, which is a uniquely human trait . . . The negation [of man's animal nature] is so completely successful on this point that merely to note and affirm that something is there is deemed less than human.[2]

Excrement is not a construction; it is both material reality and necessity. Yet the body capable of creating waste is viewed as immoral, unethical, horrifying, unhealthy, distasteful. Even talking about it can be perceived as unseemly.[3]

Hence our concern for our children's bowel habits. In *Till Eulenspiegel*, the sixteenth-century German compendium of trickster stories, a mother asks Till to help her constipated child. Till obliges by shitting and then placing the child's toilet over it so the mother thinks the child evacuated his bowels.[4] In another tale at a country inn, the children of the innkeeper "täten ihr Gemach hinter die Hustür" [defecated behind the door of house]. Till then defecates in the fireplace in a fecal commentary.[5] Once children are of a certain age, when we assume that rationality can prevail over instinct, we shame them to find excretions disgusting. While children see garbage cleaners as cool, even as superheroes, as one

> enters adult life, this changes . . . We take our children out of the muck, then we do our best to wipe out the traces of that origin. We busy ourselves in terrifying them as soon as they are old enough to take part (little by little) in our disgust for excrement, for everything that emanates from warm and living flesh.[6]

Basic needs and instincts need to be controlled through societal imposition.

Growing up entails rejecting our undeniably material origin. "It is clear that we are sorry we came from life, from meat, from a whole bloody mess."[7] Bataille neatly and graphically illustrates the process by which we deform our children into finding the dirt from which we came horrific, a muck that suggests our affinity to animals, one we reject in disgust.

> We must *artificially* deform [children] in our image and, as our most precious possessions, instill in them the horror of that which is only natural. We tear them away from nature by washing them, then by dressing them. But we will not rest until they share the impulse that made us clean them and clothe them, until they share our horror of the life of the flesh, of life naked, undisguised, a horror without which we would resemble the animals.[8]

We come to internalize culture's disdain toward excrement, alienating ourselves from our own bodies. We not only condition our children to see filth as distasteful, but also enact this scrutiny on ourselves: "Catching a glimpse of the brute physicality of waste signals a kind of failure."[9]

> Nothing is more unwarranted than to believe we are further from animality, further from natural defilements . . . We look down on [primitive societies] from our sanitary installations, and we give ourselves the impression of unassailable purity.[10]

The fervently avowed distinction between humankind and animal develops into divisions among humans, as in, for instance, among classes.

Of the many ways to villainize the vulnerable body, class distinctions prove enduring. "Villain" itself etymologically springs from the Anglo-Norman *villein* for a type of serf who is a peasant or cultivator subject to a lord (OED).[11] This class-based term possesses moral or ethical overtones—the material person morphs into a (negatively) moral entity. A neutral being becomes endowed with undesirable attributes, as though merely belonging to a particular social class imbues one with the perceived favorable or harmful effects of that set. The *villein*, the peasant, becomes marked as villainous, morally dubious, or debased spiritually. Analogously, the body, which we might scientifically regard as a bit of impartial matter, becomes infused with moral properties and implications. Filth still clings to these benighted, animalistic folk, the medieval laborer surrounded by dirt. Marked off as subhuman, the peasant becomes the symbol of stupidity and filth, associated with dirt and manure.[12] While Till generally punishes those who deserve it, the association between peasant and excrement could explain Till's pointlessly cruel shitting on the plums of a peasant.[13] Neidhart von Reuenthal associates peasants with excrement. In the "Veilchenlegende" ["violet legend"], a knight finds the first violet of the season in a springtime contest. He covers the little flower with his cap and runs off to inform the duchess of Bavaria. In his absence, a peasant finds the cap, lifts it up, shits on the flower, and replaces the cap. Upon the court's return, the knight proudly lifts the cap and becomes the victim of derision. He enacts his vengeance by cutting off the left legs of 32 villagers.

The Lower Orders and Ordure

The codification distinguishing civilized from uncivilized bleeds into the metaphor of upper as opposed to lower. The lower classes are literally lower—closer to the ground with its dirt they wallow in. The desire to dissociate ourselves from excrement, filth, and waste as much as possible permeates the long history linking peasants and dung. After all, if peasants are associated with dung and we want to see ourselves as higher socially, we need to cordon ourselves off from direct contact with muck. This division produced between self and excrement facilitates our advancement up the social ladder. One way to enhance our pristine separation from peasants is through being marked as polite. Manners extend themselves to linguistic usage, euphemism being one means to evade the shame of being. In Alan of Lille's *De planctu naturae* [*The Complaint of Nature*] (8.182–190), while preparing to describe ugly vices, Natura says that she will not speak plainly but will "gild shameful deeds with golden ornaments of modest words." This, she says, "will prevent the stench of the dung-heap from spreading too far on the breeze and causing many to vomit in the nausea of indignation."[14] Euphemism increasingly comes into play in the late medieval and early modern periods. In English, excremental words situated highest, those seen as most "polite" or publicly sanctioned, are Latin-based, while the lowest is of Germanic origin, a distinction stemming from the shift from Germanic words in the Anglo-Saxon period to French and Latinate ones in the late medieval and early modern eras.[15] The two words most acceptable in public discourse, *excrement* and *feces*, are of Latin origin, *feces* only starting to be used in the first half of the fourteenth century.[16] The association between privacy and where one defecates collapses in the word *privy*, at a time when increased privatization in the home was evolving.[17] In a recent twist on the idea of euphemisms, the insidious nature of the English class system, inextricably entwined with language, can be seen in the use of the word "toilet" by the middle class, originally used by lower-class strivers in the Victorian period to gain refinement by uttering French terms.[18] Excrement should be left in its proper place—in the outhouse. Sir John Harington, in his book on privies, *A New Discourse of a Stale Subject, Called The Metamorphosis of Ajax*, writes, "To keepe your houses sweete, clense privie vaultes. / To keepe your soules as sweete, mend privie faults."[19] This alteration in attitude to dung reflects the corresponding shift from public, life-giving excrement to repressed, hidden, private shit.[20] One's own cleanliness is dependent upon the sanitary nature of those about one, hidden from view.

Linguistic change reflects this shift in perceptions of the body and privacy. *Culture,* meaning "the action or practice of cultivating the soil; tillage" (OED), is a late Middle English word, used in this sense through the nineteenth century. *Culture,* figuratively meaning "the cultivating or development (of the mind, faculties, manners, etc.); improvement by education and training" (OED), comes about in the early sixteenth century; *cultivation* associates with *civility.* As the word becomes removed from the country to the city, it is linked to the advance of society—again, that separation from rural labourers or peasants.[21]

Alienated and Alienating Food

> We love our food when it is meate, we loathe it when it is excrement. When it goes into us we desire it, when it passeth through us we despise it.
>
> Edward Reynoldes, *Treatise of the Passions and Faculties of the Soule of Man* (1640).[22]

In the Middle Ages, the whole side of an animal was placed on the table or the dead creature was exhibited with its feathers and carved before everyone. The animal origins became disguised through the cutting up of the animal before bringing it to the table slathered in piquant sauces. By concealing the animal origin of what we consume, we attempt to camouflage our own animal origin.[23] Food that was communally ingested in the early medieval period gradually becomes increasingly individual and private.[24] As food becomes an object and commodity, the result of that object and commodity becomes distanced. Alienation from the origin of where food comes from slowly evolved. As the sign of high class embraced the ability to control a fork, one became estranged from eating with the hands. Alienation through technology (the fork) mirrors alienation from the ultimate outcome of what one eats (excrement).[25] Yet any attempt at fine cuisine is a sham. As Lothario de Segni disgustedly points out in *De miseria humane conditionis* [*On the Misery of the Human Condition*], "What was foul to swallow comes out fouler, making vile gases above and below and hideous noises."[26] This is hardly a transcendent view of human endeavor. We ourselves become edible. "He who supped once on delicacies in his hall is now in his sepulcher food for worms."[27]

Till Eulenspiegel will have none of any hoity-toity fakery. He tells it like it is. Till uses excrement not only to humiliate, but also to expose the connection between eating and defecation. When Till works for a wool-weaver, he is told to beat the wool higher, slang for harder. But Till, taking it literally, goes and sits on the roof to beat it. The infuriated boss tells him, "If you want to beat wool, beat it. If you want to fool around, fool around. Climb off the roof and shit in the wool-basket." Which Till promptly does. In desperation the boss says, "Take that excrement and stick it where no one will have it," meaning, presumably, disposing of it properly. But Till carries it into the dining room, saying, "I know perfectly well you don't want it in here. Nobody would want it in here. But I'm doing what you told me."[28] No one would want excrement in the dining room or to be reminded of the link of consumption and excretion, two acts that should be kept discrete. Another time, while in Hannover, Till visits the public baths called the "House of Cleansing." Till shits in a water trough in the bathing room, arguing that one needs to be cleansed inside and out. The bathkeeper threatens him and, as Till dresses in the dining room, he "realized that he had not cleansed himself quite enough in the bathing room. He spotted a folding table, opened it, shitted a pile on it, and folded it up again."[29] Till takes the sign, "House of Cleansing," to its logical end—literally. Taking to heart Jonathan Dollimore's insight that "the civilizing process not only fails to subdue disorder, but actively produces it,"[30] Till punishes the unsympathetic bathkeeper by ruining his meal with the sign of what food mutates into. In a related tale, a cheating innkeeper gives Till farts instead of sheets and a pillow. Till reciprocates by letting a "huge shit" fall into an "elegant folding dinner-table." In this case, the innkeeper recognizes the justice in Till's punishment.[31] Most of the excremental tales involve Till exposing shit in the city rather than the country.[32] There is a revulsion toward the linking of excrement and eating, acts that should be kept separate in a cultivated—and particularly urban—society.

In *Gulliver's Travels*, the eponymous protagonist describes for his Master the excess in eating and consuming in England, where doctors come to the following conclusion: "Their Fundamental is, that all Diseases arise from *Repletion*; from whence they conclude, that a great *Evacuation* of the Body is necessary, either through the natural Passage, or upwards at the Mouth."[33] Many items are suggested for this evacuation, either through the anus or the mouth via vomiting. In a perverse topsy-turvy satire of the natural order, he scathingly considers these "Doctors" as "Artists ingeniously considering that in all Diseases Nature is forced out of her Seat; therefore to replace her in it, the Body must be treated in a Manner directly contrary, by interchanging the Use of each Orifice; forcing Solids and Liquids in at the *Anus*, and making Evacuations at the Mouth."[34] Luis Buñuel's *The Phantom of Liberty* (1974) parodies the inevitable connection between food consumption and food excretion.[35] People shamefully eat in solitary little cubicles, but lounge around a communal table publically excreting. Given the consumerism "foodie-ism" exemplifies, Andrew Martin has argued for the inevitability of a time coming when "someone produces a cookbook with follow-through: not only will there be beautifully lit photographs of the food itself, but equally artistic ones of the results."[36] No matter how elegant we may think we are, underneath our sophisticated veneer loiters filth.

The inevitable link between eating and defecation lies at the heart of the Danes's terror of Grendel's attacks. Civilization, as seen in the hall Heorot in *Beowulf*, stands in opposition to noncultivated regions. The association of Grendel with Cain suggests that he exists outside the law. Grendel proves this alien status most graphically through reabsorbing men by literally devouring them, thereby transgressing human codes of proper conduct and the cannibalism taboo. Eating is a communal activity, one that Grendel parodies by cannibalistically eating society itself. Grendel, signifying whatever is loathsome and filthy, marks and stains his moor retreat ("mearcað mōrhopu" 450). After the fight with Beowulf, he crawls to his abode, trailing bodily fluids ("feorhlāstas bær," 846b) as a sign of his filth and defeat. The sea wells with blood—"Ðǣr wæs on blōde brim weallende" (847)—and hot gore—"hāton heolfre" (849). The killing of Grendel purges and cleanses ("gefǣlsod" 825; see also 432, 1176, and 2352) the hall, the epitome of culture. Yet this hallowed space is all too frequently stained with the blood of its own warriors. Civilization is vulnerable to filth. Heorot has a body—its door is a mouth, one that Grendel enters. By showing how easily he—the alien, the other, the waste—could be incorporated into their body/society, he must die. The creation song—a song of origins—initiates Grendel's anger because the Judeo-Christian creation myth establishes man's dominance over animals. Grendel challenges this lesson by eating men—only to eventually defecate their bodies.[37]

In Norbert Elias's view in *The Civilizing Process*, shame increases as social relational density increases; that is, as classes become more interdependent, as in the urban arena, practices of self-control become more developed and regulated.[38] A class system dependent on a code of conduct enforces submission or exclusion. Socially nurtured restrictions arouse feelings of regret and shame.[39] Shame leads to taboo. Sixteenth-century books of manners and gestures show how hierarchies are established, maintained, and often upended by etiquette. The upper secular classes or bourgeoisie in the early modern period learned their manners from the court as a means to be less distinguishable from their betters. One example of the increased privatization of filth production can be seen in the Brunswick Court Regulations of 1589: "Let no one, whoever he may be, before, at, or after meals, early or late, foul the staircases, corridors or closets with urine or other filth, but go to suitable, prescribed places for such relief."[40]

Literature from various time periods exposes how class hierarchy exists, but is continually unstable. The upper classes can act just as badly as the lower. Appearances mean nothing; no one is impervious from trash. The rich continue to make messes: "They were careless people, Tom and Daisy—they smashed up things and creatures and then retreated back into their money or their vast carelessness, or whatever it was that kept them together, and let other people clean up the mess they had made."[41] The people at Gatsby's party "conducted themselves according to the rules of behavior associated with an amusement park;"[42] the fancy dress of partygoers cannot disguise the inherent low demeanor of the attendees. Gatsby aspires to a class whose worth is as fictive as his own name.

The loosening of classes, allowing for social mobility, places heavy burden on outward behavior, and increased pressure on behavior. People police each other's and their own behavior, modifying their own actions in the scramble to improve their lot.[43] As Elias argues, the poor try to become like the rich who scorn them. The names of many people—those from East and West Egg showing up at Gatsby's parties—end up being substitutable, with nothing to distinguish them. Their names pile up like garbage, so that they become meaningless, without any joyful Rabelaisian superfluity. One man keeps coming with four different girls, "but they were so identical with one another that it inevitably seemed they had been there before. I have forgotten their names."[44] Nick, ever Jay's defender, proclaims, "They're a rotten crowd... You're worth the whole damn bunch put together."[45]

Civilized behavior is especially cultivated by those who are not yet or quite in the highest class, who strive to imitate their betters, who, in reality, are not better at all. Nick is admittedly not of the same class as Daisy or Tom, but knows a corked wine when he drinks one.[46] Gatsby seems to know he needs to work on his elevation in society, picking "his words with care." The nouveau riche status forces him to cultivate gentility attentively.[47] His book from 1906 lists his intentions for self-improvement, including, "Practice elocution, poise and how to attain it . . . Read one improving book or magazine per week." Gatsby, whose "parents were shiftless and unsuccessful farm people,"[48] whitewashes his past to seem worthy of the tonier elements of Long Island society. Gatsby vents his shame at his lowly origins by insulting his father, who informs Nick, "He told me I et like a hog once, and I beat him for it."[49] Jay, ashamed of his humble origins on first meeting Daisy, "let her believe that he was a person from much the same stratum as herself."[50] As Wolfsheim admits, "I raised him up out of nothing, right out of the gutter."[51] Tom excoriates how Gatsby got his money—with Wolfsheim—bootlegger, drugstore owner, and swindler.[52] "I suppose the latest thing is to sit back and let Mr. Nobody from Nowhere make love to your wife . . . I'll be damned if I see how you got within a mile of her unless you brought the groceries to the back door."[53]

The need to carry sway over others comes home in Dorothy Allison's frighteningly horrific "Meanest Woman Ever Left Tennessee." Shirley, a mill worker, considers herself quality, unlike her miner husband. "'A machinist is a higher class of man,' Shirley told the children."[54] Creating industry is better than working in nature. Table manners and even presentation serve to signify class, not trash. "At the dinner table, Mattie watched how her mama spooned rice out of the bowl, all the while talking about how only trash served food out of a cooking pot. 'Quality people use serving dishes.' Shirley slapped Bo's hand. 'Quality people don't come to the table with grease under their nails . . . If you'd really washed, you would be clean . . . Nobody in my family ever came to the table with dirt under their nails.'"[55] When Mattie brings up the labor union, her mother is furious. She buys into the capitalistic dream where the worker should be oppressed. She is low, wanting and needing someone even lower than herself to oppress. By participating in oppressing others, Shirley resists becoming the lowest form of trash herself. Her cruel disdain comes home to roost. When she dies, no one wanted to wash her body, thus turning her into the very filth she had accused her family of embodying.

Read in the context of Norbert Elias's discussion of the civilizing process, we can see how the European Yahoos of Jonathan Swift's *Gulliver's Travels* try to cover their animal selves with short claws [trimmed nails] and the use of perfume. Swift was highly skeptical of ideas of transcendence; excrement becomes the surest way to deflate and mock such notions. The civilizing process can never disguise innate animal nature. In Gulliver's first view of a Yahoo, he perceives a hirsute creature, whose hair grows prodigiously on an anus surrounded by bare tailless buttocks.[56] His initial reaction is one of "antipathy." Hitting one, others chase him and "began to discharge their Excrements on my head," whereby he is almost "stifled with the Filth, which fell about me on every Side."[57] While the Houyhnhnms are "orderly and rational,"[58] "elegant,"[59] the epitome of what Elias would call *civilitas*, the Yahoos are the epitome of what is "detestable."[60] Indeed, Gulliver wonders at "their strong Disposition to Nastiness and Dirt; whereas there appears to be a natural Love of Cleanliness in all other Animals."[61] Imagine Gulliver's horror when he discovers "in this abominable Animal, a perfect human Figure."[62]

Despite his Master's amazement at Gulliver's "Civility and Cleanliness,"[63] "it was plain I must be a perfect Yahoo."[64] The noble horse sets up a binary between clean and dirty *Yahoos*, a division that Gulliver himself picks up on. By explaining mankind to Master, he becomes misanthropic. Because of his very affinity, Gulliver's repulsion exceeds that of his Master. "The more I came near them, the more hateful they grew."[65] Excrement is the ideal signifier for all that is disgust-inducing in culture: luxurious excess and bodily filth work on moral and literal levels. Gulliver's self-loathing accompanies self-introspection: "I expressed my uneasiness at his giving me so often the Appellation of *Yahoo*, an odious Animal, for which I had so utter an Hatred and Contempt. I begged he would forbear applying that Word to me, and take the same Order in his Family."[66] Gulliver diligently learns to speak their language.

The human has to become animal to become most civilized.

CHAPTER 4

Memory and Narrative: Ruins, Nostalgia, and Ghosts

I have kept a journal since the age of 15. Every night I write in it the activities of the day. "It's cheaper than a therapist," I joke. Occasionally, my husband and I will dispute about an event from the past, perhaps about an argument that had transpired over some piddling issue. When we try to reconstruct this past quibble, I proudly declare I can *prove* my point by whipping out my diary dating from the period the dispute took place. Then my spouse rolls his eyes, suggesting that my scribbles are no more reliable than our memories. After all, I only wrote *my* point of view of the quarrel in the first place; I never included his take on the row. My authoritative writing on the past event is, in his view, hardly a trustworthy source since I was subjective in my initial transcription. Our mutual history, then, is doomed to incongruity.

My diary is a solitary fragment from which any attempt to reconstruct what "actually" took place is inevitably flawed.

Finding and Inventing Trash

Garbage does not lie.

William Rathje and Cullen Murphy[1]

In German, the word for *to find* is *finden*. Building on this root, *erfinden* signifies *to invent*. Archeological labor uncovers detritus from the past. *Finds*, the results of human innovation and invention, in turn aid scholars to *invent* theories of history. History is nothing more than the examination of the garbage of human presence. Our earliest ancestors left little garbage, since most organic material has since become recycled,[2] though artifacts like Acheulian hand axes from 1.4 million years ago have been retrieved as detritus from the past. In archeology, *curated* as opposed to *expedient* objects are kept for a later use. Archeologists find not only big triumphant remains like pyramids or humble homes precious—but also trash becomes *curated.* As archeologist Charles Britt Bousman has said, "We're basically saving the trash of other societies. We're the trash keepers of other societies."[3] Moreover, "trash has a history."[4] According to Michael Brian Schiffer, artifacts and residues exist in both their "systemic" (original use and position) and "archeological" (how items are found at their discovery and uncovering) contexts. Schiffer has written at great length on "formation processes," how artifacts become part of the archeological record: through discard, loss, caching, and ritual internment. When any of these artifacts is subsequently reused, it enters the reclamation process that can multiply the complications of the archeologist's job.[5] Entropy, "time's arrow," also plays a role in affecting how the artifacts will appear to the archeologist in the field.[6] Other factors affecting formation processes include the "sampling bias conception" (organic decay, for example) and ecofacts (the accumulation and effect of natural processes over time, including dirt and sediments); these might skew one's interpretation of artifacts in the field. As applied to literary interpretation, while the original author of a pre-Freudian work may have included a cigar, the post-Freudian reader inevitably senses a phallic overtone as a result of its "formation process" and not part of the de facto "refuse."

De facto, primary, and secondary refuse, concepts Schiffer invokes, are of use to our discussion of waste. *De facto refuse* is that which we find when a culture was abandoned, as in the example of Pompeii, and "consists of artifacts from the systemic inventory, often still usable, that are left behind on occupation surfaces when people abandon activity areas, structures, and settlements."[7] *Primary refuse* would be that which was already considered trash at the time of disposal, and deposited close to the site of its creation or function. In other words, it is contextual refuse or artifacts "discarded at their location of use."[8] *Secondary detritus* was likewise considered waste or trash at the original time of deposit, and consists of "artifacts *discarded* away from their use location,"[9] generally a landfill or dump expressly created for this purpose. Thus, various types of primary refuse might be in the secondary refuse site. Additionally, as Schiffer points out, "The ubiquitous dirt we labor so hard to remove is itself an artifact that has much information to disclose."[10] In fact, he almost calls one of his articles "archaeology as sedimentology," whereby we should "view deposits themselves as peculiar artifacts, the characteristics of which must be studied in their own right."[11]

The contents of archeological sites teach us how to approach waste in historical texts. While the archeologist attempts to find material remains and then interpret them, historians construct narratives that subsequent historians troll for refuse. Deposits—intended and accidental—teach us about the construction of the past. After all, "If you want to know what is really going on in a community, look at its garbage."[12] Cultural and non-cultural formation processes can alter a site. We might see this applicable to a manuscript, which may have been repurposed after its initial purpose as poetry hoard. The *Exeter* manuscript, which contains such jewels as "The Wanderer" and "The Seafarer," was notoriously used as a "hot-plate, beer-mat" and cutting board[13]—poetry put to practical use.

History as Garbage

> To *know* is to *choose*... The factory of knowledge is incomplete without waste disposal sites... Knowledge is inconceivable without ignorance, memory without forgetting.
>
> Zygmunt Bauman[14]

We could say that history not only *contains* trash, but also history *is* trash. History can be read not as collection of fragments, but "as an unclassifiable scrap-book"[15] that constructs the past "as a ruin."[16] Walter Benjamin criticizes the fetish of traditional historiography for a coherent unified narrative, one that necessitates the denial of the "'refuse' and 'detritus' of history."[17] In the search for a narrative, we inevitably create waste. As Benjamin in "Theses on the Philosophy of History" writes, "For every image of the past that is not recognized by the present as one of its own concerns threatens to disappear irretrievably."[18] To structure order, memory, story, and history, we need to designate what is valuable and that which should be forgotten. As John Scanlan writes, the "imaginary graveyard of progress... buries the past as if it was simply useless rubbish."[19] An act of codification, history values what to remember and memorialize and condemns what is to be junked. Orderly and tidy linear progression crumbles when jettisoned scraps return to prowl through narrative sequence. Repudiated, crumbs from the ruined and spectral past insist on skulking around the edges of our framed remembrance.

We are always looking into the past. When we look at the sun, it's actually eight minutes away. The stars we see may already be dead. We can only perceive light echoes. The poignant impossibility of trying to recover the past lies at the heart of the plight of history, making us look back to a longed-for state we can never be in again. In perhaps the grimmest example of utter decay, Beckett's *Endgame* shows how for Hamm and Clov "history is cancelled out."[20] Nagg and Nell try to kiss, but they are trapped in their ashbins and only have stumps so can't move. Nell: "Why this farce, day after day?"[21] Only the past is tangibly happy. Nell "(*elegaic*): Ah, yesterday! (*They turn painfully towards each other*)."[22] The present condition remains, we in our ashbins, disabled physically and spiritually, with nothing to look forward to but the end. Hamm wishes he could find oblivion: "If I could sleep I might make love. I'd go into the woods. My eyes would see... the sky, the earth. I'd run, run, they wouldn't catch me."[23] But this can only be in the imagination of the self etherized upon the table, as T. S. Eliot's Prufrock might say. Nagg and Nell are not utterly bereft of some hope. Although they are wretched, their partial bodies in ashcans foretelling their deaths, cognizant how they are being trashed and soon will be nothing but trash—despite it all they have evocative memories. Rowing on Lake Como, Nell still sees the bottom of the lake: "So white. So clean."[24] Unlike the increase in misery for the Old English Wanderer, whose past joys intensify his present misery, this reminder of the happy erotic promise of youth makes the present misery almost bearable. The present must be tolerable, despite everything.

What happens when you cannot forget the past? In Borges's "Funes, His Memory," the protagonist tells the narrator, "*I, myself, alone, have more memories than all men have had since the world began... My memory, sir, is like a garbage heap.*" After his fall from a horse, Funes cannot forget. His memory, unable to shed a single detail, was complete and "perfect."[25] "To think is to ignore (or forget) differences, to generalize, to abstract. In the teeming world of Ireneo Funes there was nothing but particulars—and they were virtually *immediate* particulars."[26] Each moment has become unique and unlike another so that no (conventional) codification can express each instant. "The truth was, Funes remembered not only every leaf or every tree in every patch of forest, but every time he had perceived or imagined that leaf... He saw that by the time he died he would still not have finished classifying all the memories of his childhood."[27] The narrator realizes his own culpability in adding to the dung heap of Funes's non-composting memory. "I was struck by the thought that every word I spoke, every expression of my face or motion of my hand would endure in his implacable memory; I was rendered clumsy by the fear of making pointless gestures."[28] Humans have to codify and forget—or else go mad. "(I repeat, the least important of his recollections was more minutely precise and more lively than our perception of a physical pleasure or a physical torment)."[29] As we see in the dire plight of Borges's Funes, memory as a topless garbage heap can only drive us beyond the brink. Funes embodies the psychotic results of what happens when the human compulsion to disremember no longer functions.

The process of history ritualizes garbage making. Two key literary texts for the Anglophone canon exemplify how "Western culture... can also be seen as a history of garbage."[30] Familiar to both the *Beowulf* poet and Shakespeare, the land constituting present-day Denmark is the site of origin for their Germanic ancestors—the Jutes and Angles—distant enough to allow sufficient freedom to impose meanings of origins. The time and space of the Angles and Jutes, distant from the poet's location, can be manipulated and exploited to signify anything, a fiction of national identity. Both works are rooted in a Denmark that works as a cipher and displacement for England itself. While the manuscript of *Beowulf* dates from about the year 1000, the orally formulated poem itself certainly emerges in the centuries when Viking incursions and settlements were endemic. *Hamlet* was written after the tumultuous sixteenth century when internal foes threatened national security. The works echo one another in uncanny ways. Both works grapple with Cain from whom the monsters in *Beowulf* are descended and whose crime Claudius and Hamlet invoke (III.iii.37–38; V.i.73). Hamlet could "drink hot blood" (III.ii.380), not unlike Grendel (742), whose existence is likewise entwined closely with his mother. Ophelia is torn between Hamlet's identities as both her lover and the man who killed her father, not unlike the grieving Hildeburh, tragically caught between her husband's family and her blood kin—son, uncle, and husband killed by violent warfare.

Beowulf, famously structured according to the battles with the three monsters (Grendel and Grendel's mother in the land of the Scyldings and the dragon in Beowulf's native land of the Geats), at its core concerns the brutality that accompanies political turmoil. Violence likewise facilitates the disruption of the reign of old Hamlet: the brother Claudius has poisoned the rightful king before the play even opens. Within each work, figures are discarded by the political victors: Grendel and his mother are aggressively defeated by the mercenary Beowulf; Claudius sends his nephew and rival Hamlet to be executed. When this plan fails, he schemes a plot that concludes with virtually universal carnage. The ultimate conquerors in each work come from the north: the Swedes in *Beowulf*, the Norwegians in *Hamlet*. As the site of origins grounding our Anglophone literary tradition, Denmark, the locus of the action, is, as it were, a space of waste and destruction, given presence in the "rubble literature" of *Beowulf* and *Hamlet*.

The process of history creates waste by disposing of inconvenient moments from the past. Both *Beowulf* and *Hamlet* emerge from periods concerned with the establishment of a new religious order, Christianity after paganism and Protestantism after Catholicism. Yet traces from the original period exist, littering society and culture. Waste stalks *Beowulf*—in the many deaths of living beings and in the decay and destruction of culture and civilization. The vexed question of just how Christian *Beowulf* is continues to plague readers of the poem. The pagan elements are amply evident in references to sacrifice made at heathen temples in an attempt to stave off Grendel's violent attacks and in the thick tapestry of Germanic mythology decorating the poem. Hrothgar's eloquent speech before Beowulf's ultimate departure invoking God suggests a diachronic positive progression—*from* paganism *to* monotheism—one the original text recipients were meant to approve of. The culture's origin, the poem seems to suggest, is grounded in paganism, an impotent and corrupt system, one that is to be rejected, so that it can be triumphant in monotheism.[31]

Hamlet was written in the wake of the enormous religious convulsions of the sixteenth century. The Reformation left "debris" and "wreckage" from English Catholicism in its wake.[32] Manuscripts were reduced to the level of "waste paper—or waste parchment,"[33] being used for everything from scarecrows, fire-lighters, mending material, wrapping paper, and even toilet paper.[34] Yet the "symbolic residue" of medieval Catholicism, its vestiges impossible to utterly trash, was still present in Protestant England.[35] In the linear or teleological view of history, the past is perceived as waste since it is not seen in the present. Yet, as Raymond Williams[36] notes about the cultural process, a residual element formed in the past still has an effective role in the present. The residual can manifest itself by being incorporated into the dominant culture or can stand in an "alternative" or "oppositional" relationship to the dominant culture.[37] The residual represents "areas of human experience, aspiration, and achievement which the dominant culture neglects, undervalues, opposes, represses, or even cannot recognize."[38] Even at the time, Reformers recognized the wastage of their Catholic heritage. "And therefore let every godly man ceasse with me from henceforth to marvail, why Canterbury, Walsingham, and sundry suche like, are nowe in these our dayes become in manner waste, since God in times paste was in them blasphemed most."[39] James Pilkington, bishop of Durham, criticized those against the cleansing of churches including those who reputedly lewdly mutter sayings such as "What shall I do at Church? I may not have my beads; the church is like a waste barn; there is no images nor saints to worship and make curtsey to." By implication, medieval churches' clutter *is* waste—these wrongheaded individuals read incorrectly.[40] John Leland's notes on the Worcestershire record, "There was a place of nunnes at the very northe side of the cemiteri of St. Oswald. It was caulyd Whitestan, now suppressyd, the churche clene rasyd downe, and a ferme place made of the resydewe of the buildings."[41] This residue, a trace or mark of the past, persists, incapable of being utterly erased.[42]

This debris and wreckage of the Reformation functions on both a literal level—the ruins of abbeys, burnt manuscripts, and disfigured and destroyed artifacts—and a figurative level—what Stephen Greenblatt in his discussion of Purgatory traces in *Hamlet* and Alison Chapman in her discussion of Catholic debris observes concerning Ophelia's songs. Iconoclastic destruction is the deliberate act of forgetting, one that is "active, purposeful, and collectively performed."[43] Margaret Aston contends that an aching nostalgia for what was destroyed and plundered in the Reformation by iconoclasts was present in sixteenth- and seventeenth-century English literature. Possible traces of Reformation detritus appear in Shakespeare's Sonnet 73, with its reference to "bare ruin'd choirs" and in *Titus Andronicus*, when the Second Goth describes how he strays upon a "ruinous monastery," a "wasted building" (V.i.2153, 2155).[44]

Shakespeare doubles words, using two opposite meanings simultaneously. This artificial conjunction creates meaning out of place, an analogue for matter out of place as in Douglas's paradigm for dirt. Claudius's doubling language reflects how he is two-faced, saying, "Therefore our sometime sister, now our Queen . . . With mirth in funeral and with dirge in marriage" (I.ii.8, 12), and later, "But now, my cousin Hamlet, and my son—" (I.ii.64). The multiplication of roles and shifting of family position suggests familial and political chaos. Doubled words multiply with Hamlet's disdainful reference to his "uncle-father and aunt-mother" (II.ii.313). The doubling, an unnatural border crossing of categories, besmirches the "incestuous sheets" (I.ii.157) of the court. Clear distinctions are muddied; the viscous yoking troubles the convulsed court. The "both/and" of doubled words reminds us that the past is never wholly eliminated; history is not teleological and progressive, but simultaneous, hybrid, sedimented, multiple, and heterogeneous.[45]

In other words, waste.

Ghostly Hauntings

Remember me.

The Ghost [Hamlet the Elder], *Hamlet* (I.v.91)

When historical breaks take place, such as the revolutions in Eastern Europe of 1989 and 1990, the past becomes purged, along with heads of state, both literally and figuratively. The idea of progress demands, as Scanlan argues, that "the past must also be seen as garbage too."[46] The massive stone statues commemorating Marx and Engels or Saddam Hussein were toppled down and trashed in an eagerness to erase recent history. "Every act of differentiation—every 'clean break' with the past, creates garbage; results in leftovers."[47] The leftovers in the wake of the regime change in *Hamlet* are literal: "The funeral baked meats / Did coldly furnish forth the marriage tables" (I.ii.179–180). Dispossessed, Hamlet seeks to understand this rewriting of history. The trashing of the past is an integral strategy of progress; yet the "debris of a life [somehow] continues as the ghost of the present."[48] Indeed, the debris from the past *literally* [re]appears as a ghost in the present. "Ghosts . . . are figures of the unresolved, often repressed violence of the past appearing in the present."[49] The Ghost describes his "murder most foul" (I.v.27). Horatio tells Hamlet how Francisco and Barnardo encountered a ghost "in the dead waste and middle of the night" (I.ii.197), who must be "purged" (I.v.13). So must Claudius and his sins—all foul corruption must be sluiced from individual bodies, souls, and the body politic. The new regime erases Hamlet *père* most effectively by marrying the new king to the queen of the former leader. This link both legitimates a new illegitimate leader and acts as a trace of the past. Hamlet himself is detritus—the son of the dead king—who threatens the present order and must be disposed of.

Prince Hamlet, the uncanny trace from the past, disturbs the economy of Claudius's body politic. Claudius's court can only exist if it rejects what has come before it—the garbage and waste of the previous regime. Claudius outsources this dumping of Hamlet to the English king via his designated sanitation workers, Rosencrantz and Guildenstern. They, in turn, evoke the spectre of Wittenberg, and hence act as traces of the origins of the Reformation. Claudius's plan to cleanse Denmark of the Hamlet regime maddens young Hamlet and sickens the realm, rendering it weak and open to conquest. As in the attempt of the Danes to "fǣlsian" (432) or cleanse Heorot, Claudius ultimately succeeds only in destroying everyone, including himself.

The fear of the past embodied by corpses bleeds over into the terror induced by ghosts in *Eyrbyggja Saga*. Evidence of both pagan origins and lived bodies haunt the living in a visceral way. "One night Thorir Wood-Leg went out to the privy to ease himself, and when he was on his way back to the house, he saw the [once bewitched, now deceased] shepherd standing in front of the door."[50] They tussle and Thorir ultimately wastes away and dies. At one point, a seal's head, presumably the manifestation of the dead Thorgunna whose final wishes had not been carried out, rises up through the floor. Drowned men come to their own funeral feast. "At that time a good many heathen beliefs still prevailed, though people were baptized and supposed to be Christians."[51] The residue of pagan beliefs is evident in these ghostly hauntings. Only the law can control these manifestations. The dead are indicted for "trespassing on the home and robbing people of life and health. All the dead ones at the fire were summonsed the same way. Then the door-court was held and charges made, the proper procedure of ordinary law courts being observed throughout."[52] All sentenced ghosts depart; the living return inside, as "the priest carried holy water and sacred relics to every corner of the house."[53] The legal code succeeds in controlling even the dead. The pagan past is put to rest by civilized law, at least temporarily, until a monster calf containing the evil spirit of Thorolf, grows into an insane bull.

The past cannot remain not wholly past.

Unless even ghosts are forbidden. In *The Journey*, a law is created that makes Jews no longer exist legally.[54] People don't see Jews, who have become "ghosts."[55] Memory and waste are integrally tied in to one another. The nadir is the trash situation in Ruhenthal, the Valley of Peace, where the Jews are quartered, "the town of visible ghosts, who also have a ton of rubbish. But since everything there is in shambles, it's difficult to distinguish normal rubbish from abnormal."[56] The trash is thrown into "an out-of-service funeral wagon,"[57] foreshadowing the imminent deaths of the other trash and rubbish, the "ghosts" who pull the wagon along. Leopold Lustig, once the doctor, becomes the rubbish collector.[58] Yes, "everything in Ruhenthal is filthy,"[59] even the doomed ghosts.

Leopold's son, Paul, imagines a time of no time, when only obituaries exist in the paper. There is no memory of the past if no time is left, "nothing remaining but rubble broken to bits and pulverized."[60] Rubble cannot exist in linear time; anyone associated with rubble forgets and is forgotten. Paul sees the people of Leitenberg as dead masks: "All towns and cities are corpses that will soon be reduced to the rubble that they will bury."[61] At the end of the war, Paul, now a displaced person, just wants to be inside a nice apartment. As he tells a lady who allows him into her flat, "We were no longer allowed to exist, and now my dearest ones are dead!"[62] Worst of all, the dead cannot even be mourned since they have been forgotten. Not even worthy of being ghosts, utterly and completely annihilated, they leave no trace behind.

Scanlan argues for the uncanny emanation we sense from trash and garbage. Waste is a form of the uncanny,[63] threatening "the very integrity of our body politic."[64] A weird disturbance animates the sense of doom pervading *The Great Gatsby*. Moved by Gatsby's attempt to recreate the past, Nick is "reminded of something—an elusive rhythm, a fragment of lost words, that I had heard somewhere a long time ago." But whatever it was "made no sound, and what I had almost remembered was uncommunicable forever."[65] "Uncanny" remains "somehow [continue] as the ghost of the present."[66] Gatsby tells Nick about how he lived in Europe collecting jewels, "trying to forget something very sad that had happened to me long ago."[67] Objects substitute for meaning in his meaningless life. Gatsby drifts, "trying to forget the sad thing that happened to me."[68]

Phantoms burden Gatsby, himself perceived of as a "ghostly celebrity."[69] "The most grotesque and fantastic conceits haunted him in his bed at night."[70] Leaving on their ill-fated trip, Jay and Daisy were "gone, without a word, snapped out, made accidental, isolated, like ghosts, even from our pity."[71] After the accident, Nick visits Gatsby: "Once I tumbled with a sort of splash upon the keys of ghostly piano. There was an inexplicable amount of dust everywhere, and the rooms were musty."[72] This decay portends death. Even his wealth, "phantom millions," is intangible.[73] Nick imagines

Gatsby's thoughts: "A new world, material without being real, where poor ghosts, breathing dreams like air, drifted fortuitously about."[74]

The French poet Charles Baudelaire describes "the ragpicker as an archivist, a cataloger" sorting through "everything that the big city has cast off, everything it lost, everything it disdained, everything it broke."[75] The historian, like Baudelaire's poet, is a version of the *Lumpensammler*, *chiffonnier*, or rag picker: a "collector of the 'refuse of history.'"[76] This *Lumpensammler* "salvage[s] the unsalvageable."[77] Irving Wohlfarth categorizes Walter Benjamin as "bricoleur"[78] who argues that the historicist "*re*constructs some bygone era out of a tell-tale need to *forget* the present," while the materialist historian "*constructs* a particular past according to the dictates of the hour."[79] Modern life and modern art creatively reuse the waste tidbits from history.

George, Jay, and Nick are all *bricoleur*. For Nick, after Jay's death, "the East was haunted for me like that, distorted beyond my eyes' power of correction."[80] Jazz age ghosts haunt Nick's narrative that attempts to recoup and tidy up an always elusive past. Visiting Jordan Baker before he leaves the East, "I wanted to leave things in order and not just trust that obliging and indifferent sea to sweep my refuse away."[81] Jordan was said to have cheated once on the golf course by moving a ball from a "bad lie." Could Nick be moving things around too, as the *bricoleur* of a history he claims to have witnessed? Contrasting himself with Jordan, Nick comments, "I am one of the few honest people that I have ever known."[82] Yet her "bad lie" might also be his. How can we know what is true?

George B. Wilson, husband of Tom's mistress, has a car garage with a sign: "Repairs." His job consists of fixing things, the ultimate fix being murder. When Nick first sees George, he is "wiping his hands on a piece of waste"[83]—the waste he remedies, the waste he embodies, the waste he creates. Nick tells Gatsby, "'You can't repeat the past.' 'Can't repeat the past?' he cried incredulously. 'Why of course you can!' He looked around him wildly, as if the past were lurking here in the shadow of his house, just out of reach of his hand. 'I'm going to fix everything just the way it was before.'"[84] Jay fixes things too, not material objects like George, but financial rackets. Jay tells the shocked Nick that Wolfsheim "fixed the World's Series back in 1919."[85] Gatsby, a bricoleur like George Wilson, needs to create order from chaos, the disorder of the smashed, trashed past. George and Jay, the fixer of an impotent, empty system, both die when they cannot fix anything any more.

"So we beat on, boats against the current, borne back ceaselessly into the past."[86]

CHAPTER 5

Failed Source Reduction: Conspicuous Consumption and the Inability to Minimize

In 1990, upon my return after a two-year sojourn from West Berlin, my fiancé drove me to IGA, the local supermarket in Providence, Rhode Island. It was nothing special. Just an ordinary, sanitary store, with common items for normal folks to purchase all their weekly needs: from toilet paper to gallons of milk, yogurt to meat, lettuce to salsa. Trancelike, I walked through the aisles. A hundred brands of cereal gaudily beckoned in garishly colored boxes. Multiple cans of beans—pinto, kidney, garbanzo, black, cannellini—were stacked like steeples to a legume God.

I had lived in the so-called showplace of West Germany, designed to mock the restricted goods available in their socialist twin to the east. Having taught in East Germany, I happily had survived on a lack of selection that sufficed my needs. Confronted by endless choice upon returning to the States, I became immobile. Frozen, I said I had to leave. I could not buy anything, because I had been given too much choice.

Consumption and Excess

> Humanity recognizes the right to acquire, to conserve, and to consume rationally, but it excludes in principle *nonproductive expenditure.*
>
> Georges Bataille, *Visions of Excess*[1]

Objects left in wills in the Middle Ages—things we would discard today (who would take Granny's old mattress or a dung heap?)—would be cherished not only out of nostalgia and desire, but also out of need or financial purpose.[2] After the use value of objects was replaced by exchange value and "people no longer made the objects they lived with," alienation was the response to objects.[3] Now, "mass production and mass distribution literally generated more stuff, and more trash."[4] One result of consumerism is the rapid disposal of items for the always new.

The Brundtland Report from the 1987 World Commission on Environment and Development's *Our Common Future* articulates that Sustainable Development (SD) is that which "meets the needs of the present without compromising the ability of future generations to meet their own needs."[5] But the "interpretation of human needs itself often varies between generations and between cultures."[6] What are *needs*? One person's need is another's excessive indulgence. As Thorstein Veblen has famously argued, the consumption of excellent goods indicates honor, merit, and superiority. It is the very uselessness of luxury items that confers prestige on the owners. As Michel Serres has argued, "money and waste define one as an owner."[7] Taste develops in consumable goods (as a "connoisseur") and massages the "cultivation of the aesthetic faculty."[8] "Conspicuous consumption of valuable goods is a means to reputability to the gentleman of leisure." Leisure and the conspicuous consumption of goods show the "pecuniary strength" of a class member striving to get to the class above it. Waste—of time and effort or of goods—signals wealth.[9] Gulliver assures his master, "that this whole Globe of Earth must be at least three Times gone round, before one of our better Female Yahoos could get her breakfast, or a Cup to put it in."[10] The excessive expenditure of time, travel, and money to procure a cup or its contents lends respectability to her consumption.

Country gossip in a village where everyone knows everyone else mitigates the need for extreme conspicuous consumption, while urban living with its many "transient observers" provokes the showing of one's wealth through goods. "Consumption [exists] as a means to repute." While Veblen regrets the negative connotation of the word "waste," "an article may be useful and wasteful both":

> Throughout the entire evolution of conspicuous expenditure, whether of goods or of services of human life, runs the obvious implication that in order to effectually mend the consumer's good fame it must be an expenditure of superfluities. *In order to be reputable it must be wasteful.* No merit would accrue from the consumption of the bare necessaries of life, except by comparison with the abjectly poor who fall short even of the subsistence minimum [my emphasis].

The Great Gatsby embodies Veblen's views. The song sung by Klipspringer, "Ain't we got fun," sums up conspicuous consumption and its flip side—conspicuous lack: "One thing's sure and nothing's surer / The rich get richer and the poor get—children."[11] Both Myrtle and Gatsby try to appear upper class, which they interpret as the ability to consume conspicuously. Jay, despite his bootlegger past, has the appearance of class, yet his magnificent library contains uncut books, wasted for the mere pretence of cultural capital.

The rich do not remain waste-free. Michael Thompson jokes about how the rich man puts snot in his pocket, while the poor man does not. The solution—pointing to how the rich man possesses a handkerchief, while the poor man cannot—sets "out a relationship between status, the possession of objects, and the ability to discard objects . . . The condition of richness or poorness is determined by the quantity of objects one possesses."[12] Status and wealth lie in the quantity of objects one can *reject*.[13] Yet shedding objects never satiates the owner.[14] The more one tosses out, the more one craves new objects to (ultimately, in turn) throw away. After all, objects are meant to "'serve us,' literally, 'like slaves.'"[15]

Georges Bataille sneers, "It takes a lot of money to be refined,"[16] the objects constituting refinement merely garbage.[17] Sadness pervades the famous scene of discarded shirts in *The Great Gatsby*:

> [Gatsby] took out a pile of shirts and began throwing them, one by one, before us, shirts of sheer linen and thick silk and fine flannel, which lost their folds as they fell and covered the table in many-colored disarray. While we admired he brought more and the soft rich heap mounted higher—shirts with stripes and scrolls and plaids in coral and apple-green and lavender and faint orange, with monograms of Indian blue. Suddenly, with a strained sound, Daisy bent her head into the shirts and began to cry stormily. "They're such beautiful shirts," she sobbed, her voice muffled in the thick folds. "It makes me sad because I've never seen such—such beautiful shirts before."[18]

The ecstasy of the piles of stuff instills a kind of post-chemise *tristesse* in Daisy.[19] The colors of the shirts echo food that decays, including "orange," like the oranges sent away in the garbage at the end of the party weekend.[20] The lack of distinction between objects worth something and those that are valueless is shown in one scene where Daisy drinks: "She groped around in a waste-basket she had with her on the bed and pulled out the string of pearls [worth three hundred and fifty thousand dollars]."[21] The expensive necklace had found a true place in the wastebasket—no more than trash. Pearls themselves are created by a bit of detritus getting inside of a mollusk and then growing. Like the false Pilgrim in *Piers Plowman* and Chaucer's Pardoner selling fake relics, Gatsby has a medal from "little Montenegro" and shows it to Nick. Is it real? Is it fake? Does it matter? All that counts is that it gives Gatsby a—ultimately false—sense of security in his social standing. The very uselessness of luxury items confers an imagined prestige on the owners.

One way to trump a competitor is to have others bear witness to one's consumption; for example, by throwing a party. Daisy's husband, Tom, points out: "I don't give big parties. I suppose you've got to make your house into a pigsty in order to have any friends—in the modern world."[22] The consumer class, epitomized by Jay, is "more than buying a product, in the process, they bought themselves a self."[23] Luxury requires excessive consumption. The objects necessary for Gatsby's parties at the start of the weekend become trash by the end: "And on Monday eight servants, including an extra gardener, toiled all day with mops and scrubbing-brushes and hammers and garden-shears, repairing the ravages of the night before. Every Friday five crates of oranges and lemons arrived from a fruiterer in New York—every Monday these same oranges and lemons left his back door in a pyramid of pulpless halves."[24] The ideal commodity is Gatsby's brand name. But with commodities comes disposal. "[Gatsby] broke off and began to walk up and down a desolate path of fruit rinds and discarded favors and crushed flowers."[25] Despite the addictive spiral of acquiring and disposing goods in a frenzied cycle of procuring and sloughing, people want to have the option to waste goods. "Wasting" and "disposability" make "people feel rich."[26] Even the names of people at the party are piled up like trash, people disposable to each other. And such friends! No one comes to Gatsby's funeral, himself disposed of like trash. "The party was over."[27]

Those who cannot acquire goods inevitably find that they will be perceived of as waste themselves. Those called "white trash" threaten from within, "simply in terms of their failure to rise to a socially acceptable level of behaviour . . . and because of the disorder they threaten to spread." On the margins of society, "social trash" like "'bums', 'gypsies', 'hobos', 'loiterers', 'floppers', 'moochers', and so on)" threaten social stability from without.[28] To sign herself as higher class, aspiring to the culture of East or West Egg to which she can never belong, Myrtle rejects her husband, George. She rejects his frugality because it is *not* wasteful. In referring to George, Myrtle confesses, "I married him because I thought he was a gentleman . . . I thought he knew something about breeding, but he wasn't fit to lick my shoe." His mistake was in borrowing a suit to get married in.[29] Ultimately discarding her husband, she claims she needs to make a list of things to buy—including a dog collar—one that ultimately gives away her affair and indirectly causes her death, as well as those of her husband and Gatsby himself.[30] Henry Gatz, Jay's father, comes and shows Nick "a photograph of the house, cracked in the corners and dirty with many hands . . . He had shown it so often that I think it was more real to him now than the house itself."[31] The photo turns into a fetishized metonym for his son, the image more authentic than the material building. Jay collects newspaper articles about Daisy, fixating on her, the articles substituting for the flesh and blood woman.[32] Daisy herself is a valuable object, increasing in worth for Jay because other men desire her.[33] His vain quest for her takes on religious and sacred overtones, as though in pursuit "of a grail."[34] When

Gatsby shows Daisy his house for the first time, he tells her how he can see her house. "You always have a green light that burns all night at the end of your dock."[35] Nick observes that

> Possibly it had occurred to [Gatsby] that the colossal significance of that light had now vanished forever... Now it was again a green light on a dock. His count of enchanted objects had diminished by one.[36]

Objects ultimately become meaningless. "All objects, begins from nothing (garbage) and eventually return to nothing (garbage again)."[37] Commodities are merely "deferred trash."[38]

Two kinds of human consumption exist, according to Georges Bataille. Some are "necessary" for the continuation of life, such as food, shelter, and clothing. Then there are those "unproductive expenditures: luxury, mourning, war, cults, the construction of sumptuary monuments, games, spectacles, arts, perverse sexual activity (i.e., deflected from genital finality)—all these represent activities which, at least in primitive circumstances, have no end beyond themselves." These are "unproductive" forms Bataille calls "expenditure." Owners, in turn, shed goods in a process designed to enhance their economic vitality and reputation. Bataille compares jewels to excrement; both are "useless."[39] "Ostentatious loss remains universally linked to wealth, as its ultimate function."[40] Wasted goods lend caché to their owners. Waste, itself an excess, proliferates, creating even more waste in an uncontrollable spiralling process. Money, identified metaphorically as excess, Freudian excrement, and invisible signifier, ironically makes those who possess it active subjects, while those without this economic waste are perceived of as passive objects. The rich waste, the poor get wasted.

In Dorothy Allison's *Trash,* the narrator of "Steal Away" tells how, as a poor college student, she steals books, marks them up, and returns them. Other objects she steals to sell or destroy. Her sociology professor makes a move on her, saying, "Your family is very poor, aren't they?"[41] Apparently, the poverty of the narrator's family justifies the privileged and powerful hitting on her. Waste or be considered waste.

Avaricious Agency

Excess lies at the root of all the sins—for example, lust is an extreme perversion of love—and is particularly morally suspect in the sin of avarice. William Langland's *Piers Plowman* sets up a contrast between those who fulfil their necessary economic role and those who exploit others' work in the opening lines of the *Prologue*: "Somme putten hem to the plough, pleiden ful selde, / In settynge and sowynge swonken ful harde, / And wonnen that thise wastours with glotonye destruyeth" ["Some applied themselves to plowing, played very rarely, / Sowing seeds and setting plants worked very hard, / Won what wasters gluttonously consume"] (*Prologue* 20–22).[42] Religious figures are shown to be generally greedy: "I fond there freres, alle the foure ordres, / Prechynge the peple for profit of [the wombe] . . . Manye of thise maistres mowe clothen hem at liking / For hire moneie and hire marchaundise marchen togideres" ["I found friars there—all four of the orders— / Preaching to the people for their own paunches' welfare . . . Many of these Masters may clothe themselves richly, / For their money and their merchandise march hand in hand"] (*Prologue* 58–59; 62–63). Even parish priests and parsons, longing for London, are not immune from greed, "syngen ther for symonie, for silver is swete" ["sing Masses there for simony, for silver is sweet"] (*Prologue* 86). Neither Langland nor God condone such ultimately doomed actions that "Arn doone undevoutliche" ["Are done undevoutly"] (*Prologue* 98). Intermediaries, like mayors between the King and the Commons, shamelessly exploit the poor and powerless:

> For thise are men on this molde that moost harm wercheth
> To the povere peple that parcelmele buggen.
> For thei poisone the peple pryveliche and ofte,
> Thei richen thorugh regratrie and rentes hem biggen
> With that the povere peple sholde putte in hire wombe.
>
> [For these are the ones in the world who work most harm / To the poor people that purchase small portions; / Because they poison the people privily and often / They get rich through their retail-sales and buy rental property / With what the poor people should be putting in their bellies] (III.80–84).

A salutary work for our own time with the 1 percent versus 99 percent, Occupy Wall Street, and housing bubbles, Langland's poem excoriates the vicious appetite of the privileged whose prodigality harms the helpless.

Lady Meed allegorizes the concepts of reward, recompense, and the profit motive. Those with legal and political power—coupled with greed—most intimately cavort with the slippery and seductive Meed: "Ac Symonie and Cyvylle and sisours of courtes / Were moost pryvee with Mede of any men, me thoughte" ["But Simony and Civil and assizers of courts / Were most intimate with Meed of any men, I thought"] (II.63–64). Meed's fellows, including Liar and Guile, are especially friendly with merchants: "Ac marchaunts metten with hym and made hym abyde, / And bishetten hym in hire shoppes to shewen hire ware, / Apparailed hym as a prentice the peple to serve" ["But merchants met with him and made him stay / And shut him up in their shop to show their wares, / Appareled him as an apprentice to wait on purchasers"] (II.213–215). Meed bribes all the justices at Westminster with "Coupes of clene gold and coppes of silver, / Rynges with rubies and richesses manye, / The leeste man of hire meynee a moton of golde" ["Cups of clean gold and copes sewn with silver, / Rings with rubies and riches aplenty, / The lowest lad of the household a Lamb of gold"] (III.22–24). A friar even absolves Meed for his own personal gain,

> Theigh lewed men and lered men hadde leyen by thee bothe.
> And Falshede hadde yfolwed thee alle thise fifty wynter,
> I shal assoille thee myself for a seem of whete,
> And also be thi bedeman, and bere wel thyn er[ende],
> Amonges knyghtes and clerkes, Conscience to torne.
>
> [Though learned men and unlearned had both lain with you, / And though Falsehood had followed you all these fifteen winters, / I shall absolve you myself for a seam of wheat, / And also be your go-between and bear your message well / To ecclesiasts and knights at court, to chop Conscience down] (III.38–42).

The vilified Meed defends herself for giving hope to soldiers in despair, paying for mercenaries to defend the nation, and supporting apprentices with their masters. "Marchaundise and mede mote nede go togideres: / No wight, es I wene, withouten Mede may libbe!" ["Meed and merchandise must go together; / No life, as I believe, can last without meed"] (III.226–227).[43] In her opinion, money greases the wheels of a successful political and economic state.

Langland compactly suggests the corruption of a money-based economy through the adroit gendering of Meed, epitomizing the fears expressed in misogynistic writings of grasping, highly sexual, and sexualized females. Meed is false: "And now worth this Mede ymaried to a mansed sherewe, / To oon Fals Fikel-tonge, a fendes biyete" ["And now this Meed is being married to a most accursed wretch, / To one False Fickle-Tongue—a fiend begot him" (II.40–41). Holy Church warns Will not to desire Meed, who is put in opposition to the Virgin Mary. "'Now I bikenne thee Crist,' quod she, 'and his clene moder, / And lat no conscience acombre thee for coveitise of Mede'" ["'Now I commend you to Christ,' said she, 'and to Christ's pure mother, / And don't let your conscience be overcome by coveting Meed'"] (II.50–51). Conscience rejects Meed as a bride because of her "wantounnesse" ["wantonness"] (III.125); she is "commune" ["common"] (III.132), corrupting the church "For clergie and coveitise she coupleth togidres" ["Clergy and covetousness she couples together"] (III.165). Langland castigates in the late fourteenth century what has sprouted full-blown in late capitalism: free-floating money corrupts.

Edward Hopper's *Drug Store* (1927; MFA Boston) depicts a pharmacy at a street corner at night. The show window is lit up like the stage in a darkened theater. Festooned with blue and red curtains, hanging vases containing jeweled red and green liquids glimmer like a shrine's reliquary. Over this rich and gaudy stage dressing stands the painted sign: "EX-LAX." The large capital letters of the laxative remedy seem highlighted beyond more generic treatments displayed, such as "Prescriptions" and "Drugs." Barely visible above these signs is painted the name of the store: Silbers Pharmacy. Alchemical preparations would use excrement, the result from laxatives, to concoct potions that would miraculously turn into precious metals such as gold and silver (*Silber* in German). The store stages the link between medicine, excrement, and money, the alchemical transformation of disease that purges into literally filthy lucre.[44] Waste gives life, as in excrement alchemically transforming into gold, itself a kind of waste. As Bataille points out, decay is procreative. Although "we have no greater aversion than the aversion we feel toward those unstable, fetid and lukewarm substances where life ferments ignobly,"[45] life depends on death and decay. Decay makes new life.[46] Waste creates spiritually and materially.

Gold or money is no more than a version of trash or, from Freud's perspective, excrement. As Alan Dundes points out, manure and wealth were long associated with one another in German folklore.[47] Till Eulenspiegel, near death, offers the priest who confesses him a piece of silver. Till fills a jug halfway with excrement and sprinkles some money on top of it. He offers it to the priest, warning him not to be greedy and just to take a bit. The priest promises to control himself, but, inevitably, tries to get a "good handful." His hands get covered in shit and he walks out furiously.[48] While this tale, reminiscent of Chaucer's "Summoner's Tale," aligns feces and silver, it likewise casts the priest's sacred role in society into question by "filthing" his solemn duties already undermined by his greed. By eighteenth-century colonial America, little had changed since Langland's time. Excess, greed, and filth swirl in a financially and morally suspect stank that literally stinks. For Benjamin Franklin in *The Way to Wealth,* monetary precepts are only ignored.

> *Poor Richard says, buy what thou has no need of, and ere long thou shalt sell thy necessaries*... The artificial wants of mankind thus become more numerous than the natural... But what madness must it be to run in debt for these superfluities... *you give to another power over your liberty [when in debt].*

Excess creates debt, therefore you lose liberty to your creditor. "*The borrower is a slave to the lender, and the debtor to the creditor.*" Franklin's narrator ultimately decides to keep his old coat and not buy a new one. Though people listen to the admonitions of Poor Richard and approve of them, they continue to buy! A waste of words, Poor Richard's maxims are commodities themselves to be disposed of.[49] A foundational idea of the American Dream, the aphorism, "God helps those who help themselves," concerns self-responsibility. Just like garbage, proverbs are everywhere and cannot be pinned down, "the all-purpose utterance of proverbial expression is protean, as opposed to the fixity of written meaning."[50] These words are as much garbage as Franklin's own maxims.[51] Rather than his listeners gleaning wisdom, the defied maxims are themselves waste, ignored by listeners.

Greed destroys communities, shrivelling altruism. A grim shadow eclipses the success of the urban farm in Paul Fleischman's young adult novel *Seedfolks*, in which a motley crew of residents works to create a green space in a gritty part of Cleveland. Florence, whose ancestors were freed slaves, describes the first year of the garden, "before there were spigots and hoses, and the toolshed, and new soil. And before the landlords started charging more for apartments that look on the garden."[52] This gentrification nearly scuttles the Edenic community. Well-meaning but hapless individuals cannot help themselves getting caught up in the tentacles of greed. The father of a Haitian boy Virgil is determined to make money from the garden, selling Bibb lettuce to fancy restaurants. His sole motive is avarice, for which he is punished. The seeds he plants in the heat of the summer die.

Wayne Anderson speculates that the origin of the term "bullshit" comes from a George Cruikshank cartoon depicting Napoleon holding a pan underneath John Bull's behind as it defecates coins, this at a time when England was promoting paper currency.[53] Excrement and money are both excess—feces are expelled from the body as useless waste. Money may be symbolically endowed with financial worth, but coins are mere trinkets and bills only flammable signs, without our imagination imbuing them with economic value. In 2009, a spate of vandalism in San Francisco resulted in the arson of over a dozen portable toilets, mainly in affluent areas. These arsonists openly expose and mock the long-established association between gold and feces. "Potty pyromaniacs" send a sobering message to the rich who have these toilets set up while their homes are being renovated.[54] Money and excrement, both forms of excess, indict the wealthy addicted to their dirty rapaciousness.

The flip side of greed and wasteful expenditure is hoarding. Roger, the garage sale fanatic in Miles Wilson's wildly amusing short story, "Everything," "briefly entertained the notion that his interests were political. Such an appraisal displayed him to advantage: he was simply helping to mend the tattered safety net beneath the flying trapeze of runaway capitalism. On examination, however, it turned out that he rarely bought from the truly poor and bargained aggressively with them when he did."[55] While this story wryly pokes fun at modern consumerism, the protagonist has enough self-awareness to call his obsessive hoarding his "cumulative behavior."[56] The protagonist is an obsessive collector. As his wife leaves him she points out, "'Look at it this way... You'll have more room when I'm gone.' Roger had to admit that she was right."[57] Roger is so far gone that it logically makes sense for his wife to leave and his marriage be destroyed just so he can put more stuff in his house. The garage sale fanatic-genius protagonist starts to lose his touch. His lover finally leaves him, cleaning him out of all his stuff, and partially burning the house.

> Roger looked away wistfully at the charred and sodden mattress still smouldering in the driveway.
> "I'm not sure I was cut out for this."
> [His neighbor asks,] "Cut out for what?"
> Roger gestured at the steaming ruins.[58]

What emptiness is Roger trying to fill? A technical writer not writing, he collects stuff to give fulfilment to his life. When Roger's wife leaves him because of his obsession with collecting stuff, Amanda, grieving that he can't have some "normal" aberration like drinking, gambling, or adultery, says, "Nobody talks about what you're doing; there isn't even a name for it."[59] Some stuff is truly useless. Roger never misses his "acquisitions" once they are gone.[60] The very meaningless of these things contains the heart of their meaning. The collector of stuff prides himself on the uselessness as a tactic against capitalist utilitarianism.[61] Yet Calvino points to the constipated hoarder. "Alas the unhappy retentive (or the miser) who, fearing to lose something of his own, is unable to separate himself from anything, hoards his faeces and ends up identifying with his own detritus and losing himself in it."[62]

What is the fine line between artifacts that are dismissed as trash and those worthy of conservatorship? Hoarding plays itself out in tragic ways in *Beowulf.* The most successful adaptive culture is that of the monsters, who reproduce to populate marginal areas with their offspring. Unlike the relatively infertile humans, monsters are prolific. Cain, Grendel's kin, was commonly thought to father monstrous races. Grendel's mother has spawned the infamous "monster" and, although Grendel dies along with his mother, there is the one tantalizing hint that Grendel himself may have had offspring (line 2006). Hrothgar fails to produce a son of Beowulf's caliber or might. Bodily survival is necessary for successful reproduction, something Beowulf himself fails to do at his nation's cost. Because Beowulf does not reproduce, his style of action will die out; indeed, societies structured around lauded heroes such as Beowulf are doomed from the start if those heroes do not reproduce.[63] The dragon, jealous of his treasure, is a shadowy figure for Beowulf himself; Beowulf is likewise a treasure miser, hoarding his genes and thereby becoming a natural selection loser. Beowulf should have fathered a son. The last line of the poem, calling Beowulf the most gracious of men and most eager for fame or praise, is not triumphant, but indicates how the hero, entrapped by the cultural code that endorses fame-seeking, brings about the ultimate destruction of his people.

CHAPTER 6

Urban Myths: The Civilized and Pristine City-Body

The principles of "'new' waste management discourse" in the 1990s focused on material recovery through recycling and composting and energy recovery through incineration. As Simin Davoudi writes, "the waste hierarchy puts energy recovery through incineration on a par with material recovery through recycling, on the grounds that both options reduce the amount of waste being landfilled."[1] In Davoudi's specific example, the use of shires as landfill sites for large urban areas in northern England, urban waste was tipped into rural mines. These mines and quarries were perceived as empty "holes in the ground."[2] Urban areas had not really worried about waste production, since waste was exported to rural areas. There was no "immediate environmental or political problems for them." But by the mid-1990s, "it became evident that a waste management policy based on a continuing and unchecked flow of the waste stream from conurbation to the shires was no longer sustainable."[3] The tension between the metropolitan and the shires created an antagonism,[4] leading to this system unravelling.

For the city to maintain its identity as a well-functioning organism, filth needs to be hidden. Public filth undermines the authority of urban infrastructure. Like those bodies hiding their vulnerable and exuding margins and orifices, the city, metaphorically imagined as a body, sets itself up against the country, home to dungy fields and garbaged landfills out of view.

The city and the country have long had a mutually profitable, yet vexed, relationship. Lewis Mumford points out how the countryside surrounding urban centers profited from city dung, increasing the amount of arable land in Europe from the eleventh through thirteenth centuries.[5] Human waste, produced in the city to a potentially noxious extent, was nonetheless vital to both urban and rural populations.[6] Dung was a form of profit and even common profit in the sense that fertilizing crops (food) would help society.[7] The medieval attitude toward excrement varied depending on context. Within a city, it would be excess filth; without the city, it could be utilized as fertilizer. Tanners would typically be located at urban edges, utilizing urine to productive effect. Dung produced by farm animals constituted an item to be manipulated by humans for their own amelioration. "Cottagers' beasts produce dung, which is required to fertilize farmers' corn (wheat, oats, barley, not maize), which yields straw, and straw is necessary for keeping beasts sheltered in winter where they can produce more dung, and so on."[8] This mucky product generated more than crops. In a probate inventory from 1456, one man was said to have dung in the cowshed worth 1 shilling 8 pence and another in a similar document from 1464 was noted as having a "parcel of dung 5 shillings."[9] The medieval use of waste echoes more recently with attempts to make money from recycled waste. In Jan Karon's novel, *These High, Green Hills*,

> Lank Pitts drove a pickup load of rotted manure into town and parked it in front of Dora's hardware, where he sold it by the pound in garbage bags. "Most people give that away," grumbled a customer, who nonetheless purchased two sacks full. "I pay f'r th' feed that goes in m' horses," said Lank. "Seems fair t' charge f'r what comes out."[10]

We can glimpse a medieval view of garbage as akin to our own, such as those that use animal manure as power. It is consoling to reflect on the Texas Panhandle, described as the "Saudi Arabia of cow manure . . . there's always going to be more," one cattleman says.[11]

While we may have impressions of medieval cities as being ubiquitously filthy, "older city models were more environmentally conscious—the smaller scale that allowed for walking rather than driving, or buildings that used natural ventilation provided by opening and closing windows." In fact, recent architects suggest that they should "incorporate some of these elements from older urban planning."[12] We could do worse than to look to medieval cities for guidance. In her extensive work on medieval filth control, Dolly Jørgensen argues that the individual in late medieval English and Scandinavian cities was expected to have certain responsibilities concerning cleaning the city. Her work, based on legal documents and archeological findings, argues for a well-functioning infrastructure for addressing issues of waste management. Needless to say, if individuals are expected to have responsibilities pertaining to urban cleaning, some residents will refuse to comply, leading to infractions noted in legal cases and evident in material cultural evidence.[13] While urban waste could be used in the rural economy, getting it there was an issue. City workers would use dung-boats to carry excrement from London to the south bank where Southwark was situated.[14] Though individuals hired privy cleaners, "too much was left to the initiative of the individual and too little responsibility was taken in the way of close and constant supervision by the city."[15] Civilizations learned to use detritus. Archeological findings suggest that there was increasingly "less waste disposal in the city and/or more extensive cleaning than in earlier medieval eras."[16] Cities have always incorporated trash. Animal bones and pottery would be repurposed as fill below paved streets. Waste was reused and through the reuse transformed into a commodity.[17] Solutions to our present environmental crisis include "living responsibly" and utilizing "rubbish as resource,"[18] though some "countercultural" activists "emphatically deny the terms 'garbage,' 'trash,' 'waste,' and 'refuse.' They argue that 'resources' is the proper term for materials found on the street in a black plastic bag or found on a store shelf in transparent shrink-wrap."[19]

Tom Loback, a sculptor, uses flotsam and jetsam from the Hudson River. The artist laughed when a woman on a bicycle twice reprimanded him as he was yanking and propping pieces from the water. "She said I was ruining the city's 'pristine' nature... *Pristine* in this city? I had to tell her the shoreline we stood on was a landfill from railroad days."[20] Despite well-functioning infrastructure that allows us to forget garbage, cities are actually "haunted by garbage."[21] Only the ideal city is blank.[22]

City as a Human Body

The human body has been recognized as the source of cultural symbols and the basis of social relationships. Society itself was conceived of in terms of the body, with God or the king at the head. Such a body at its densest and most populated form is to be found in the city.[23] If the filth of human bodies had to be disciplined, the garbage of urban bodies ratchets up the dire nature of guarding the borders and orifices exuding waste.

Given the multitude of legal records revealing the interest civil and secular authorities had in controlling filth, it is not difficult to see how the city as a body could be read and figured as a site of bodily illness. Eustache Deschamps (c. 1340–1406) writes, "Contaminated air, poisoned ground, / Infected bodies in the cemetery, and the dead / Scattered in the fields in grievous wars, / Latrines piled with infectious dung, / Stench from garbage dumped / In fields, towns, and castles, / Great filth and heaps of excrement, / Refuse which is foolhardy to make, / Stale, polluted air, and swine: / All of these cause plague in many places."[24] The miasma theory conjectured that bad smells caused illness. A sick city infected by contagion could blight its inhabitants and environs. An increased emphasis on the privatization of domestic space occurs in the late medieval period, paralleling a corresponding increase in the privatization of defecation. A roll from about 1485, including a genealogy and images of Troy and Eneas, depicts various cities, including London and York, where one can just make out the water systems. Little rivers flowing into the Thames suggest a circulation of waters, clean and filthy.[25]

In Book II, Canto ix of Edmund Spenser's *Faerie Queene*, Guyon and Arthur pass through the body realized as a physical structure. In the Hall, a "iolly yeoman" (28) named Appetite is the Marshall. Then, in a *tour de force* description of digestion, Spenser follows standard physiological ideas of the late sixteenth century in his description of elimination:

> But all the liquour, which was fowle and waste,
> Not good nor seruiceable elles for ought,
> They in another great rownd vessell plaste,
> Till by a conduit pipe it thence were brought:
> And all the rest, that noyous was, and nought,
> By secret wayes, that none might it espy,
> Was close conuaid, and to the backgate brought,
> That cleped was *Port Esquiline*, whereby
> It was auoided quite, and throwne out priuily. (II.ix.32)

The Port Esquiline was the gate in ancient Rome through which garbage was thrown, thus the "anus" of the city. Spenser embodies the late sixteenth-century city in this accurate rendition of digestion and disposal.

We learn in the sixteenth-century *Till Eulenspiegel* how excrement was disposed of in early modern German cities. In the winter in Wismar, "die Schinder die heimlichen Gemach reinigten . . . sie . . . pflegen in das Wasser zu führen" ["harnessmakers were cleaning out the outhouses," which they planned to "dispose of in the water"]. Till buys the excrement from them to fill barrels almost full, topped with tallow with which he tricks a shoemaker. Upon discovering what is really in the barrels, the shoemaker disposes of the filth "uff die Schelmengrub" ["the slaughterers' graveyard"].[26] When Till shits in a bathhouse, the bathkeeper angrily says, "So dann hie pflegt man uff dem Scheißhus abreinigen, und der Schelmenschinder pflegt das uszeführen uff die Schelmengrub" ["One cleanses oneself of this sort of stuff in the shithouse, and the horseslaughterer takes it out to the slaughterers' burial place"].[27] When Till shits on a pile of plums, "Also müßt der frumm Mann die Pflaumen hinwegführen uff die Schelmengrub" [the owner "had to drive his plums off to the slaughterers' graveyard"].[28] Excrement was evidently either thrown into a water system (river) or taken to the animal corpse dump. Another tale suggests that there was filth of some sort in the streets as well. In Hildesheim, Till is told by his merchant boss to clear out the house, whereupon Till puts all the furniture and pots in the street. The furious merchant cries, "Es hat mehr gekostet, dann daß man das in Dreck werfen sollt" ["These things cost too much to be thrown into filth like this!"].[29] His outrage suggests that there is at least garbage, if not excrement, already mouldering in the street. Not all cities have systems—however faulty—in place. Michael Thompson, calling Jonathan Swift the "founder of Rubbish Theory," quotes "A Digression concerning Criticks" (1696). One species of Mortal Man who proceeds "with the Caution of a Man that walks thro' Edenborough Streets in a Morning, who is indeed as careful as he can, to watch diligently, and spy out the Filth in his Way, not that he is curious to observe the Colour and Complexion of the Ordure, or take its Dimensions, much less to be padling in, or tasting it: but only with a Design to come out as cleanly as he may."[30] Here Swift mocks the ability of pedestrians to make their way through the byways of the Athens of the North.

The tension between filth and urban centers has only been exacerbated by the growth of cities where the virtual impossibility to integrate one's excrement in an urban environment makes the disposal of such matter of paramount importance not only aesthetically but also in terms of health.[31] Unlike biological ecosystems that are closed and cyclical, households and cities are open systems, disrupted by industrialization, making "flow . . . one-way."[32] Excremental practices became more regulated with increased urbanization; concurrent with growth was the privatization of shit.[33] Dominique Laporte points out how in France the purification of the language was yoked with the elimination of excrement from city space.[34] An edict from 1539 insists on the evacuation of ordure: ["*We enjoin such persons [who leave waste in public spaces] to collect droppings and wastes and to gather them inside their homes, where they shall pack them into receptacles and wicker baskets to be then carried outside the aforementioned city and its surrounding areas*"].[35] The danger inherent in private filth marks the problem of waste disposal. At its most dire in the nineteenth century, "shit [became] thinkable as a public problem and *un*thinkable as a private secret."[36] Naturalizing the concealment of systems of waste processing, the state enhances its own power through maintaining the surveillance of our filth by sewage management. Gay Hawkins proposes that the "hint of shit in a public space doesn't just call the self into question, but technologies of governance, *faith in infrastructure*."[37] When shit enters the public sphere, it destroys the illusion of the public secret.[38] "Publics don't shit" since shitting happens only in private; public shit scandalizes and shocks.[39] Polluting, dangerous shit, which should be secret, makes it all the more powerful and threatening. The capacity of the state to remove this filth shores up the tentacles of the state's power.[40]

Separation or "differentiation establishes culture," which is itself a "retreat from an *undifferentiated* mass of things (which may also be called 'nature')."[41] Separation from garbage makes culture possible.[42] Matter *in* place would be, for example, trash in its appropriate bin. Taken to the margins of our property, we properly dispose of it, thus adhering to codified societal expectations and law. In his meditation on trash collection and disposal in the modern city of Paris, Italo Calvino poetically meditates on how he must put out the trash at the proper times and in the proper manner. In response, the city, to whom he pays an annual "taxe de'enlèvement des ordures ménagères,"[43] picks up what he leaves. This orderly process reflects well on the garbage producer and city, all functioning smoothly and according to the ordered regime. If not for dustbin men, "the city is transformed into a corrupted dunghill."[44] The city as a body metaphor controls Calvino's image of the incinerator, "that great purifying instrument, the essential guts of the city."[45]

The desire to keep the filth of one's life private suggests that there are certain public standards for the *poubelle* so that it is not easily observed or smelt. Calvino extols the

> poubelle agréée, which is to say a pleasing dustbin, something approved and acceptable (with the implication: approved by the prefectorial regulations and by the authority that is made outwardly manifest in them and that is inwardly present in the individual consciences of the citizens, thus founding the basis of our social contract and of the expediencies of good living).

The person who throws out the garbage has a social role: "I am confirming my dependence on the institutions without which I would die buried under my own rubbish in the snail shell of my individual existence."[46] The urban dweller agrees, whether she realizes it or not, to adhere to the public contract with regard to the proper way and times to take out the trash.[47] This silent contract among citizens exists, in which we are expected to take care of our own waste, or at least aid the public services that take away our trash. It is a contract so pervasive that public rules silently become inwardly obeyed. Emptying the trash is a "contract [that] concerns no one but myself, it is a mutual agreement between me and myself, me and my interior law, or Kantian imperative, or superego."[48] As Calvino points out concerning emptying the garbage,

> That transfer from one container to another, which for most inhabitants of the metropolis takes on the significance of a passage from private to public, for me in our house, in the garage where we keep the big *poubelle* during the day, is only the last gesture of the ceremonial upon which the private is founded—and as such is accomplished by myself as paterfamilias—my taking leave of the leftovers of things confirming their complete and irreversible appropriation.[49]

As a dutiful fulfiller of this contract, Calvino's narrator reconfirms his own proper position, both as an urban city dweller and as the patriarchal head of the household.

Ragpicking as Essential Urban Function

Waste authors specifically target the urban arena as a filth-ridden space due to the density of waste-producing bodies and the very real danger and problem of what to do with all the detritus inhabitants produce. As Jane Bennett points out, "the sheer volume of products, and the necessity of junking them to make room for new ones, devalues the thing."[50] For her there exists an

> affinity between thing-power materialism and ecological thinking: both advocate the cultivation of an enhanced sense of the extent to which all things are spun together in a dense web, and both warn of the self-destructive character of human actions that are reckless with regard to the other nodes of the web.[51]

The more we value the thing—even waste—the more attuned we can be to what we waste, throw out, and discard. This, in turn, might lead to "less thoughtless waste, and so perhaps less waste overall."[52] Of course, reduction can have unintended negative consequences. Trash pickers or *pepenadores* in Mexico City suffer because of government plans to increase composting and recycling efforts, thus decreasing the robustness of their livelihood.[53]

Not "marginal" to urban economy, the scavenging by ragpickers has been a form of "recuperation" that was key to the vitality of the city's economy.[54] Their toil, absolutely necessary for their very survival, adheres to "one of the number of survival strategies for the very poor."[55] In the developing world (developing into what? more wasteful cultures like ours?), "the bidonvilles, bustees, favellas, and villas miserias" offer, "along with the insanitary conditions, more room for initiative, for community, for culture."[56] Barrie Ratcliffe projects a reinvigoration of marginalized areas of disposal. "If this were done, the physical margin would then become a measure of dynamism rather than a place of exclusion, its vacant lands the object of a discrete, if anarchic, colonization by cowsheds, recycling activities and market gardens, warehouses and space-hungry industries, the shanty towns of workers and chiffonniers."[57] What if the uncanny spot of the valley of ashes in *The Great Gatsby* were made central and not abject? The novel forestalls this option; doomed figures inhabiting the desolate valley weigh down the region, sinking its potential for vitalism.

We must not romanticize the life of slum-dwelling recyclers.[58] The places they live in literally shift—like the shifty[59] waste they dwell upon—and are unpredictable. Payatas, on the outskirts of Manila, is "said to support up to 100,000 people, about one-fifth of whom actually lived in shacks built on the garbage."[60] Some of these unfortunates died in a 2001 disaster and continue to die periodically in landslides of rubbish. In a garbage dump in Phnom Penh, Cambodia, "you come across a child ambling barefoot, searching for old plastic cups that recyclers will buy for five cents a pound. Many families actually live in shacks on this smoking garbage." Nicholas Kristof ultimately defends sweatshop jobs for these people as a step up. He even promotes bringing manufacturing industries there. "But take it from 13-year-old Neuo Chanthou, who earns a bit less than $1 a day scavenging in the dump. She's wearing a 'Playboy' shirt and hat that she found amid the filth, and she worries about her sister, who lost part of her hand when a garbage truck ran over her. 'It's dirty, hot and smelly here,' she said wistfully. 'A factory is better.'"[61] Yet the factory would produce more goods doomed, in turn, to become waste destined for the rubbished domiciles of marginal and marginalized landfill dwellers.

These garbage cities stink, while the wasted caverns of the bombed cities in Adler's *The Journey* echo emptily as the war ends. Yet signs of life emerge. People move from one ravaged apartment to another slightly less bombed out one. People

> were just too weak to carry off all the rubbish, which overpowered them. They gave up and simply crept away from one heap to another in a different street... Out of the rubble treasures left behind appear, even fragments seeming precious to whomever wants them. It doesn't take much effort, just bend down and grab hold, or just use a stick to scrape away the rubble, for there's much that's there to take.[62]

Paul, now a displaced person (a peripatetic wasted soul, as Bauman would say), talks to someone disturbed by a ruined theater, which morphs into a symbol for both life and the crematorium. Paul wonders in the wake of this conversation: "Or the people, the many, many people? If it has to do with a crematorium?... The difference, perhaps one can only flee the specter, no rest... the journey... taken away... rubbish, rubble everywhere... Forgive me!"[63] Paul's interlocutor proclaims one can rebuild from the rubble; but Paul objects, "My friend, the rubble, the rubble! You will not make a dent in it with picks and shovels. It stands in the way and will trip you up... But I don't want to quarrel with you. And the crematorium? I mean, the theater?... Theater or rubble, there's no longer any difference."[64] Waste disguises the essential nature distinguishing art (theatre), rubble (waste), and cremation (death house). In the wake of devastation, Paul no longer sees the viability of art.

While Adler disagreed with Adorno in their correspondence as to whether or not there could be poetry after Auschwitz, trash, as well as "trashing," demands being recognized for what it is.[65] Recycling and ragpicking are central to culture. When Jewish rag dealers were no longer allowed to deal with rags, new people were allotted this job. The insidious nature of Nazi thought strips any vibrancy from the recycling process that remains necessary when all goods are useful to the state. "Everyone tosses it all away for the sake of the state; everyone tosses away what is worthless and sees the state transform it into something of worth once again."[66]

Just as in medieval understandings of the city as a body, *The Journey* describes the bombed out urban landscape as though it were a human cadaver torn open:

> The streets have been recently ripped open, their naked entrails exposed, thick, bulging cables wind their way and are operated on . . . Anything scandalous is covered over and sealed off, the water mains and the telephone lines, everything banished and placed under the earth in order that the healing can fully begin.[67]

In the old days there was the ceremonial hauling away of trash. As in Böll's description of washdays, which were like a communal festival, joy was found in rubbish: "Clouds of dust rose from the wagon, as well as ashes and tiny grains of dirt, now and then a bright piece of paper dancing impishly on the wind."[68] But then it all changed. Just as fascism tainted recycling, garbage removal acquires a sinister stench. "No longer is the whole process carried out in such good spirits, nor is the glee involved in getting rid of the rubbish either enjoyed or looked forward to or perhaps even fathomed." Impersonal, anonymous, disposed of in a sanitary can, trash disposal loses it potential for catalyzing conviviality:

> A huge wagon appears with two men in overalls with rubber gloves that look like fins. The men walk into the courtyard, confident and indifferently cool, lifting one can after another, emptying each coldly and in a professional manner so that no dust rises, not a word is said, the empty cans rolling back into the courtyard as if it all meant nothing . . . Sighing and gasping, the rubbish sinks into the field, then the wagon rushes back to the city in order to clean out more streets and suck up more victims into its monstrous stomach.[69]

This sanitized public cleaning eliminates the personal touch of the ragpicker, known, perhaps, as a character integral to city life. The more anonymous the person cleaning up, the more unidentified and unidentifiable your own detritus becomes.[70]

CHAPTER 7

Interiorized Waste: Sin and Metaphysical Meaninglessness

Purity, materially and spiritually, constitutes generative meaning, while impurity signals sin and metaphysical crisis. Moral hygiene precludes transgressions, ranging from venial sins to obscene offenses like sodomy. Since no one can exist in a wholly pristine condition, doubts inexorably invade the soul and psyche, spoiling them with sullied qualms, tarnished trust, and blemished beliefs. If termites infest the strictures scaffolding our societal assumptions, perhaps nothing has significance.

The Old Law clearly rejects filth that is "externalized" in the form of excrement, lepers, and dead bodies. The New Law *internalizes* filth. As Kristeva argues, "Abjection is no longer exterior. It is permanent and comes from within."[1] The Christian "interiorization of impurity"[2] blends with guilt to create the concept of sin. We see this internalization all through Christian medieval literature. The insistence upon cleanliness—literal and symbolic, individual and social, bodily and spiritual—is well documented in *De miseria humane conditionis.* Lothario writes in the opening passage how he intends to comment on "the vileness of the human condition,"[3] pointing out how we were conceived not only "from the itch of the flesh, in the heat of passion and the stench of lust," but also, and yet worse, "with the stain of sin."[4] Regarding this "primeval taint," he comments, "Just as drink is polluted by a soiled vessel, anything that touches something polluted becomes polluted."[5] He delights in the visceral filth of fecal matter. "No one," Lothario claims, "can boast of the cleanness of his heart . . . Almost the whole life of mortals is full of mortal sin, so that one can scarcely find anyone who does not go astray, does not return to his own vomit and rot in his own dung."[6] Even nocturnal emissions confirm one's foulness, physically and spiritually. "For impure images appear in dreams, and by such nocturnal illusions not only is the flesh soiled, but the soul is spotted."[7] He readily recognizes our affinity with beasts as a source of humiliation, revulsion, and disgust. You, humans, "produce nits, lice, and tapeworms . . . you give off spit, urine, and dung . . . you give off a dreadful stench."[8] We can never deny "the body is the soul's prison."[9] As Lothario, filled with revulsion, relates, "And yet what is a man adorned with gems but a whited sepulchre outside and full of rottenness within? Blue and purple cloth, scarlet and fine linen rot in the ground; gold and silver, precious stones and gems tarnish and grow dull in the mire."[10] Those useless jewels cannot prevent the body from being on the menu of a worm's banquet.

Gluttony brought about the Fall through the act of eating. As the future Pope Innocent III writes: "Gluttony demands a heavy tribute but gives the meanest returns: the more delicate the food, the more stinking the dung."[11] The result of gluttony, excrement, becomes a sign of corruption. Indeed, one religious figure, the fourteenth-century Franciscan, Francesc Eiximenis of Valencia, said that before the Fall, excrement did not even stink.[12] Ultimately, God will determine who is saved and who is damned according to his own binary, a system of codification that uses shepherding as the vehicle. Citing Matthew 25: 32–33, Lothario writes, "All nations shall be gathered together before him, and he shall separate them one from another, as the shepherd separates the sheep from the goats. And he shall set the sheep on his right hand, but the goats on his left."[13] God, who originated codification in the division of darkness from light, will adhere to it at the Final Judgment.

Filth and sin are inextricably united throughout medieval works. As Venus says, in Guillaume de Deguileville's fourteenth-century *The Pilgrimage of Human Life*, "So I am foul in abstracto, but I am even more foul in concreto, and therefore I carry a false painted face to cover my own face, which is covered with filth. This false face is called Make-up."[14] Chaucer's Parson identifies sin with the excretory parts of the body:

> Of the hyndre part of hir buttokes, it is ful horrible for to see. For certes, in that partie of hir body ther as they purgen hir stynkynge ordure, / that foule partie shewe they to the peple prowdly in despit of honestitee, which honestitee that Jhesu Crist and his freendes observede to shewen in hir lyve. (*Canterbury Tales* X.427–428)[15]

He continues in this vein against adultery (*Canterbury Tales* X.848, 850). In a slightly later work, the fifteenth-century English visionary and pilgrim Margery Kempe relates the parable of the shitting bear, equating feces and inner sin.[16] Excrement wipes itself off on sinful condemnation.

The epitome of inward sin is externalized in an imagined excrement-laden hell.[17] The most famous depiction of this is in Dante's *Inferno*:

> This was the place we reached; the ditch beneath
> held people plunged in excrement that seemed
> as if it had been poured from human privies.
> And while my eyes searched that abysmal sight,
> I saw one with a head so smeared with shit,
> one could not see if he were lay or cleric. (*Inferno* 18.112–117)

Here in the *Malebolge*, the evil pouches of chaos structuring the *Inferno*, Dante the pilgrim descends ever lower. In the Catholic sense, sin *is* chaos, a misunderstanding of behavior appropriate in God's order.[18] Excremental ordure manifests the interior sin in a compacted turd never to be released or expelled.

As we saw in the previous chapter, historical records recognize the importance of filth control for the societal functioning of the urban body. This quotidian reality reflected in literature is, in turn, imposed on the material human body. Urban excrement, viewed as superfluous since it was not as readily utilized as it was in the sustainable rural culture,[19] was increasingly associated with moral filth in the Church Fathers' tradition. Sanitation was more than the physical act of recycling or disposal of pollutants. It carried an allegorical nuance and reflected on the cities' "good and godly rule."[20] In addition to its importance to mere material concerns, regulating urban waterways gained metaphysical and moral resonance.[21] Literary works illustrate the tension between the city and filth production, fusing the urban and the moral. Filth in its material form transmutes to stern prohibitions and ultimately vile sin. Possibly the most abhorrent depravity to come under condemnation by religious authorities was sodomy, a deviance associated with urban density, filth, and perversion.

In *Leviticus*, rules are stipulated concerning whom not to sleep with or see naked, along with prohibitions against sodomy: "Defile not ye yourselves in any of these things: for in all these the nations are defiled which I cast out before you: And the land is defiled: therefore I do visit the iniquity thereof upon it, and the land itself vomiteth out her inhabitants." (*Lev.* 18: 24–25). Pope Innocent III, clearly and explicitly a huge fan of *Leviticus*, cites it in *On the Misery of the Human Condition*, where he refers to "unnatural intercourse."[22] Sodomites were thought to come into contact with one another at all sorts of places, including public latrines, further associating sodomy with literal filth and with the anus, the site both of sexual deviance and the production of excrement.[23]

Excremental elements are present throughout the fourteenth-century Middle English poem, *Cleanness*, a poem devoted to establishing proper behavior and clean conduct, with extended passages on the Flood, the destruction of Sodom and Gomorrah, and the tale of the kings of Babylon—Nebuchadnezzar and Belshazzar—and the virtuous Daniel. Contrasting its remarkable passage praising God's order in creating heterosexual intercourse, the poem strenuously condemns sodomy.[24] The voiding of evil parallels the voiding of physical filth. The anal Dead Sea functions as "the primal locus . . . of evil . . . the grave as opposed to the birth portal."[25] In the section on Sodom and Gomorrah, an extended paean extols heterosexual sex as positive intercourse (695–708), while condemning sodomy as unclean (709–712). God hates lewdness more than anything, especially unnatural intercourse. In the stable where Christ is born, the "animals' waste is transformed into the sweetest of smells, while the filth of sodomy, unnatural and sinful, will stink forever."[26] Even the words of the Sodomites are "equated with defecation" (see ll. 842–848).[27] When Sodom and Gomorrah are destroyed, the earth again opens up to a hell filled with filth:

Gorde to Gomorra that the grounde laused . . .
The grete barres of the abyme he barst up at ones,
That alle the regioun torof in riftes ful grete,
And cloven alle in lyttel cloutes that clyffes aywhere . . .
The brethe of the brynston bi that hit blende were,
Al tho citees and her sydes sunkken to helle.

[Bombarded Gomorrah, broke open the ground . . . / The great bars of the abyss burst up all at once, / Rent the entire region in rifts gaping wide / And the cliffs and crags everywhere all cloven to pieces . . . By the time the smoke of the brimstone had ceased, / All those cities and their surroundings had sunk into hell] (957, 963–965, 967–968).[28]

The cities are voided:

This was a vengaunce violent that voyded thise places,
That foundered has so fayr a folk and the folde sonkken.
There fyve citees wern set nou is a see called,
That ay is drovy and dym, and ded in hit kynde;
Blo, blubrande, and blak, unblythe to neghe,
As a stynkande stanc that stryed synne,
That ever of synne and of smach smart is to fele;
Forthy the derk dede see hit is demed evermore.

[It was a violent vengeance that devastated those places, / That swallowed such fair citizens and submerged the land. / Where five cities stood a sea is now found, / That is always dreary and dim, and dead in nature, / Bubbling

> and black, baleful to approach, / Like a stinking cesspool that destroyed sin, / Ever acrid in aroma and taste. / Thus it is deemed the Dead Sea, dark forever] (1013–1020).

God's vengeance annuls these places. Sodom and Gomorrah represent what is most threatening about cities and their potential for filth production. Displaced onto the unforgivable sin of sodomy, those who cavort or associate with sodomites embody urban waste.

Two of God's messengers come to visit Lot. The men of Sodom, in turn, demand the angels to teach them of love (841ff.). Lot goes out and offers his daughters instead. Refusing, the angry men accuse Lot: "Wost thou not wel that thou wones here a wyye strange, / An outcomlyng, a carle?—we kylle of thyn heved!" ["Have you fully forgotten that you are a foreigner here, / A stranger, a serf? We will strike off your head!"] (875–876). His angels warn Lot before Sodom's destruction, assuring him a safe place, "For thou art oddely thyn one out of this fylthe, / And als Abraham thyn eme hit at himself asked" ["Since you are singled out as separate from this filth, / And also Abraham your uncle asked Him for it"] (923–924). The destruction of Sodom is horrific and makes such a horrible noise "that Kryst myght haf rawthe" ["For Christ to take pity"] (972). Indeed, Christ is more likely to find pity and affinity to these dirty souls than the vengeful God of the Torah.[29]

Cleanness adheres to Gospel medicine; Pentateuchal moments display gruesome scenes of God's incision and surgical amputation of filth.[30] Indeed, in the wake of the 1215 Fourth Lateran Council's establishment of the bifurcation between surgery and medicine, surgery was cast down as a lower form of medical attention. Canon 18 specifically refers to medicine: "No subdeacon, deacon, or priest shall practice that part of surgery involving burning and cutting." Medicine functions as a metaphor for spiritual cleansing; no surgery is suggested:

> Let the priest be discreet and cautious that he may pour wine and oil into the wounds of the one injured after the manner of a skilful physician, carefully inquiring into the circumstances of the sinner and the sin, from the nature of which he may understand what kind of advice to give and what remedy to apply, making use of different experiments to heal the sick one. (Canon 21)

We are assured that confession cleans us. *Mankind* likewise invokes the notions of surgery and medicine. Mercy advises the sinner to ask for mercy and not continue the sin: "While a wound is fresh it is proved curable by surgery, / That, if it proceed overlong, it is cause of great grievance" (857–858).[31] The bodily presence of the sinner has some efficacy.[32] Mercy says,

"Ask mercy and have, while the body with the soul hath his annexion; / If ye tarry till your decease, ye may hap of your desire to miss" (863–864). In other words, ask for mercy while your body and soul are together. Your body asks for salvation; once you are dead, it is too late when body and soul are divided. Only *with* and *in* your body can you ask for forgiveness, and thereby get mercy. So the body can be the source of goodness and possible mercy. Only through Mary and Jesus does a new kind of cleanness come, one that is internally, not externally, determined.

Just as excrement is only useful once it has decomposed, filthy or corrupted souls need penance to decompose sin and once again become clean and useful. As Dame Penance defines herself in Deguileville's *Pilgrimage of Human Life*,

> in order to chastise them well I strike them with my stinging switches, inflicting pain and suffering upon them for their benefit and amendment. At one time I make them remember their old sins and say, "Oh God! why did you consent to that, only to be miserable now?" At another time I make them say: "Dear Lord God! Sweet Lord! I promise you I will amend my life"...Sometimes I make them endure the difficulties of a pilgrimage or some other long journey.[33]

Penance, through her beating of the corrupted soul, facilitates a process of amendment, integral to pilgrimage and the act of cleansing, both figurative and literal. An allegorical passage in *Cleanness* suggests that, one day, we will all be summoned to a splendid feast, but those who have committed filthy deeds, clad in dirty clothes, will go to hell. "Bot war the wel, if thou wylt, thy wedes ben clene" ["But be careful, if you come, that your clothes are clean enough"] (165). We are urged to always be careful concerning our behavior:

> Forthy war the now, wyye, that worschyp desyres
> In his comlych courte that kyng is of blysse,
> In the fylthe of the flesch that thou be founden never,
> Tyl any water in the worlde to wasche the fayly.
> For is no segge under sunne so seme of his craftes,
> If he be sulped in synne that syttes unclene,
> On spec of a spote may spede to mysse
> Of the syghte of the Soverayn that syttes so hyghe.
>
> [Let this be a lesson, if you would like to be honored / In His glorious court who is King of bliss, / That the filth of the flesh never be found on you, / So that all the water in the world would fail to wash you. / For nobody here below behaves so well, / That if he is soiled by sin, he sits unclean. / One speck of a spot may ensure that he misses / The sight of the Sovereign who sits so high] (545–552).

As Jeremy J. Citrome has shown, John Audelay's early fifteenth-century writings delineate the internalization of the rhetoric of sin as filth and the need to purge it through confession.[34] We see this process of spiritual cleansing inwardly most movingly in *Piers Plowman*, in the Everyman figure of Haukyn, who is literally soiled in dirty clothes that symbolize his internal tainted sin. Likewise, *Cleanness* teaches us the most important lesson: "Thenne confourme the to Kryst, and the clene make" ["Then keep yourself clean, conform yourself to Christ"] (1067). Christ was born "lyght withinne the lel mayden" ["alit in that loyal Maiden"] (1069). Mary herself was cleansed by bearing Jesus. "Bot much clener was hir corse God kynned therinne" ["Her body all the cleaner for bearing God within"] (1072):

Thenne was her blythe barne burnyst so clene
That bothe the ox and the asse hym hered at ones;
Thay knewe hym by his clannes for kyng of nature,
For non so clene of such a clos com never er thenne.
And yif clanly he thenne com, ful cortays therafter,
That alle that longed to luther ful lodly he hated;
By nobleye of his norture he nolde never towche
Oght that was ungoderly other ordure was inne.

[Then her cheerful Child came so cleanly polished / That the ox and the ass honored Him at once; / They recognized Him by His cleanness as the King of nature, / For none had come so cleanly from such an enclosure before. / And if He came cleanly, He was most courteous afterward, / And hotly hated all that harbored evil; / The nobility of His nature ensured He would never touch / Anything that was ungodly or infected with filth] (1085–1092).

Yet Jesus cannot help attracting filthy outcasts:

Yet comen lodly to that lede, as lazares monye,
Summe lepre, summe lome, and lomerande blynde,
Poysened and parlatyk and pyned in fyres,
Drye folk and ydropike, and dede at the laste.

[Yet the Lord attracted the loathsome, lazars and more, / The leprous, the lame, and the lumbering blind, / Those poisoned, paralytic, and pained by inflammations, / The dessicated and dropsical, the dead ultimately] (1093–1096).

Christ has the ability to clean filthy things of their foulness: "For what-so he towched also tyd tourned to hele, / Wel clanner then any crafte cowthe devyse" ["For whatever He touched immediately turned healthy, / Far cleaner than any medical craft could achieve"] (1099–1100). The court of the one we need to seek out is "kyryous and clene" (1109) ["careful and clean"]; the way we can get there is through confession. Penance purifies us.

Fear and Self-Loathing

Something is rotten in the state of Denmark.

Marcellus, *Hamlet* (I.iv.90)

The German word *Angst*, literally meaning *fear*, has come to signify a spiritual malaise that corrodes one from within. A poster child for *Angst*, *Hamlet* is a play replete with literal and metaphorical anguish. Fragments, belaying wholeness and wholesomeness, lie strewn throughout the halls of Elsinore. In response to Barnardo's question, "Is Horatio there?," Horatio replies, "A piece of him" (1.i.18). Hamlet quotes the speech about Pyrrhus (II.ii.446ff.), scraps of words from works outside the hermetic seal of the play. In informing the players how to perform, he defames hearing an actor "tear a passion to / tatters, to very rags, to split the ears of the groundlings" (III.ii.9–10). Hamlet calls Claudius himself a "king of shreds and patches" (III.iv.99). There are countless examples of Mary Douglas's "matter out of place" in *Hamlet*, a play larded with waste. Not stable in Purgatory, Hamlet's father appears on the battlements. Funeral meats are to be used for a wedding celebration. Rosencrantz and Guildenstern are famously out of place—killed for Hamlet. Claudius himself is out of place—he is the wrong man as king, in an illegitimate bed. Gertrude, likewise, lies between the incestuous sheets with the murderer of her husband. Polonius, not Claudius, is behind the arras. Claudius is praying, not fornicating, so Hamlet cannot kill in hopes of sending his soul to hell. Hamlet and Laertes are *in* Ophelia's grave, not above it. The poisoned wine is drunk by the wrong person; meant for Hamlet, Gertrude consumes it. The wine itself is not that of the Eucharist, suggesting eternal life; it is the wine of death, with no hope.[35] The final scene is one of consummate wastage, the many bodies on the stage, dead and bloodied. The very land fought over in Poland is "a little patch of ground / That hath in it no profit but the name . . . I would not farm it" (IV.iv.17–19). Hamlet comments, "Two thousand souls and twenty thousand ducats / Will not debate the question of this straw. / This is the th'impostume of much wealth and peace / That inward breaks" (IV.iv.24–26).[36] The "impostume" is an abscess filled with corrupted matter or pus; so, too, is the very venture of colonization. The land in Poland, acknowledged to be useless waste, is scrapped over until soldiers die in defense of it. All these moments signal the rot, filth, and detritus that constitute Denmark, its politics, and its unfortunate inhabitants. Denmark itself is a passageway, the intestine, in the play's faulty geography, between Norway and Poland.

Hamlet himself is filled with malaise, nausea, and metaphysical despair. If "I am to be distinguished *from* my trash, not *through* it,"[37] what does it mean for Hamlet to be so closely aligned with filth and trash? This waste transmutes into or is a sign of despair seeping into—and out of—the body politic.[38] Marcellus observes, "Something is rotten in the state of Denmark" (I.iv.90). If the body politic functions metaphorically in *Hamlet*, who constitutes its excrement? Hamlet himself, a malingerer who does and contributes nothing, not even revenge, is valued as nothing. Hamlet is, himself, the rot at the heart of Denmark, the rot we all contain.

Metaphysical or spiritual waste—that one is spiritually wasted, meaningless, or void—comes about when the division between self and waste becomes impossible.[39] Garbage, Scanlan's overarching noun, suggests "waste, imperfection, disorder, and ruin," all indeterminate,[40] all ambiguous. Ambiguity or non-codification causes psychological dis-ease. Melancholy "is considered semi-excremental."[41] Interior "dirt" stains so that rational, clean, or objective thought flounders. Waste defies model or system; whatever waste enters *itself* becomes sullied and infected. Whenever the same/self and other are not distinct, we react with horror (as with the recognition that the wasted other could *be/come* us), compassion (the other is *like* us), humor (something is odd or prohibited), or disgust (again, something prohibited that is mixed with the self).[42] Hamlet, feigning madness, discovers the difficulty of conjuring non sequiturs. Sane, Hamlet needs system for uttering his self into a discourse and cannot maintain a fey demeanour. Even Polonius detects the "method" in his "madness" (2.ii.202–203).

Greta Garbology

I vant to be alone.

Greta Garbo in *Grand Hotel* (1932)

Isolated when we do not want to be, we feel oppressed when, desiring solitude, we are surrounded by others. That other Hamlet—Hamm, whose truncated name suggests a crippled conscience—likewise struggles with his rubbished soul and feels the paralysis of inaction. In Samuel Beckett's *Endgame*, all is waste, existing on multiple levels.[43] Adorno's analysis of *Endgame* focuses on the physical manifestation of existential waste.[44] We are confronted by waste from the very opening of the play: two ashbins signifying the waste of creation or life sit on the stage. The lack of rain, no more bicycle wheels, no more pap, no seeds coming up suggest the wasteland into which nature has devolved.[45] One of the most corrosive metaphoric uses of waste is the transfer from that materiality of waste to that experienced psychologically and spiritually. The internalization of waste renders one's inward spiritual being as a waste—all is wasted and meaningless. This wasteland of the self, the interior landscape,[46] is often evoked, as when Hamm asks Clov, "Have you not had enough?" Clov: "Yes! (*Pause.*)

Of what?" Hamm: "Of this... this... thing."[47] Waste creeps inexorably into our mental landfills.

Hamm reminisces about a mad engraver and painter, a man he used to visit in the asylum. Hamm would take the madman to the window to show him loveliness, whereupon the madman would "snatch away his hand and go back into his corner. Appalled. All he had seen was ashes. (*Pause.*) He alone had been spared. (*Pause.*) Forgotten. (*Pause.*) It appears the case is... was not so... so unusual."[48] If you see the truth, the end of the world hidden behind the apparent fecundity of nature, then you are considered mad. And what do you see? The ashes—those of incinerated Jews, of nuclear war—what we all inevitably must become: ashes to ashes, dust to dust. The madman sees reality and recognizes death. Was the madman Hamm himself? Clov points out to Hamm that, if Clov died in the kitchen, his body would soon begin to stink. In fact, says Hamm, "You stink already. The whole place stinks of corpses." Clov: "The whole universe." Hamm: "To hell with the universe."[49]

Clov pushes Hamm in his wheelchair up to the wall, which Hamm lays his hand against: "Beyond is the... other hell."[50] We are all in the center of our own existence(s) and existential beings; all around us lie borderland and waste. Hamm tells Clov, "One day you'll be blind, like me. You'll be sitting there, a speck in the void, in the dark, for ever, like me... Infinite emptiness will be all around you, all the resurrected dead of all the ages wouldn't fill it, and there you'll be like a little bit of grit in the middle of the steppe."[51] One day this is what you will be: alone, a solitary bit of waste.

A space void of sun and night, all "gray,"[52] as *Endgame* reverts to the moment before God divided light from dark, before God codified, dividing Night from Day, a time and "state of waste," when all was a "formless, primeval vacuum."[53] Order was created from waste,[54] in an alchemical process of divine composting. We can see the transfer from the materiality of waste to the metaphoric in the paradigm of codification as articulated by Douglas. Metaphorical order emerges out of the designation of categories like dirty and clean. Clov "love[s] order. It's my dream. A world where all would be silent and still and each thing in its last place, under the last dust." As he starts picking up again, Hamm asks in exasperation: "What in God's name do you think you are doing?" Clov replies, "I'm doing my best to create a little order."[55] If he can make order, then civilization, rather than the apocalypse, prevails. With order and codification come meaning, something Hamm half hopes, half dreads. He asks Clov, "We're not beginning to... to... mean something?,"[56] a suggestion Clov mocks. Hamm: "To think perhaps it won't all have been for nothing!" Through telling stories, we make our lives meaningful. Finally Clov pleads: "Let's stop playing!" Hamm agrees, "Then let it end." But first, he needs to know, "Did anyone ever have pity on me?"[57]

Nell dies; Nagg cries, a sign, as Hamm points out, that Nagg is "living."[58] But outside, there is neither sea nor light. We are flung back to a time before creation. This is life: "You weep, and weep, for nothing, so as not to laugh, and little by little... you being to grieve." Such is the human and divine condition. "You're on earth, there's no cure for that!"[59] cries out Hamm. The only process is death. "The end is in the beginning," says Hamm, with echoes of T. S. Eliot, and circling us back to before God created us. "Perhaps I could go on with my story, end it and begin another."[60] Another story, another creation. God as the playwright *extraordinaire.* In *Endgame*, the eternal question is asked. Hamm: "Why do you stay with me?" Clov: "Why do you keep me?" Hamm: "There's no one else." Clov: "There's nowhere else."[61] We stave off the annihilation through the existence or presence of the Other. If Hamm is God, then Clov is man, the waste of creation itself and its Creator. What could be bigger than God and his play: Creation itself?

Even God has metaphysical despair: "Old endgame lost of old, play and lose and have done with losing."[62] At one point, Clov, Hamm, and Nagg try to pray to God but fail. Hamm: "The bastard! He doesn't exist!" Clov: "Not yet."[63] Time is circling back to before God again. God doesn't exist, Hamm doesn't exist, we don't exist. But Nagg tells Hamm, "Yes, I hope I'll live till then, to hear you calling me like when you were a tiny boy, and were frightened, in the dark, and I was your only hope." The Father does not seem to be believed in.

The metatextual, metaphysical, cosmic/divine, and existential all fuse in Hamm's final speech: "Moments for nothing, now as always, time was never and time is over, reckoning closed and story ended."[64] He calls, "Father!" And there is no answer. "Good." "And to end up with? (Pause.) Discard." Ending with waste and destruction, he tears apart the toy dog. "Since that's the way we're playing it... let's play it that way... and speak no more about it... speak no more... Old stancher! You... remain."[65]

What remains? God? The author? Man? Your remains are all that remain. You are the remainder after God is dead.

The rest is silence.

CHAPTER 8

The Toxic Metaphor of Wasted Humans: Those Filthy Cleaners Who Scrub Us Spotless

> And in your sleep you sense that the dustcart isn't just grinding refuse, but human lives too and social roles and privileges and it won't stop until it has done the whole round.
>
> Italo Calvino, "La Poubelle Agréée"[1]

In every chapter thus far, wasted humans have appeared, deemed as such for some deficiency: the wrong gender, not enough money, incorrect lineage, lower heritage, something "off" about them, sinful, immoral, poor, undifferentiated, or hybrid. Material garbage collection intimately mingles with the stockpiling of human existence. Those who literally pick up our filth become filthy in turn. The rhetoric of othering constructs unprivileged races, religions, and ethnicities as unclean or inhuman. Wasted humans can be anyone—no one is immune. The rich become poor, the powerful weak, the young old.

Humans feel the compulsion to point out difference. In our worse moments, we take that difference to vilify those other than ourselves. We seem to need hierarchy and the consequent inflicting of misery on someone other than ourselves. Once that othering is recognized, we set ourselves up as better than them, in a process that creates what Polish sociologist Zygmunt Bauman argues produces "wasted lives," those wasted humans rendered invisible as non-entities. Waste itself threatens to overpower us, so we respond by ignoring it, rendering it invisible, powerless, and non-existent. We myopically avoid that which disturbs us. One way that we make wasted humans invisible is to make them cognate to waste; waste is something we take all means to avoid. Wasted humans—disdained, ignored, and made invisible—are ontologically non-existent.

Embodied Metaphor

> Regarded as "'impure" and therefore contaminating, manual scavengers are condemned to a life of filthy, dangerous work—their job, ironically, to remove the most obvious proof of each and every person's humanity.
>
> Wellcome Collection, "Dirt: The Filthy Reality of Everyday Life."[2]

Metaphor lies at the base of communication. George Lakoff and Mark Johnson's paradigm of standard metaphors stems from the view that our world emerges from our embodied state. In spatialization metaphors, up is high and good, low is down and bad.[3] Our bodies are bounded by a surface (skin). We are oriented both out and in, hence a plethora of *container* metaphors, references to "in" and "out." "We're *in* a pickle." "How are we going to get *out* of this?"[4] The body shapes the mind. Recent work on embodied cognition suggests that even "the experience of washing your hands is associated with moral purity and ethical judgement."[5] As Lakoff and Johnson conclude, "How we think metaphorically matters."[6]

Martha Nussbaum explores the corrosive aspects of what she calls the "politics of disgust," a "deep aversion akin to that inspired by bodily wastes, slimy insects, and spoiled food." This disgust expresses itself in "a fundamental refusal of another person's full humanity."[7] "Culture [may feel] like nature,"[8] since disgust "is taught by parents and society"[9] and "is an especially powerful vehicle of social teaching."[10] The disgust explored most famously by Mary Douglas—the bodily emissions propelled from the bodily borders—"expresses a universal human discomfort with bodily reality, but then uses that discomfort to target and subordinate vulnerable minorities."[11]

> Thus, throughout history, certain disgust properties—sliminess, bad smell, stickiness, decay, foulness—have repeatedly and monotonously been associated with, indeed projected onto, groups by reference to whom privileged groups seek to define their superior human status. Jews, women, homosexuals, untouchables, lower-class people—all of these are imagined as tainted by the dirt of the body.[12]

The rhetoric of othering those of a different race, religion, ethnicity, or gender constructs them as unclean or inhuman. The classes of wasted peoples vary over the course of history: immigrants, poor white trash, Jews, untouchables, women, old people, ragpickers, Roma, and so on. All these groups have something in common—they are the wasted peoples of society. Wasted people are seen as carriers of filth and dirt and "are those required to perform social functions equivalent to the excretory functions of the body."[13]

Waste cannot be easily limited, hence the desire to restrict it, such as in a container (our old friend, the container metaphor) signified as a waste can. Once located there, anyone who touches the waste becomes, in turn, contaminated as refuse, to be thrown out socially, geographically, and morally. The television show "Undercover Boss" places the CEOs of various corporations anonymously in their own companies in the role of janitors. This playacting allows for the viewer's Schadenfreude (those rich bastards get taken down a peg or two and have to wallow in their own filth), yet simultaneously reinforces the association we make between metaphorical lowness and cleaners of filth.[14] Within the French economy, Calvino points out, "Refuse collection is a precarious occupation, performed by those who have not yet managed to put down roots in the city."[15] Met with resistance and tolerated only because "within two days the whole city would be stinking and filthy," India's ragpickers live on donations and sell discarded items they find as a kind of gleaning.[16]

People throw things out to clean up. In turn, people who are associated with trash as cleaners or processors become trash in our minds. It is no problem for *us* to waste *them*. Often, poor people of color—particularly women or immigrants—clean up filth in hotels or public restrooms. Marginal outsiders to patriarchal capitalism become made into metaphors for waste. Unclean things contaminate. We waste them for their low work since they literally touch what we dispose of. As early as H. Mayhew's *London Labour and the London Poor*, 1851–1852, we see an association between "the dust-yard sifters" and "moral impoverishment and contagion."[17] Ragpickers were considered untouchables, seen as stemming from a peripatetic and rootless class.[18] The nineteenth-century urban equivalent of the gleaner is the ragpicker, *chiffonier*, or *Lumpensammler*, a figure of both "all-embracing negativity [and] all-encompassing positivity."[19] In a passage from Baudelaire's essay *Du Vin et Du Haschisch* (written in 1851, before his poem on the same subject), the narrator urges the reader:

> Let us descend a little lower and consider one of those mysterious creatures who lives, as it were, off the leavings (*déjections*) of the big city...Here we have a man whose task is to gather the day's rubbish in the capital. Everything that the big city has cast off, everything it lost, everything it disdained, everything it broke, he catalogues and collects. He combs through the archives of debauchery, the stockpile of waste. He sorts things out and makes intelligent choices; like a miser assembling his treasure, he gathers the trash that, after being regurgitated by the goddess of Industry, will assume the shape of useful or gratifying objects.[20]

Eerily, like those medieval Jews and Saracens who were forced to wear distinctive and distinguishing dress under Canon 68 of the Fourth Lateran

Council, like Jews under the Third Reich with saffron yellow patches depicting the Star of David sewn onto their clothes, ragpickers had to wear metal identification badges.[21] Like Adler's Jews, who intertwine with waste, becoming forgotten, "rag pickers have no voice in history, leave no direct traces in the sources. They are merely talked about and down to in middle-class writings and in administrative archives."[22]

Calvino comments, "My contemplation of the heap of peels, shells, packaging and plastic containers brings with it the satisfaction of having consumed their contents, while for the man who unloads the *poubelle* into the rotating crater of the dustcart it offers only an idea of the amount of goods which are denied to him, which reach him only as useless detritus."[23] The ragpicker or garbage collector becomes, in Calvino's view, depressed at the sight of what is not permitted to him to possess (and, then, throw away) himself. He who unloads the dust cart is cognate to "useless detritus" himself. Invisible to an eye attuned only to what capitalism deems visible—the useful commodity—the one who picks up detritus cannot be seen.[24] Rubbish carrying in *The Journey* is a "despised task, they let the unfortunate people cart it all away, away, away so that it is out of sight, out of mind . . . Thus near the camp, the rubbish lies there and rots, unobserved by most, despised by those who must look at it or who see it by chance."[25] The rubbish and its unfortunate carters are equally ignored and rendered nonexistent. Even Jay Gatsby worked at St. Olaf's College "despising the janitor's work with which he was to pay his way through."[26] Paid to eliminate trash, Gatsby ultimately turns into the waste he managed. Like William Blake's chimney sweepers, anyone who touches dirt or waste is contaminated socially as waste himself or herself.

In reflecting on the dustbin man of his childhood days, Calvino observes that "our genteel lifestyle seemed guaranteed for all eternity by the availability of cheap labour."[27] Will "we" always dominate "them"?[28] Perhaps we will all become our own trash collectors, mutually tainted by the filth we engender.

Perhaps we already have.

SD discourse values reuse and lack of waste, since waste is conventionally viewed as ethically or physically bad or immoral. The human who *produces* waste in SD discourse would then be, you would think, *unethical.* Yet that isn't the case. "Wastelands" were key to rural economies before the fifteenth century. They were communal areas for grazing, gathering wood, and collecting herbs for medicine. It was the space, Myra Hird says, of the "productive remainder," where poorer members of society could find sustenance in this "social safety net."[29] While it was uncultivated, it was not unused. But the process of enclosure from the fifteenth through the eighteenth centuries argued that these wastelands were "*wasted* lands" since they weren't being exploited for profit. And those who survived from these lands were, in turn, considered wasted humans.[30] More recently, in the case of the proposed Brazilian dam

> Bishop Kräutler, president of the Indigenous Missionary Council (CIMI), claims that indigenous groups are subjected to hostility from other Brazilians because they are seen to live off the land without *producing* anything: "The idea of the white dominant society, is that only those who produce are entitled to live, the rest are superfluous and disposable. And the Indians fall into this category."

While Indigenous Peoples are positioned as "unproductive and idle" within the dominant SD story line, within the alternative story line they and their cultures are seen as having "served to defend the environment in contrast to destructive development projects."[31] In SD discourse, one can ignore "the many indigenous communities who already live sustainably, meeting the basic human needs of their societies while preserving land and nature, yet who are associated with backwardness."[32] The government agency endorsing dam building is producing waste but sets up the indigenous peoples as wasters. The rich people call the shots.

Perhaps we need to rethink our model or understanding of waste for an innovative metaphor to emerge. If waste were vibrant, our metaphors would change. Wasted humans are, in fact, key to sustainability, "our" sustainability—that is, we rich First World people. Waste goes to marginal areas in developed countries. After all, "the waste products of human productive and consumptive activity need to go someplace."[33] Poor countries integrate waste, rendering those areas "marginal."[34]

The marginalized pick through old cell phones and salvage metals, they load trash into landfills, they are ragpickers and aluminum can recyclers. Wasted humans may be the most "sustainable" acting humans around. It is with them that shifty[35] "objects pass back and forth between the realms of useful objects and wastes with greater ease."[36] Their sustainability work allows "us" who produce waste to feel as though *we* are acting sustainably. Bauman quotes a study arguing that "the main population problem is in wealthy countries. There are, in fact, too many rich people."[37]

Wasted Humans

> We see that disgust, which always serves the purpose of setting us at a distance from our own animality and mortality, easily takes as its object other persons and groups, who come to represent what is avoided in the self.
>
> Martha Nussbaum[38]

Bauman's *Wasted Lives: Modernity and Its Outcasts* parallels Nussbaum's exploration of the history of the "politics of disgust." He writes how our planet has become too full to absorb both human waste and the wasted bodies of rejected peoples.[39] Who are "wasted" humans? Those who are unemployed or redundant. They are unnecessary; they have "been disposed of *because of being disposable*." "'Redundancy' shares its semantic space with 'rejects', 'wastrels', 'garbage', 'refuse'—with *waste*."[40] The wasted, "untied to any place, shifty, unpredictable,"[41] can turn up anywhere and threaten the pristine borders of our constructed inviolability. There have always been some humans who have been considered better than other humans. "Societies have felt strong disgust toward many people and practices, including members of lower castes and classes, foreigners, people with disabilities, people with physical deformities, Jews, and people who contract interracial marriages," as well as to homosexuals, as Nussbaum's book attests.[42] This is a *projective* disgust: "the contamination is not literal and physical, but imaginary."[43] The identification of some humans in a society as disgusting is "a stratagem adopted to cordon off the dominant group more securely from its own feared animality."[44] To take Timothy Morton's paradigm, "There is a Western habit of thinking things as *reduceds*."[45] In the case of wasted lives, those in power are reducers, degrading wasted humans into reduced things without agency, identity, subjectivity, or worth. Bauman characterizes "marginalized" or "supernumerary" people as

> "declassés individuals, possessing no defined social status, deemed redundant from the point of view of material and intellectual production and regarding themselves as such." "Organized society" treats them as "scroungers and intruders, charges them at best with unwarranted pretences or indolence, often with all sorts of wickedness, like scheming, swindling, living a life hovering on the brink of criminality, but in each case with feeding parasitically on the social body."[46]

Ethical Dangers Loom with Metaphor

> It is always the excess of *them* that worries *us*.[47]

The trope of "humans as waste" ultimately goes back to the inherently metaphorical way humans think. It tells us not just that humans can conquer other humans, but how we think in general and how we think poetically. Naomi Quinn argues that metaphor maps "from one domain of experience to another." Source domains (the familiar and everyday seen in throwing out trash) become mapped onto target domains (the abstract and conceptual whereby people are trash).[48] But while for Lakoff and Johnson metaphors structure understanding, Quinn says, "particular metaphors are selected by speakers, and are favored by these speakers, just because they provide satisfying mappings onto already existing cultural understandings . . . Metaphor, far from being productive of understanding, is actually highly constrained by understanding."[49] Whether or not understanding or metaphor comes first, we enter into an associative process to make knowledge. The cultural model of waste is mapped onto humans.

Literature mercilessly exploits the filthing of others to excoriate. Filth, specifically excrement, has long been used to mock and satirize others.[50] The "slinging of excrement and drenching in urine are traditional debasing gestures" in grotesque realism and elsewhere.[51] The narrator in Catullus 23 insults Furius, saying:

> Add to this elegance something more elegant,
> that your "fundament" is purer than a salt cellar,
> nor do you shit even ten times in the whole year;
> and, when you do shit, it's harder than a bean or pebbles,
> which—if you grind and break it up in your hands—
> you're not able—ever—to befoul a finger.[52]

In the late medieval period, newcomers to even an exclusive club constitute the "filth" of that privileged group. New university students were symbolically associated with rusticity; hazing rituals involved excrement.[53] Even among the Yahoos a hierarchy exists, the most "deformed" being the ruler. Once his Favourite, "hated by the whole Herd," loses his office, all the other Yahoos "discharge their Excrements upon him from Head to Foot."[54]

Filth is linked to humiliation in *Isaiah* 36: 12 (also in 2 *Kings* 18: 27) when it is suggested that men may one day consume dung and piss. The chief officer answered, "Is it to your master and to you that my master has sent me to say this? Is it not to the people sitting on the wall who, like you, will have to eat their own dung and drink their own urine?"[55] This rebuke carries the figurative color of humiliation, made even more explicit when God utters a curse: "If you will not listen to me and pay heed to the honouring of my name, says the Lord of Hosts, then I will lay a curse upon you. I will turn your blessings into a curse; yes, into a curse, because you pay no heed. I will cut off your arm, fling offal in your faces, the offal of your pilgrim-feasts, and I will banish you from my presence" (*Malachi* 2: 2–3). The flinging of offal, itself a form of waste, in turn transforms the recipient in a transmission of wastage. Those who commit outrages against God are likened to filth: "I will bring dire distress upon men; they shall walk like blind men for their sin against the Lord. Their blood shall be spilt like dust and bowels like dung" (*Zephaniah* 1: 17). Ordure signals a humiliating downfall: "Those who once fed delicately are desolate in the streets, and those nurtured in purple now grovel on dunghills" (*Lamentations* 4: 5).

We associate the lowest ranks of any society with filth. Unclean creatures must be separated from ourselves, until they are made clean or remain forever banished. *Leviticus* 15 instructs us how the man who becomes unclean must purify himself to integrate himself back into society. During this process, he must be separated from the community before he can be allowed back in. The stranger, on the other hand, is naturally, inherently, irrevocably, and ontologically tainted: "Neither from a stranger's hand shall ye offer the bread of your God of any of these; because their corruption is in them, and blemishes be in them: they shall not be accepted for you" (*Lev.* 22: 5). Seen as filthy waste, those who are not like us must be kept separate. Seen as transgressing basic laws, wasted people become taboo themselves.[56] In many societies, one important taboo is "the provision against the dangers incurred by handling or coming in contact with corpses, by eating certain foods, etc." Additionally, taboo includes "the sanctity (or uncleanness) which results from the violation of the prohibition."[57] Uncleanliness affixes itself onto the violator. One becomes contaminated by what one touches, as though infected by the material one encounters:

> Or if a soul touch any unclean thing, whether it be a carcass of an unclean beast, or a carcass of unclean cattle, or the carcass of unclean creeping things, and if it be hidden from him; he also shall be unclean, and guilty. Or if he touch the uncleanness of man, whatsoever uncleanness it be that a man shall be defiled withal, and it be hid from him; when he knoweth of it, then he shall be guilty. (*Lev.* 5: 2–3)

Those who touch corpses become polluted, just as corpses themselves are suspected to be. Hence, washing dead bodies becomes a corporeal work of mercy, for who would want to *voluntarily* elect to enter the ranks of the wasted?

Linguistic difference constitutes a means for distinguishing between *us* and *them*. The *shibboleth*, now a figurative term for a catchword distinctive to a group's followers, originates from a moment in the Hebrew Bible. Jephthah forces those suspected of those not belonging to his own Gileadites to say the word "Sibboleth" (*Judges* 12: 4–6). The enemy Gileadites "could not pronounce the word properly," whereupon they would be executed. The word enters English in 1382 in the Wycliffite translation of the Bible as "Sebolech" (see OED).[58] In the following decade, Chaucer references Flemish immigrants with their proverb "sooth pley, quaad pley" (I.4357). This could be his subtle reference to the tradition that rebels in 1381 marked and killed Flemish immigrants by making them say, "Bread and cheese," which Flemings could only say in their distinctive accent.[59]

Distinctions among living human bodies reflect spatial distinctions. Marginal peoples live, generally speaking, on the margins. These peripheral areas, populated by the waste of society, literally come to be associated with waste.[60] Not an innovation of the modern era, walled medieval towns were sometimes marked by the presence of hospitals, particularly on the margins, forming a boundary around towns that heightened the distinction between urban and rural and fostered a sense of urban identity. Hospitals treating the diseased and sick often appeared at the major gates of cities.[61] Stigmatized groups, such as lepers, would be placed at the boundaries of towns.[62] The South Bank in London, for example, remains the site of rejection—filthy in its literal and material manifestations—even today, in the form of immigrants newly arrived to the city.[63] Projective filthing extends to geographical and topological liminality. While West Egg is perceived of as inferior to East Egg, there is a worse place in which to reside. Between West and East Egg is an abject marginal area, "a certain desolate area of land. This is a valley of ashes—a fantastic farm where ashes grow like wheat into ridges and hills and grotesque gardens; where ashes take the forms of houses and chimneys and rising smoke and, finally, with a transcendent effort, of men who move dimly and already crumbling through the powdery air." Tended to by "ash-gray men,"[64] whose mysterious work lingers ominously throughout the novel, the ash-piles in this wastescape "become a metaphor . . . rehabilitated into the world of ideal reason."[65]

Wasted humans, not seen as "real," are uncanny figments to be ignored and shunted aside. In *Seedfolks*, Leona decides that an urban blight has to be cleaned up by the city:

> Then I studied all the trash on the ground. Don't know why anyone called that lot "vacant." The garbage was piled high as your waist, some of it from the neighborhood and some dropped off by outside people. The ones who don't want to pay at the dump, or got dangerous chemicals, or think we're such slobs down here we won't mind another load of junk. We can't get City Hall to *pick up* our trash, but we got it *delivered* just fine.[66]

Given the runaround over the phone, from city office to another city office, Leona looks for "the key that would make that trash disappear."[67] Finally she goes in person to the Public Health Department "to make myself real to 'em." The "dolled-up receptionist" pretty much ignores her until Leona opens up a garbage bag. "I was *definitely* real to them now."[68] The next thing we read, an even more wasted group than the lower-class neighbors of this urban area—prisoners from the local jail—are cleaning up the trash-filled lot.

Marginal status does not transfigure wasted humans into necessarily more compassionate human beings, celebrating mutual affection *among* wasted humans. As Katherine Boo writes in *Behind the Beautiful Forevers: Life, Death and Hope in a Mumbai Undercity*, being poor does not confer automatic virtue on a soul. In a garbage-filled slum of Mumbai, the impoverished "improved their lots by beggaring the life chances of other poor people."[69] Even among filth cleaners, hierarchies insinuate themselves. Richard Jago in his pastoral poem, "The Scavengers: A Town Eclogue," presents two street scavengers chatting about their work, lending humanity to workers most often ignored or passed over: "Last of the toiling race there lived a pair, / Bred up in labour, and inured to care; / To sweep the streets their task from sun to sun, / And seek the nastiness which others shun." The wife happily recalls how the husband made money from the dung as fertilizer: "Again you sold its well digested store, / To dung the garden where it grew before." The husband observes that some folks thought themselves too fine for this scavenging pair: "What though the beaux and powder'd coxcombs jeer'd, / And at the scavenger's employment sneer'd, / Yet then at night content I told my gains, / And thought well paid their malice, and my pains." After all, "Sweet is the scent that from advantage springs, / And nothing dirty which good interest brings." They clearly earned a decent living at this toil. The woman observes cattle and remarks, "Their fat the butchers pleased, but me their dung." The wife reminisces, "Ah! Gaffer Pestel, what brave days were those, / When higher than our house our muckhill rose! / The growing mount I view'd with joyful eyes, / And mark'd what each load added to its size. / Wrapp'd in its fragrant steam we often sat, / And to its praises held delightful chat." The rising dunghill is described in terms of community—how she and her husband enjoyed discourse with one another in its sight. Although they worked *in* filth, they *were* not filthy. As the wife points out, "For sure 'twas dirtiness that kept me clean: / Clean was my gown on Sundays, if not fine, / Nor Mrs ***'s cap so white as mine. / A slut in silk or kersey is the same, / Nor sweetest always is the finest dame." Although they were seen as dirty because of their labor, in fact, they were cleaner than supposedly finer folk.[70] They proudly distinguish themselves from others they disdain for their lowliness and dirt.

In *Seedfolks*, even among wasted humans, there exists the desire to stay with one's own kind. Sam notices the gardeners self-segregate, "the blacks on one side, the whites on another, the Central Americans and Asians toward the back."[71] People still dump trash; a "crazy homeless man"[72] rips out plants; people put up fences. "God, who made Eden, also wrecked the Tower of Babel, by dividing people. From Paradise, the garden was turning back into Cleveland."[73] The idealized garden threatens to return to abjectness by replicating the antagonizing divisions of the city surrounding it.

Even Calvino, in his lyrical meditation on refuse, admits that his garbage collector will one day become a garbage *producer* in the climb up the social ladder: "His having been taken on as a dustbin man is the first step up a social ladder that will eventually make today's pariah another member of the consumer society and like everybody else a producer of refuse."[74]

The dustbin man is "the entry point into a system that swallows up men and remakes them in its own likeness and image."[75] Immigrants and people of color typically carry away the trash of those with deeper roots in the city. Unlike his counterpart in Italy where "the position of dustbin man is defended as a stable occupation," in Calvino's Paris, "Loading the dustcart, the immigrant in his first job discovers the city as one might look at the wrong side of a carpet: he judges the wealth or the poverty of the different areas from the quality of their refuse, which then stimulates his dreams of the consumer's destiny that awaits him."[76] Eventually that immigrant dustbin man will move up and become a *consumer* of detritus rather than a collector of it. He, in turn, can be replaced by a newer arrival "escaping from the deserts of the 'developing countries.'"[77] The wasted do not consume; in late capitalism, one must consume and waste—in order not to become wasted oneself.[78] Slipping out of the tentacles of the capitalistic system, waste "cannot, or can no longer, be commodified," hence its dangerous aura. Similarly, wasted lives, those people who cannot consume, escape capitalism's confines.[79]

In *The Journey*, Johann the street cleaner sees the ghost train go by, not understanding the "symbol on their breast." He sees that

> there's dirt everywhere, and it all has to be cleaned up. It's the same in France as it is in Russia. But then why do people go to war? Everyone should just take care of their own streets at home. Street sweepers can get along wherever and work together to clean things up... Everyone has his own broom to sweep.

Johann imagines when the war is over and all can celebrate with him to clean up the dirt:

> Then all of the streets will be full of people, one next to the other, the people will leave behind a ton of trash, though it won't matter, for that's also part of the victory and Johann will happily put in the overtime so that Leitenberg is clean once again, if only it means peace, finally peace.[80]

He realizes that if he showed sympathy to Jews, he would go to jail. You have to think of yourself first. Even as a street sweeper, Johann has privileges. Among undifferentiated rubbish, differences materialize.

Aliens, Foreigners, and Immigrants

> The one by whom the abject exists is thus a *deject* who place (himself), *separates* (himself), situates (himself), and therefore *strays* instead of getting his bearings, desiring, belonging, or refusing... The deject is in short a *stray*.
>
> Julia Kristeva[81]

Wonders of the East, found in the late tenth-century *Beowulf* manuscript,[82] contains descriptions of marvels and otherworldly creatures in far-off lands. The description of 13-foot-tall women, with "boar's tusks and hair down to their heels and oxen's tails from their loins," amuses us at first by this unlikely hodgepodge of characteristics. Foreigners are not really human; their bodily emissions, for example, must be fouler and more infectious than our own. The Other must be rendered dissimilar in language. If this difference succeeds, then the Other really *is* different. "They have camel's feet and ass's teeth. Because of their uncleanliness they were slain by the great Macedonian, Alexander. When he could not take them alive, he killed them because they are foul in body and contemptible."[83] The undeniable greatness of Alexander's legacy validates his slaughter of these (literally) animalistic women. Their difference makes them unclean, foul, and contemptible. The physical lack of unity—the hodgepodge of haphazard body parts—renders them available for slaughter. Foreigners, too far off to actually observe, become monstrous hybrids, readily expendable. While these impossible creatures mystify and attract, their total otherness offers them up for dissection. Not truly human, they lie on the operating table of analysis, ready to be sacrificed for their otherness, to make *us* whole, complete, and defiantly *not* Other.

The American myth that all immigrants were welcome does not hold true. The United States has never really welcomed everyone. Emma Lazarus's "The New Colossus" (1883), famously displayed at the Statue of Liberty, calls this totemic bronze icon the "Mother of Exiles." Lazarus credits her with saying, "with silent lips. 'Give me your tired, your poor / Your huddled masses yearning to breathe free, / The wretched refuse of your teeming shore, / Send these, the homeless, tempest-tossed to me.'" Though intended sympathetically, immigrants are referred to as "refuse." Is the "container" the United States, whereby they are discarded only to be recycled as Americans? Or is the container old Europe, rejecting this flotsam from its ghettoes, liberating it from disease and poverty? Immigrants were usually physically filthy upon arrival in ships. If we "other" a group successfully enough, ultimately they then turn into the very filth we accuse them of embodying. This makes us innocent of victimizing others—see, they really *are* filthy![84] Literally soiled, they prove their affinity to filth and waste. Rejecting immigrants rhetorically made cognate with waste,[85] we physically designate them to waste dumps created by us, in a pathetic fallacy of association. Al Young, a California poet laureate, in his poem "Watsonville after the Quake" refers to the invisibility of devastated immigrants:

For all their huff & puff & blow your house down,
the mediators of disaster and distress
didn't find this sickly devastation sexy.

> Besides, who's going to cry or lose sleep
> over a spaced out, tar papered, toppled down town
> by the sea, brown now with alien debris?

These others, Mexicans in their makeshift tents, "alien" and "brown" themselves, constitute "debris" to be abandoned, forgotten by the media and those living around them.

Colonized Peoples

> Where the European has trod, death seems to pursue the aboriginal.
>
> Charles Darwin[86]

As Alexa Bingham points out, SD discourse has been criticized for being paternalistic and ethnocentric. "Development" is a problematic word, vague and "vulnerable to manipulation."[87] In her case study concerning a dam project in Brazil, we can see the dam project is promoted as an example of "clean energy," since "clean" is judged in terms of emissions. But an "eco-populist" campaign booklet argues otherwise: "Clean energy can be there in the receiving companies, but in the Tapajós basin there will be dirt and pollution."[88] What does "clean" signify? Modernity requires developing or colonized countries to become addicted to waste creation, consumption, and disposal.[89] Ironically, while rich countries try to persuade "developing" countries to have birth control because the birth rate is out of control, in fact it is those rich countries that use up the most resources proportionately in the world. "Overpopulation" is a loaded term, imbued by the hegemony of capitalism. The poor are suspect because they are "flawed consumers."[90] Although all those rich people drain the world's resources, the First World encourages the Third World to "develop."

Among countries, a lower one becomes the waste dump for another and provides "waste" workers for higher caste consumers. In the US colonialization of the Philippines in the late nineteeth and early twentieth centuries, Americans figured and constructed Filipinos as lower, animalistic, and feces-ridden in opposition to a medicalized, sanitized civilization embodied by their own "'classical' bodies: a completed body, isolated and closed off from other bodies, something individual and expressive."[91] Part of the "colonial project" was to somatically discipline the locals.[92] In contrast to the life-affirming lower bodily stratum celebrated by Bakhtin, in this instance

> the medical grotesque is more destructive than regenerating, more debasing than liberating. Physicians thus construed the Filipino body as an essential negation: of health, of discipline, of civilization. The colonizing process must therefore be a "civilizing" process.[93]

In the Philippines, starting in 1914, Americans, in an attempt to impose "civilized" habits of sanitation and cleanliness on the local people whom they figured as filthy and debased, created "Clean-Up Week" to beautify one's town. This included privy day, when one was expected to attend to the building or repairing of one's privy.[94] Excremental colonial sites threaten the integrity of European identity,[95] but simultaneously confer the smug sanctimoniousness of Western superiority.

Once something has lost its usefulness and become trash, it is convenient to send it to poor parts of the world and then those parts of the world materially "become" trash. As Michel Serres points out, "the rich readily discharge waste—another case of *dumping*—where the very poorest live."[96] The gruesome and stomach-churning inducing story by Marcus Heeresma, "Dumping Ground," set in the massive rubbish tip on the outskirts of Lima, Peru, features a Dutch protagonist, whose house in his native land sits on a canal.[97] This voyeur of third-world countries, or "dumping grounds" as he calls them, finds the filth of poverty sexually exciting.[98] This compact story epitomizes many aspects of wasted peoples, in terms of class, gender, race, and global politics. He exploits a 15-year-old girl, part of their staff, who receives his wife's cast-off panties. "The man thinks it rather exciting and *piquant*, actually, to have the little one in his bed in the reduced underwear of his spouse."[99] Listening to the Vienna Choir Boys, he drives in his car to the rubbish heap where "thousands on the tip are busily at it, spread in among the dirt. Silently."[100]

Some of the "rag-wearers"

> press themselves up against the window again. A primitive life-form, the man thinks. A barely viable variant of shit, dung and garbage creepers and gorgers, of dumpground-shuffling rustlers, that has raised itself up on its hind legs. The only solution, even so, would be, never mind the politics: wire netting round the plain and set the flame throwers on them.[101]

Little better than insects, they deserve to be exterminated. And what is wrong with insects? Throughout the story, told from the man's consciousness, cockroaches who continually ease their way into their car via holes by the pedals are periodically exterminated with a hammer he has brought along for the purpose. As he hammers each bug to a squashed mess, he gazes and despises the "waste products" he sees laboring away.[102] He sees a naked Cholo girl, "still Indian, primarily," perhaps 13 or 14 years old. Gazing at her almost naked body, her legs spread apart in the dirt, he starts to masturbate. While he fondles himself, he sees a "strong negro" pushing a garbage cart that he tips over the girl. Our shock and revulsion turns to disbelief when she laughs and, seemingly willingly and with delight, gambols with him on rubbish.[103]

Meanwhile, the *cucarachas* have not been exterminated as thoroughly as before owing to his onomastic diversion. They cover his belly and thighs, and this excites him. Then the car, "that sinks like the immense garbage tip," itself starts to move. Suddenly the "dazed man" realizes "that he is covered in those sticky waste and excrement eaters." The rubbish collectors have taken the car and attack it. It goes down "one of the sloping garbage-sides of a deep, black-scorched pit. In a moment or two, the car is covered with a slow, insect-like layer of crawling people. The four doors are pulled open. Almost simultaneously, the car fills up."[104] The man, who "wasted" these people, this culture, this land, becomes waste himself. Inevitably, the activity of trashing others—in this case making them cognate to material waste—comes home to roost in an excrement-laden nest, in this phantasmagorical, allegorical nightmare of colonialism and waste.

As colonial allegory, the idealized horses in *Gulliver's Travels* treat the barbaric Yahoos as Europeans treated the benighted peoples they came to conquer. The *Houyhnhnms* at their Grand Assembly acknowledge that Yahoos "were the most filthy, noisome, and deformed Animal which Nature ever produced."[105] Indeed, they have a creation myth about the Yahoos in which they might have been produced by "Mud and Slime, or from the Ooze and Froth of the Sea,"[106] somehow arriving on the island of the *Houyhnhnms* and since degenerating. While the Master's view is that

> although he hated the *Yahoos* of this Country, yet he no more blamed them for their odious Qualities, than he did a *Gnnayh* (a Bird of Prey) for its Cruelty, or a sharp Stone for cutting his Hoof.[107]

The *Houyhnhnms* are not perfect, using the Yahoos as slaves. The color-based hierarchy in their own breeding, Gulliver argues, brings about a perfect balance in behavior and appearance among horses. In the class hierarchy among humans, the nobler one is, the weaker and more vile one becomes, until the nobleman's wife is reduced to sleeping with her "neighbours, or Domesticks, in order to improve and continue the breed."[108] Gulliver rails against the perfidy of those who colonize under the false flag of introducing civilization, whereby natives "[are] driven out or destroyed, their Princes tortured to discover their Gold; a free Licence given to all Acts of Inhumanity and Lust: the Earth reeking with the Blood of the Inhabitants: And this execrable Crew of Butchers employed in so pious an Expedition, is a *modern Colony* sent to convert and civilize an idolatrous and barbarous People."[109]

Ironically, he contributes to the further degradation of his fellow *Yahoos*. Telling his Master of the convention of castrating young male horses, Gulliver spurs the virtuous *Houyhnhnms* to consider, in their great reason, that this is an excellent way to rid themselves of *Yahoos* through their gradual demise.[110]

Self-hatred divides Gulliver from the *Yahoos*. Because of his very affinity, Gulliver's repulsion induces him to identify himself with the horses. Attempting to make distinctions between himself and his hated self/other, he sees his own language as "barbarous."[111] Yet the *Houyhnhnms*, in their fear that Gulliver through his advanced state of being could become the Yahoos' leader and lead an insurrection, decide he must leave. The grieving Gulliver, nauseated by his own kind, can barely allow himself to wear the humane Portuguese captain's shirts until they are washed for fear of being "defile[d]."[112] He loathes his family:

> As soon as I entered the House, my Wife took me in her Arms, and kissed me; at which, having not been used to the touch of that odious Animal for so many Years, I fell in a swoon for almost an Hour... During the first Year I could not endure my Wife or Children in my Presence, the very Smell of them was intolerable; much less could I suffer them to eat in the same Room. To this Hour they dare not presume to touch my Bread, or drink out of the same Cup.[113]

Gulliver thinks he has distinguished himself from his fellow *Yahoos*, yet he sails out using *Yahoo* children's skins,[114] indicting him in the very perversity he claims to scorn. In this vile parody, Gulliver distances himself from the *Yahoos* by exploiting them as the *Houyhnhnms* do, even utilizing Yahoo tallow and finding horror at own children: "And when I began to consider, that by copulating with one of the Yahoo-Species, I had become a Parent of more; it struck me with the utmost Shame, Confusion and Horror."[115]

He cannot get off his high horse.

Sanitary Slaughter

> This cursed Jew.
>
> The Prioress, Geoffrey Chaucer's *The Canterbury Tales* (VII.570)

It is no accident that Chaucer's tale for his anti-Jewish Prioress should associate the filth of a privy with the "Hebrayk peple" (VII.560).[116] Peter the Venerable of Cluny, in his 1143 treatise against Jews, describes them as "the excrement of the human species" (*humani generis feces*).[117] The link between Jews and excrement, embedded as "part of a much larger iconography of sinners and moral corruption,"[118] rendered them as so literally other as to constitute another race.[119] Lothario Dei Segni writes of a Jewish siege and a "wretched" Jewish mother, who, driven mad with hunger, eats her own baby. Lothario adds that, since she had been a wealthy noblewoman brought low, "she became insane with resentment." After roasting and eating part of her son, the robbers, smelling the delectable roasted flesh, rush in and are horrified. Describing her as "more cruel than the robbers themselves"[120] for urging to share in the meal, it is hard to know whether Lothario's disgust is aimed at her for her Jewishness or her perversity as a mother, but he seems to intermingle these two aspects, reflecting one another. The eating of her boy child functions as a perversion of the Christian mass, echoing blood libel cases rife in the later Middle Ages.

While Martin Luther early on was pro-Judaic, he later turns against Jews, even going so far as to recommend committing violence against them. In linking Jews and excrement, Luther relates a tale about a "malicious rabbi [who] does not call the dear mother of Christ *Maria* but *haria*—i.e., *sterquilinium*, a dung heap."[121] Written in Germany during the time of Luther's activity, *Till Eulenspiegel*'s eponymous protagonist uses excrement to humiliate or commit vengeance; shit works as a form of "satire and aggression"[122] in anti-Semitic rhetoric. In one tale, Till notices that a young man is selling expensive musk at fairtime in Frankfurt. Coincidentally, Till is bitten on the rear by a flea, and, while attending to it, discovers some "knots on his behind,"[123] what are colloquially referred to as "dingleberries." Till wraps them up in material and sets himself up as a seller. Eventually three "rich Jews" come by. Till tells them "they were prophet's berries, and that whoever put one of them into his mouth, and then stuck one in his nose, would speak the truth forever after." The Jews, wanting to be able to know when their Messiah will come, readily decide to buy them. Till says, "If you don't wish to pay them, you dogs, take off and leave the junk where it is." A Jew named Issac then eats it after ritual fasting and prayer. "Then the man had it in his mouth, Moses asked him, 'Well, dear Isaac, how does it taste?' 'Servant of God! We've been cheated by Goi! This is nothing other than human excrement.' So they all tasted the prophet's berry—until they saw the 'tree' on which the berry had grown."[124] By *ingesting* excrement, these Jews are physically identified with and even *become* it. The taboo of excrement ingestion plays with anxieties about the Eucharist transforming into excrement in human bowels.[125] Till clearly sees having Jews consume fecal filth as a good joke.

Trashed as a group, the specter of Jewish infection wafts through *The Great Gatsby* like a contaminated miasma. Merely associating with a Jew could taint the non-Jew. The photographer's wife, Lucille McKee, talks about a mistake she made in almost marrying "a little kike...I knew he was below me. Everybody kept saying to me: 'Lucille, that man's way below you!'"[126] With ominous prescience, owning a company called "The Swastica Holding Company," Wolfsheim personifies anti-Semitic views about Jews and money.[127]

Within National Socialist logic, it was the utmost extension of logic to eliminate human beings who were judged to be waste.[128] Did waste systems and treatment in Germany come to *influence* or simply *reflect* the National Socialist treatment of wasted humans? Did Germany burn trash in the 1920s and 1930s, making it easy later to incinerate Jews? Was trash put into landfills so that Jews, who were seen as trash, could be dumped into ghettos? Was recycling a valued endeavor, so gold could be extracted from teeth for reuse? Did Germany practice source reduction, in which case it is unproblematic to eliminate Jews, cognate to trash? The Nazis literally trashed Jews by ruthlessly garbaging their possessions. Recently found were "piles of looted Jewish possessions" in Klandorf, 30 miles north of Berlin,[129] placed there after the savagery of Kristallnacht (November 9, 1938). This dump signified not only how the Nazis considered Jews and all they possessed as mere trash, but also how the perpetrators wished to keep their own trash-making expertise hidden and peripheral.

Division among the wasted in camps likewise was divided between those to be immediately incinerated and those whose inevitable disposal was postponed. In *Seedfolks*, the Indian, Amir, notices one Polish lady can't bring herself to thin her seedlings. "She looked down at them and said she knew she ought to do it, but that this task reminded her too closely of her concentration camp, where the prisoners were inspected each morning and divided into two lines—the healthy to live and the others to die."[130] This survivor remains haunted by the disposal of her fellow Jews.

H. G. Adler's novel, *The Journey*, recurrently uses waste as a theme—rubbish, detritus, filth, and garbage.[131] Aspects of memory, history, urban sanitation, and literary structure collude to express the wasting of entire populations of people. The filthing of one group of people seeps into the cultural groundwater. People come to abandon others easily. "Those who live on create special memorials to the dead, whom they call the departed, wishing still to possess them or to appear to possess them, though they also turn them into rubbish or refuse, which is easier to abandon when necessary."[132] A mocking voice describes how people disdained in life are given tombs and stones in death as though that makes it all right. "Existence is all that remains, an almost incomprehensible collection of detritus left over."[133] Jews are held in barracks:

> The sinister is buried or, without anyone noticing, is thrown out with the rubbish that the town hauls away once a week and dumps onto open ground near the Scharnhorst barracks in Leitenberg. There all despair is tossed away and forgotten, for so much intense disgust is associated with the useless that the community has decided as one to block it out of their immediate view.[134]

As is typical for all wasted peoples, Jews are understood to be rubbish and, as such, disposable:

> For that's how a fiend spreads trouble, who then is only satisfied when it leads to an orgy of destruction that swallows up everything in its reach. That, however, is when the rubbish blows about! And it blows around as well what is not yet rubbish, but will be! It has to be stepped on and kicked out of the way, its memory left to rot![135]

Once forgotten, they can be disposed of like waste; once wasted, they can be disposed of and forgotten. The camp of Ruhenthal is itself described as "useless, completely useless, nothing but rubbish."[136] Rubbish, and those constituting it, should be eliminated and banished from existence and even memory. Here, there is no codification, no distinction between clean and dirty, because everything is dirty, including, just as the Nazis argued, Jews.[137]

One narrative voice delivers an anonymous address to Jews; rubbish reflects inner or metaphysical rubbish:

> The stinking rubble that is the mirror of your own unquestionable hideousness, what you yourselves are and what you still don't wish to recognize, though it's the despair within yourselves that makes you long for the help of the Beautification Association... You are rubbish, but the kind that is not allowed between table and bed, between chair and cupboard. Rubbish mixes with rubbish, and sin with sin, all of it a disgusting gruel that is only good for the vermin that help it to rot even further.

The narrator dismisses hopeful ideas in the minds of Jews:

> You are your own graveyards... Yet you still long to get away from the rubbish; you still long to be elsewhere, which only demonstrates how disingenuous all those ideas are about beautification floating around inside your head.[138]

In imagining a time of no destination, just journey, Paul contemplates a language without verbs, whereupon language would be "destroyed."[139] Language *has* lost "all meaning"; words "are barked out"; life "can no longer even be picked out of the wasteland of rubbish."[140] Animals would take over; Mutsch the cat would roar, "Everything is rubbish!" Then the street sweepers would be overwhelmed: "The endless stream of rubble overwhelms the street sweepers,"[141] just as it has overtaken language itself.

Time affects social exclusion. We do not want to be associated with dying people or the elderly of the workforce.[142] Likened to rubbish, the elderly, along with the very young, wear diapers. Neither contributes economically to society, but children have at least the potential to one day be useful. Since we all fear becoming trash ourselves and in order not to be reminded that we ourselves will one day be seen as trash, we sequester the elderly, those visible ghosts who we refuse to remember, into nursing homes. In *Endgame*, Nagg and Nell are relegated to garbage, literally confined to trashcans.[143] Both the old (Nagg and Nell) are like babies—needing constant attention, their "pap,"[144] and to be cleaned up.[145] Their wasted bodies are just stumps because of an accident they nostalgically allude to.[146] Hamm says to chuck this muck into the sea, wanting to remove any reminder of death. While it is hard to imagine a grimmer scene than the Beckettian world of annihilation, at least a grim humor resonates.

In *The Journey*, on the other hand, language excludes hope. Leopold embodies the waste of waste, the double whammy: an old Jew.

> The old ones should be happy with their lot, a teeming heap of rubbish, for there they can wait patiently until they are dragged off, though if they don't want that, then, indeed, there are plenty of small jobs by which such creatures can make themselves useful. Get away from the rubbish! You've had it too good for too long!... It was a clever idea to gather together thousands of old folks in Ruhenthal... They cleaned the toilets and swept the streets.[147]

Anything—anyone—useless is rubbish and must be disposed of. The living dispose rubbish, including people once valued, until they themselves are stripped of value. In the film *Judgment at Nuremberg* (1961), Burt Lancaster's character mirrors the association between Jews and filth to indict himself and his fellow Nazi judges who upheld laws of the Third Reich as corrupt butchers. "Ernst Janning," he says, referring to himself, "who made his life excrement because he walked with them [his fellow judges]." At least he *chose* this path to the [im]moral privy, while his Jewish victims wielded no agency.

In a brilliant tour de force, Adler enters us into the voice of ultimate Evil and perversity—where he welcomes us to see the slaughter of people as a sanitary solution to societal pollution. "The crematorium is practical and hygienic. It's one of the nicest and most useful inventions of the modern era."[148] The sarcastic narrator invokes how tasteful the crematorium process is. "How wonderful! And cheap!"[149]—an advertisement for the Holocaust and mass murder:

> Also, our executions are carried out in the quickest manner in our crematoriums. After disrobing, the patients are shot from behind on marble tiles, everything done with the utmost consideration in order to avoid any undesirable mess-ups. The corpse is then placed immediately on a conveyor that

> feeds into the fire of the furnace such that the lifeless corpse is never touched by human hands. As a result the danger of infection is reduced to a minimum. The perfect diet! Success guaranteed! Other methods of execution that are supposedly as good can hardly compare!

The final product of the Final Solution is neatly wrapped up: "At reasonable cost the ashes can be sent to your house in a simple mail packet that holds an urn carefully wrapped in wastepaper." Fortunately, there are sanitary regulations. "Spitting within the crematorium, and especially during executions, is strictly forbidden!"[150] Adler's tragically sardonic mockery of mass execution plays off of the repressed bodily fluids that Douglas and Kristeva so vehemently assert repulse us.

Genocide is fine—just don't spit!

PART II

Energy Recovery and the Dynamic Power of Things

CHAPTER 9

The Secret Life of Objects: The Audacity of Thingness and the Poignancy of Materiality

After my parents died within two months of each other in the spring of 2010, my brothers and I were left with emptying the family house we had called home for 47 years. As both my parents were writers, the amount of paper and folders crammed into every file cabinet and container was daunting. I remember venturing into the attic one dreary day, depressed and downhearted. This top floor held my father's study, replete with hundreds of chemistry journals and books, as well as files chock-full of everything from book contracts to chapter versions from the 1960s. Then I braved one of the many closets. I carried masters' theses from my dad's students at New York University written in the 1950s and 1960s. Catching a corner awkwardly, I dropped these weighty tomes on my feet. Dispiritedly, I finally managed to clear a path to the file cabinet in the corner, where my mother had long hoped to retrieve her journals from the 1930s and 1940s. The metal drawer creaked open. Imagine my joy in finding income tax returns from the 1980s, packing lists for trips taken in the 1970s, and bills for items purchased decades before.

But—what's this? An envelope: "Bobby's baby curls." My father's curls from 1919. Then—my parents' college and graduate school diplomas. Their wedding certificate.

The treasure in the dross.

"You've Got That Thing, That Certain Thing."[1]

Well, here's another nice mesh you've gotten me into.

With apologizes to Oliver Hardy[2]

This book furthers emerging and overlapping areas of inquiry called by a variety of names: Discard Studies, New Materialism[s],[3] "ecomaterialism,"[4] material ecocriticism, the "material turn,"[5] vitalism, vibrant materialism, and Neomaterialism.[6] Other adjacent influences include Object-Oriented Ontology (OOO),[7] contemporary materialisms, stuff theory,[8] ecological posthumanism,[9] post/humanism,[10] speculative realism, actor-network theory, "flat ontology,"[11] Deleuzian "assemblage," thing theory, "relational materialities,"[12] "'material ethics' . . . based on the co-extensive materiality of human and nonhuman subjects,"[13] and "*alien phenomenology.*"[14] These approaches, situated against "the hubris of human exceptionalism,"[15] recognize objects as "full-fledged actors."[16] By "reconsider[ing] the human as just another object," humans gain "equal ontological status with all other things."[17] These approaches concur that "*things matter* . . . they possess integrity, power, independence and vibrancy."[18] This levelling of ontological status flattens the hierarchy privileging and centering on the human.

Jane Bennett's work on "thing-power"[19] proves useful for the study of waste. She asks if trash ("garbage, litter, dirt, debris, filth, refuse, detritus, rubbish, junk"—cognate to my waste) is "stuff whose power to move, speak, or make a difference has become dormant or dead." Have these materialities lost their thing-power or integrity?[20] The catalyst for Bennett's enterprise was provoked by the power she recognized in trash. One day spying some debris—including a glove, pollen, rat, bottle cap, and wood—Bennett claims its constituents spoke to her. Rather than being "irreducible to the culture of objects," the stuff she witnessed was comprised of "vivid entities." "Dead stuff" became "live presence . . . junk, then claimant."[21] Even waste has thing-power in Bennett's schema, where she argues, "Inanimate things have a life of their own, that deep within them is an inexplicable vitality or energy, a moment of independence from and resistance to us and other things. A kind of thing-power."[22] While we react to waste with horror, aversion, and avoidance, it manifests the potential to resist us.

The Garbage Project is an archeological-sociological attempt to figure out consumer patterns according to neighborhoods. William Rathje and Cullen Murphy have argued through their garbage archeological work that "historians are understandably drawn to written evidence of this kind." After all, "garbage has often served as a kind of tattle-tale, setting the record straight . . . Garbage, then, represents physical fact, not mythology."[23] As Paul Reyes describes the harrowing, disgusting work of helping out his father in the "trashing out" of foreclosed homes—the process whereby a team enters a now defunct domicile to clean out the remaining junk so that it can be torn down or resold—"It is a guilt-ridden literary forensics, because to confront the junk is to confront the individuality being purged from a place."[24] He cannot get inured to the variety of items left over and abandoned: "The squalor is a shock every time, each excavation a peek into a state of mind, like dismantling some diorama of dejection."[25] Objects provoke, cause, act upon us. Stacy Alaimo's concept of "transcorporeality" suggests the linkage of "humans to global networks of consumption, waste, and pollution, capturing the strange agencies of the ordinary stuff of our lives."[26] Things are not mere objects, but possess an independence from us. W. J. T. Mitchell memorably puts it this way: "Things . . . [signal] the moment when the object becomes Other, when the sardine can looks back, when the mute idol speaks, when the subject experiences the object as uncanny."[27] Possessing agency, matter itself is not static but "of a generative becoming."[28] Converging approaches focus on live matter, the virtue of stuff, the material life. Flattening the differences between the human and nonhuman, they aspire to help ecology. The more horizontal to one another humans and nonhumans are (as opposed to a hierarchy privileging humans), the better that will be for the earth.[29] By "raising the status of objects to the level of people,"[30] we can unsettle the centrality of the human whose intelligence has succeeded in fashioning "hyperobjects" like nuclear waste, things that "do not rot in our lifetimes."[31]

Anthropocentrism lies at the heart of unsustainable actions in ecology. I take Timothy Morton's point that Nature—a pristine impossibility, with its aura as a "holistic, healthy, real thing"—functions as a concept that perpetuates the division between human and nature.[32] Indeed, the very concept of Nature may lead to its very degradation by "allowing us to see ourselves as *other* than our environment, rather than embedded within it."[33] If we witness Nature as this inviolate/inviolable entity, we can only succeed in violating it. We appropriate our surroundings through ownership. Rejecting the term Nature as artificial, Morton suggests "ecology" as opposed to "environmentalism" as a less proprietary word. Michel Serres concurs, "Don't we actually admit as much when we say *environment*? That which surrounds man makes him into the center."[34]

Waste risks being seen, as Nature has been, as out there, not part of us, leading to the possible destruction of the planet. In castigating the human era, Myra J. Hird argues: "Waste has become *the* signifier of the Anthropocene, inaugurating the only epoch that centralizes humans."[35] Not only that, but we are not content to simply destroy the earth. "The Anthropocene marks an epoch of permanently temporary waste deposits left for imagined futures to resolve."[36] The Anthropocene, where "garbage is a sign of the human,"[37] has a lot to answer for. As Morton jokes—or is it laments?—"What to do with their poop will be the last thing on future humans' (or humanoids') minds."[38] Probing waste belongs to "dark ecology," where we can "explore all kinds of art forms as ecological . . . The ecological thought includes negativity and irony, ugliness and horror."[39] Waste, even "toxic waste,"[40] belongs here in Morton's dark ecology.

As A. R. Ammons in his National Book Award winning poem, *Garbage,* writes, "A waste of words, a flattened-down, smoothed- / over mesa of styrofoam verbiage; since words were / introduced here things have gone poorly for the / planet."[41] Eventually, "we may replace our mountains with / trash." The first human was a scavenger or chiffonier. Ammons imagines the time of the dinosaurs and then "along came the frail one, our ancestor, / scavenger, seed finder, nut cracker, fruit / picker, grubs, bulbs, etc., and here we are at / last, last, probably, behold, we have replaced / the meadows with oilslick."[42] Without that first scavenger, no civilization would have been created. Now we are only those "who define meaning / only in the blasted landfalls of no meaning."[43] As Ammons writes in *Garbage*, "stuff here is illusory."[44] New materialists shift our understanding as seeing our connection webbed and integrated into the universe "all around us, all the time. The goal here is to overcome the spatial distinction between ourselves and the rest of the biosphere."[45] The integral nature of humans with the "biosphere" constitutes the conclusion of Walt Whitman's "When I Heard the Learn'd Astronomer." The poet criticizes the "learn'd astronomer" who insists on dividing the stars into charts according to numbers, "charts and diagrams." The speaker, sickened by this triumphant master of the world around him, goes outside, "In the mystical moist night-air, and from time to time, / Look'd up in perfect silence at the stars."[46] Whitman understood the need to see ourselves *with* the world, not as separate *from* it.

Nature is an active agent, not a passive object, for "all are subjects."[47] As Bennett points out, "vital materiality can never really be thrown 'away,' for it continues its activities even as a discarded or unwanted commodity."[48] If "storied matter" has "narrative agency,"[49] what story does *waste* tell, as opposed to what story *we* tell *about* waste? Waste has "a life of its own . . . we can (and perhaps should) get into conversation and relationship with all the life and mindedness we encounter around and about us, whatever form it takes." From this, "new ways of knowing become possible." The "*productive* question"[50] might be, "What does waste *know*?"

Rather than exclusion, I concur with Morton that we should focus on "thinking of interconnectedness," the mesh,[51] which is "*the ecological thought*."[52] Bennett's insight that the "political potential" of thing materialism "resides in its ability to induce a greater sense of interconnectedness between humanity and nonhumanity."[53] The "interconnectedness of all material bodies,"[54] can help us to [re]consider the ethics of waste. For Morton, "The ecological thought . . . is realizing that we're always already responsible for the other."[55] If waste is seen as animate, it might make it more threatening. But there can be ways to tease out a more compassionate response to material and metaphorical waste.[56] What is disposable has energy; through trash meditation we recover energy from waste itself.

Bauman's wasted human is for Morton "the strange stranger [who] is not just the 'other'—the 'self' is this other."[57] No man is an island (to coin a phrase); "all beings are interconnected. This is the mesh."[58]

The Dignity of Things and Waste

> They shimmered back and forth between debris and thing—between, on the one hand, stuff to ignore . . . and, on the other hand, stuff that commanded attention in its own right, as existents in excess of their association with human meanings, habits, or projects . . . Stuff exhibited its thing-power.
>
> Jane Bennett[59]

Thing-power does not adhere simply to late capitalism and its overproduction of stuff. The dynamic agency of objects litters the literary canon, a repository of stuff and matter. *Beowulf* bears witness to how human constructions attempt to stave off the inevitable destructive forces of history. All cultural means to hold back social chaos—peace-weaving marriages, halls, *wergild* (man price—the paying off of one's criminal deeds), weapons, and gold artifacts—are ultimately futile. Women, crucial agents in the peace-weaving marriages, are "gold-adorned" ("goldhroden," 614), just as the hall Heorot likewise ornamented with gold ("goldfāh," 308, 1800). Marriage, constructed to contain violence, and the hall, constructed to exclude violence, are both intended to bring about a space of peace; yet these artifices are doomed to invasion, transgression, and destruction. As Beowulf himself points out, no matter how good the woman is, a peace-keeping marriage rarely maintains the peace. Just as men and women join in marriages that eventually disintegrate, so, too, is the hall Heorot joined but doomed (773b–775a).[60] Violence inevitably comes to halls (81b-85, 144–146a, 411–414), even Beowulf's own (2324–27a), due to incidents committed in the past that come to bear poisonous fruits. Warfare, chaos, and entropy characterize the natural state of the world; peace and progress are artificial, viable only in fantasy.

Beowulf tries to "own" his fight with Grendel when he relates his adventure to his Geatish lord, Hygelac. In this passage (2069b–2092b), he mentions something the reader did not earlier witness: the *glōf* (2085b), an intricately woven artifact, into which the monster Grendel presumably put dead body parts for later consumption or use. Like Judith's bag into which her maidservant puts Holofernes's head in another text of the *Beowulf* manuscript (Cotton Vitellius A xv), the glove functions as a garbage bag or man purse. Yet, why is it not mentioned earlier in the text? Has Beowulf made this up for Hygelac and, if so, why? In this retelling or adaptation, Beowulf mentions that his beloved companion, Hondscio, was killed by Grendel in the raid.[61] The glove is, perhaps, a perverse and displaced objectification of the hero's companion Hondscio, whose name literally signifies *glove*.[62] The killed companion as a glove, as an object, suggests the inability for constructed artifacts, even the most intricately woven or created, to survive and prevail, something the poem overtly suggests as well. Artifacts that do survive in *Beowulf* are those made by giants or gods, such as Beowulf's mail fashioned by Weland; the sword that kills Grendel's mother (1558, 1562), the hilt of which Beowulf later hands to Hroðgar (1679); the sword that Wiglaf uses to help Beowulf in his fight against the dragon (2616); and the sword with which Eofer kills his brother's murderer, Ongentheow (2979)—all are *e(o)tenisc/enta* or *giganta* (*geweorc*). Hrunting, Unferð's sword lent to Beowulf, is made by a weaponsmith (*wǣpna smið*, 1452) and is called an "ancient treasure" (*ealdġestrēona* 1458). The glory of the gods' supremacy in the past still glimmers in the glittering object capable of slicing open a human body—an object sentenced to hot gore and grim destruction.

Waste stalks *Beowulf*—in the many deaths of living beings and in the decay and destruction of culture and civilization. Just as man is doomed to die, the permanence of his art is transitory as well. Objects are material, tangible things doomed to decay. Once deteriorated and without utilitarian value, cultural artifacts become trash, insignificant in the wake of violence. Goods hoarded don't help the dying man; they are "lāf," a surviving remnant (1688); like the human body itself, they are "lǣne," transitory or on loan (1754). Once fetishized objects will eventually tarnish and decay ("Sceal se hearda helm (hyr)stedgolde / fǣtum befeallen" 2255–2256). As the Last Survivor tells us, even the iron coat of mail decays like its warrior owner ("ġē swylċe sēo herepād sīo æt hilde ġebād / ofer borda ġebræc bite īrena / brosnað æfter beorne," 2258–2260a). Objects convey poignancy. Treasures are consumed by rust (3049). Like a toxic waste dump, the concealed treasure, hoarded by the dragon, comes back to haunt Beowulf's people, poisoning their existence. Even the dragon's gold that precipitated this disaster is fated to be burned as waste (3010–3015). The burial of treasure, in a moment of *primary refuse*, underscores how it is useless ("unnyt," 3168). Even the barrow may be pillaged by the invading Swedes, its potential usefulness as a beacon to sailors and reminder of Beowulf's glory doomed to plundering.

Decaying objects retain their composure in Richard Wilbur's "Junk": "Yet the things themselves in thoughtless honor / Have kept composure like captives who would not / Talk under torture." When "material powers . . . call for our attentiveness, or even 'respect,'" we acknowledge the dignity in stuff.[63] Further, when things and humans are "partly strangers and partly kin," we free objects "from their merely instrumental status in a world of human uses."[64] By civil engagement with things, we are no longer separated from them. Yet Wilbur's poem sadly observes that eventually even "they shall waste in the weather toward what they were."

> The sun shall glory in the glitter of glass-chips,
> Foreseeing the salvage of the prisoned sand,
> And the blistering paint peel off in patches,
> That the good grain be discovered again.
> Then burnt, bulldozed, they shall all be buried
> To the depth of diamonds, in the making dark
> Where halt Hephaestus keeps his hammer
> And Wayland's work is worn away.

The "disappearance of objects" echoes sensually and elegiacally.[65]

The Utility of Uselessness

> Most wildlife species . . . are always going to be "useless," or occasionally annoying, from a human perspective . . . Wildlife is and should be useless in the same way art, music, poetry and even sports are useless.
>
> Richard Conniff, "Useless Creatures"[66]

In her work on dirt theory and material ecocriticism, Heather Sullivan cautions us against declaring stability between the categories "clean and unclean, sanitary and unsanitary, or the pure and the dirty . . . With dirt theory, we see that most of these boundaries are actually porous membranes." So why do we feel the need to discipline dirt, "an essential element in our environment"?[67] Codification with its cognate purification/cleansing expends energy in the wake of encroaching waste. Community ultimately results from codification, sometimes for better, though often for ill, as when codification can marginalize and demonize other humans. Culture attempts to control waste by giving it a use; for example, human excrement has long been cited in medical cures. Dirt's ability to purify has its basis in homeopathy. A Mayan recipe for expelling afterbirth included horse dung and herbs.[68] The dung of land crocodiles was used in cosmetics.[69] In a play by Marguerite de Navarre, *Le Mallade* [*The Sick Man*] (c. 1535), the wife suggests folk remedies to cure her husband, including the "shit of a pure white / Pigeon." In a system in which to be valued means to be useful, the useless is automatically considered to be waste. Humans insist on imposing the concept of commodity onto objects, so that even waste can become useful.

Waste as garbage has long been studied. Michael Thompson famously investigates how rubbish is socially defined.[70] Rubbish creates, making "possible transformations in use and value. Without a category of rubbish it is impossible for value to move or change."[71] All cultures have "three categories of possessable objects: valuable, valueless, and negatively valued."[72]

Objects are transient (and decrease in value over time) or durable (and increase in value over time). A third set of objects doesn't belong to these two. They constitute a "*covert*" category called *rubbish*.[73] Transient objects over time likewise become rubbish.[74] This transmogrifying penchant of rubbish suggests its thing-power, dependent on, but slipping away, from human perspective.

Historically, useful detritus has been monitored by the law. In the *Eyrbyggja Saga*, driftage, incredibly valuable in Iceland, was controlled by certain expectations, ones that Ospak, a blatant plunderer, ignores. After a terrible blizzard, the word goes out that a finback whale has come ashore. While the farmers cutting up the whale are dutifully following the expectations of shares and division, Ospak rows up with some men and brutally takes the meat. Here, fleshly flotsam in the form of a dead whale has become useful, even vital, providing the necessary calories for survival. Yet the rules for accessing that salvaged meat are disobeyed by some through violence; in turn, only hostility can resolve this transgression according to the rules of Icelandic sagas. What might normally be viewed as waste—a decaying, stinking corpse—becomes endowed with vitality necessary for the culture.

The dead cetacean has use for humans, as do many living nonhuman creatures. Anthropocentrism insists on use-value for allowing animals, for example, to thrive. In arguments for saving animals and wildlife, "the argument also implies that animals matter only because they benefit humans, or because just possibly, at some unknowable point in the future, they might benefit humans."[75] And it is not just animals that do or do not have use-value; geographical features are likewise judged. Remember those holes in the shires (chapter 6) begging to be filled with urban waste?[76] Where cities seem full and quarries seem empty, it becomes imperative to fill those empty spaces with the leftovers from the urban environment. An alternative story line would be to allow those holes to speak and not be co-opted by the cities. Seeing waste from their point of view permits this readjustment in our understanding. That something is "useless" should not automatically compel us to condemn or exploit it. For example, Marcus Hensel notes that in Anglo-Saxon poetry, an anthropocentric view of the landscape appears when human characters interact with it. Typically, especially in heroic or elegiac verse, the land comes off pretty negatively—as threatening, menacing, or malevolent. But from a deep ecology view, wherein the "use-value" of the land is immaterial, we can read the land differently, as neutrally evoked. Use-value requiring human judgment is an anthropocentric stance. An ecocentric reading would take the human, and its judgment, out of the equation.[77] Poetry can alter our view, making us see waste from its point of view.

Things are seen as waste if they are not *useful*. Yet nothing is inherently waste, if you look at it as a *thing*.[78] Perhaps redemption lies in the uselessness of stuff, in its very redundancy. Gay Hawkins analyzes the plastic bag lyrically dancing back and forth in the film, *American Beauty*.[79] While in one context a plastic bag constitutes filth and litter, in the film it is rendered beautifully and philosophically evocative, attaining layers of meaning as it reappears.[80] While celluloid aestheticizes the sublimity of a petroleum-based product, trash art on the streets is no less meaningful. A trashball costs 25 cents to buy. The creator sees in these one-inch plastic orbs "archeology, a divination of who people are from what they leave behind... 'The most important thing for me is the secret history of objects.'"[81] Still, plastic kills, leeching into marine animals diets as well as our own, "entangl[ing] us in ghastly nets of responsibility."[82]

In *Seedfolks*, the Haitian boy Virgil and his taxi driver father dig up lots of trash in their endeavor to turn the soil. One day, the boy finds a locket. "It was shaped like a heart and covered with rust, with a broken chain. I got it open. Inside was this tiny photo of a girl. She was white, with a sad-looking face. She had on this hat with flowers on it. I don't know why I kept it instead of tossing it on our trash pile."[83] Once his father's lettuce wilts and dies, he looks at the picture, remembering the Greek myths he'd learned about in fifth-grade. "In our book, the goddess of crops and the earth had a sad mouth and flowers around her, just like the girl in the locket." After cleaning the locket until it shines, he opens it and whispers, "'Save our lettuce,' to the girl."[84] The trashed locket becomes a talisman or relic with hidden deific powers.

"Waste has a kind of agency"[85] to enchant and disturb,[86] its possibilities suggesting the "networked quality of things."[87] Trash causes communication, such as between the trash collectors and those who throw away the trash in Paris. This contact "lies in the contents of the *poubelle*, a rich source of information indeed if anyone chose to consider them day by day."[88] While an object's uselessness generally renders it uncanny and taboo, at the same time, the fetishization of useless objects such as artwork make them commodities that do not signal waste, but lead to community. Even in the bleak desolation conjured up by Adler in *The Journey*, there is the acknowledgement that "perhaps someone else will need what Paul throws away."[89] Mindful of all around him, Paul takes his leave of objects. "He gets up and thanks the bed that has taken him in... Paul thanks the room that has put him up so well."[90] His acknowledgement of the vital nature of "bed" and "room" reflects the ethics of recognizing the vitality of things. We act ethically when we show "stewardship of objects."[91]

By a waterfall, I'm calling you-oo-oo-oo...

Sung by Dick Powell and Ruby Keeler in *Footlight Parade* (1933)[92]

The call from things is "more than a figure of speech, more than a projection of voice onto some inanimate stuff." Tellingly, it was trash that first called to Jane Bennett. With this "call from matter," she "caught a glimpse into a parallel world of vibrant, powerful things."[93]

Does trash call to us? Is it a thing? It *was* a thing. It is a *former* thing.

If objects are things that have agency and power, is it ethical to speak *for* them? Is it just to personify an object? It could be that "anthropomorphism... is, in short, anthropocentric."[94] *We* impose meaning on *it*. Juliet Fleming acknowledges the risks of anthropomorphizing, which smacks of "muteness envy: finding something that cannot 'speak' for itself you can speak for it, endow it with your own concerns, and, if you [are] a historian or literary critic, make speaking *its* concern."[95] Nevertheless, material ecocritics have coalesced to support anthropomorphism as a gesture of affinity, linking us *with* things. As Morton points out, "The problem resides not so much in the content as it does in the attitude that comes bundled with the accusation."[96] While thingness—taking the point of view of an object—smacks of colonizing it, we can be jolted out of our complacent perspective by acknowledging the "sentience" of the other.[97] Working to dismantle anthropocentrism, anthropomorphism "can be a narrative expedient intended to stress the agentic power of matter... reveal[ing] similarities and symmetries between the human and the nonhuman."[98] In the giving "voice to unspeakable things,"[99] we acknowledge their affinity to us, our affinity to them.

Emmanuel Levinas writes, "A thing exists in the midst of its wastes. When the kindling wood becomes smoke and ashes the identity of my table disappears. The wastes become indiscernible; the smoke drifts off anywhere."[100] While never fully formulating how "one substance in the midst of its wastes [encounters] another in the midst of its wastes," Levinas—Harman suggests—"invites us to consider" the problem. In *Totality and Infinity*, Levinas writes, "Things have no face"[101]—the face we must confront and acknowledge within his ethics. Levinas identifies the "Said" as reified knowledge and the "Saying" as the place ethics comes from, pointing out that in the Saying words are not frozen into a "literal meaning."[102]

> The *saying* is the fact that before the face I do not simply remain there contemplating it, I respond to it... It is difficult to be silent in someone's presence; this difficulty has its ultimate foundation in this signification proper to saying, whatever is the said. It is necessary to speak of something, of the rain and fine weather, no matter what, but to speak, to respond.[103]

Levinas writes, "But the face glows in the trace of the Other... His trace... *is* disturbance itself, imprinting itself... with irrefutable gravity."[104] And what leaves a sign better than waste, an "authentic trace [that] disturbs the world's order?"[105] In that profoundly spiritual atheist trilogy, *His Dark Materials*, Phillip Pullman's dust is a metaphor for consciousness. Dust, a form of waste, *makes* us conscious to act responsibly.

What if we argued that *things* have a Levinasian *face*—that we recognize their affinity to us?[106] In Calvino's "La Poubelle Agrééе," the dishes put in the dishwasher are appealingly cute. "Now the plates are all caged up in their little carriage, round faces astonished to find themselves standing upright, curved backs waiting for the storm about to break over them." Mute, their vulnerability appeals to the narrator who acknowledges their "exile" until the "cycle of cloudbursts, waterspouts and steam jets is over." Their relationship is mutual, his actions dependent on theirs: "This is the moment for me to go into action."[107] Calvino's narrator yokes himself to the "dustbin man."[108] The human is human not just because he can speak, but also because he *is speaking*: "Language as *saying* is an ethical openness to the other; as that which is *said*—reduced to fixed identity or synchronised presence—is an ontological closure to the other."[109] Only through the process of language can we acknowledge the other and ourselves. Our responsibility lies in the saying that disrupts the said or reified language. "Any reading which interrupts the said and proposes that the art work is a site of fracture makes us aware of the constant force of our responsibilities."[110] Mindfulness can enable a full, aware, non-judgmental, and loving experience of our bodies, the bodies of others—including objects—and the world.

Can we then, do we, anthropomorphize waste? Like gleaners, we see objects in terms of themselves, not in terms of something else—something useful or something culture perceives as valuable. Rabelais gives us the view of shitters from the point of view of the privy itself:

> Thy bung
> Hath flung
> Some dung
> On us.[111]

Rabelais forces us to see our shitting selves as Other and the privy as subject. The privy coheres to what Michael Uebel has called a "frontier."[112] This contact zone, "a continual realm of ambivalence," where the Other cannot be erased,[113] presents itself in the literary space where we and the Other collide, causing fissures and slippages in authoritative discourse to appear.[114] Dung leaves an ontological deposit, preventing us from seeing ourselves as wholly different.

Normally, we would see scraps of discarded paper as trash. In Leo Lionni's *little blue and little yellow* [*sic*][115], the heroes, little blue and little yellow, are depicted as torn pieces of colorful construction paper, as is everyone in their community, family, and school. One day, little blue sneaks out of his house to play and suddenly comes upon little yellow. They hug each other, eventually turning green, whereupon they run off to play. Upon returning home, their parents do not recognize them. "Little blue and little yellow were very sad. They cried big blue and yellow tears. They cried and cried until they were *all* tears." Once they had "pulled themselves together" (literally), they return home and hug; the color confusion, once clarified, results in happiness. "They all hugged each other with joy and the children played until suppertime." Possible readings could see a subtle argument for civil rights (written as it was in 1959), a comment on antimiscegenation laws, or even today an underlying defense for gay rights or a "green" reading (harmony produces a better relationship with the environment). The simple quality of the prose lends the clarity of the parable to this tale that pictures (literally) a multicolorful community of harmony. The materiality of color and torn paper forces us to confront our affinity with differences. The pun with tears/tears—in the first reading, little blue and little yellow cry "until they were all *tears*"—shows how they literally dissolve into tears that are "torn" from themselves. But they are also torn up bits of leftover paper scraps. This pun tears into us. We can become as easily fragmented as these little paper children and we tear up upon recognizing this.

Even our rubbish bin can be sentient. Heather O'Neill's trash can readies itself for the garbage dump. Her daughter, on seeing the now forsaken can, says,

> "Aww... He's all abandoned."
> "He's probably wondering what he did wrong," I said.
> "I wonder if he's wondering why we just walked by him."[116]

Filled with "guilt and sadness," O'Neill shares the history of the can. Originally a gift from her father, it followed her from fire escape to sidewalk as she moved from apartment to new abode throughout the city. She recognizes that her attachment to the can is uncanny. "Maybe because it was one of the last things left from the magical realm of my daughter's childhood, where everything was alive." Ultimately, O'Neill cannot bring herself to dump the damaged can now patched up with duct tape. "It looked as if it had been attacked by fierce pirates and come home to [my daughter] one last time to tell the marvelous tale." By anthropomorphizing the injured, elderly container, mother and daughter weave a mesh of warp and woof, human and nonhuman.

Generous Rocks

We need to "keep returning the focus to the nonhuman bodies of the hoard, considered as actants."[117] The role of humans could be "to serve as mouthpieces... for the non-human entities whose existence and whose concerns we hope to make present or knowable."[118] The riddle emerging from the Old English word-hoard allows us to do this. Its structure voices either an animal or an object from its first-person (first-object?) point of view or the speaker bears witness to the thing's inherent qualities. Since the riddles were most likely written by Christian monks, the objects appear, alongside humans, under the controlling guidance of God.

Our ethics need to be "attuned to the voices of things,"[119] including texts themselves. By listening to the "'call' from things," "insight into nonhuman agency might also be pursued via poetry."[120] Like things, we have materiality. The risk of seeing the thing-ness of humans is that we exploit other humans as things. But this might "be mitigated when the blurring of the human/nonhuman distinction is combined with the attempt to enhance the ethical standing *of things*."[121] We should not see *things* as *other*. "If matter itself is lively . . . the status of the shared materiality of all things is elevated."[122] By rendering anything or anyone—human, animal, multi- or single-cell, organic or inorganic—manifest, we lend it visibility and respect. We may even become its loving companion. As Marcel the Shell, the quirky speaking shell with pink sneakers and skis made from toenails, says about her lint-dog, "Lint is a shell's best friend."[123] And "Worthless," the heart-rending song in *Brave Little Toaster* (1987), provokes compassion for the cars retelling their life stories before they are crushed into oblivion by a terrifying machine. We feel the panic of the imminent demise of the brave little toaster and all his pals—Blanky the blanket, Lampy the lamp, the radio, and Kirby the vacuum cleaner—objects we have come to love and care for. The happy ending concludes with laughter when the morose Kirby quips, "Aah, you're all a bunch of junk." Do they laugh because they merrily accompany their "Master" to college? Or because the thought that they, beloved physical objects, are merely junk is utterly ridiculous, an insight we should take to heart.[124]

The sad little characters on the Island of Misfit Toys in *Rudolph the Red-Nosed Reindeer* (1964) tug at our heartstrings. Each misfit toy deviates from its proper category: the cowboy with the ostrich (instead of a horse), a polka-dotted elephant, and the jack-in-the-box named Charlie. Matter out of place from their appropriate class, these toys suffer, deprived of a child's love. Even a seemingly ideal doll, whose problem, producer Arthur Rankin Jr. suggested, was psychological, resides in this lonely, chilly, and remote location.[125] Yet we love these sad little creatures, anthropomorphized trains with square wheels and jelly-filled water pistols. Their thing-power transcends their pitiful state. Aware of being ostracized, they maintain their integrity, banding together in this frigid zone. The misfit toys suggest a key lesson: If things have dignity, then we cannot "waste" other humans as things, since things have inherent worth.[126]

Hildegard von Bingen's work on gemstones in her *Physica* demonstrates the thing-materialism inherent to jewels. Her gemstones "contain energy and moisture within them." They are useful for healing; "many actions can be carried out with them." It is even "the nature of precious stones to seek the honest and useful and reject the evil and false in human beings."[127] In suggesting how to activate this power, the human and inanimate together generate healing best. For a pain "in the heart, stomach, or side," simply carrying the smaragdus (emerald) will help owing to its warmth.[128] "Irrepressibly vital,"[129] rocks and stones are signs of living organisms.[130] To cure a headache, you take a smaragdus (emerald). "Place the stone in your mouth and warm it with your breath until it is moistened from your breathing." After stroking your "temples and forehead" with this moistened stone, you place it "in your mouth and keep it there for a short hour, and you will feel better,"[131] your headache dissipating. Improvement comes along with the collaboration between you and the stone, the human and inanimate, generating the power of healing, a therapeutic union of two actants in a networked mesh.[132] Bingen's onyx participates in the "interconnectedness of all material bodies"[133] by dissipating melancholia: "Focus your attention on the onyx and then put it immediately into your mouth; your mental depression will then cease."[134] Could it be this meditation on an object, a welcoming of its presence, remedies our anthropocentric malaise?[135]

Breath is vital for Hildegard—the catalyst for song and sound in her music. Saliva, though generally seen as a dirty effluence, here becomes healing. The objects are inert until activated by saliva. Just as the host is accidental bread until activated by the catalyst of the priest's words and becomes Christ's body, placing the stone in the mouth suggests a quasi-Eucharistic moment.[136] Medical ritual parallels religious ritual with the key element of vitally acting objects—the Eucharist or the gemstone—embodying or reflecting the power of divinity.

In the Middle Ages, "rock and human differ[ed] more by degree than by kind"[137] because "the line between the human and the natural world" was "porous."[138] In Chaucer's "Franklin's Tale," the rocks are varyingly seen: by Dorigen, as potential murderers of her husband; by Aurelius, as the path to Dorigen's bedchamber and sweet body; and by the magician, as the means to an ultimate source of revenue. Yet, they "cannot be said to be a merely passive reflector of competing human desires."[139] The question, "Which was the mooste fre, as thynketh yow?" (V.1622)—which the most generous—ossifies us. Perhaps, ultimately, the rocks *themselves* are the most generous, opening themselves to interpretation by others.

Anti-Anthropocentric Anthropomorphism

Bones rattled in the crypt where a tow-headed boy and his pals snuck in to play at the turn of the twentieth century. Clandestinely, the kids, frightened and exhilarated, terrified and exalted, defied religious and parental authority to creep into a subterranean space filled with the decay of their Viking forebears. The little boy, my grandfather, played with the ossified remains of his Scandinavian ancestors. Bones—the rattling discards of decayed corpses—became part of the web linking the boys to other humans and to the skeletal objects themselves.

For OOO and its related approaches like ecomaterialism and vibrant materialism, "Nothing has special status, but . . . everything exists equally."[140] As Ian Bogost explains, "To put things at the center of a new metaphysics also requires us to admit that they do not exist just *for* us."[141] While we can readily see the ocean as alive—with its fishy, mammalian, vegetative, and mineral (coral) creatures endangered—we are less able to see hard substances as vulnerable. Old English riddle 69, "Ice," poignantly observes: "A wondrous thing happened at sea: the water turned to bone."[142] By "going glacial,"[143] we can witness the transformation of water into ice quite readily. Less easily observed is the growth of bone due to its slow development.[144] An anthropocentric time frame can only perceive quick changes. That is why "'[o]bjects' *appear* [fixed] because their becoming proceeds at a speed or a level below the threshold of human discernment."[145] Vibrant materialism makes us aware of slow decay, porosity, and sympathy with the inorganic. Building on Latour's "slowciology,"[146] the "Ice" riddle jogs our long-term memory, forcing us to acknowledge that bone—a substance that appears solid, unchanging, and immutable—itself is subject to change, itself a productive and vital force. By turning to bone, the embodied sea in the Old English riddle distills how the environment is not separate from us.[147] Just as we have many bodies in us, in the form of microbiomes, the sea has many bodies in it. As a sign of a generative and becoming nature, it even, in this poem, becomes a body. This anthropomorphism works against anthropocentrism, the differences between us and the boney sea lapping ever smaller, pooling our commonalities.

Garbage is considered waste only because it once was something else, something "better." While it exists materially in space, garbage provokes reflections on time. In the Old English poem, "The Wanderer,"[148] the memory of a happy past is all the more painful when juxtaposed with the present, bereft of glee. "Thus this middle-earth / droops and decays every single day" (62–63):

> The wise man must realize how ghostly it will be
> when all the wealth of this world stands waste,
> as now here and there throughout this middle-earth
> walls stand blasted by wind,
> beaten by frost, the buildings crumbling. (73–77)

In a haunting passage known as the "Ubi sunt" ("Where have they gone?"), and famously adapted by J. R. R. Tolkien in *The Two Towers*, the poet reflects,

> Here wealth is fleeting, here friends are fleeting,
> here man is fleeting, here woman is fleeting,
> all the framework of this earth will stand empty. (108–110)

Untethered from his lord, the Wanderer sees himself as waste in this world of decay and despair. Another Anglo-Saxon poem, "The Ruin," muses on former joys. Buildings are in ruin, their architects dead; kingdoms eventually decline. Past joy, life, and glory contrast present decay:

> The roofs are ruined, the towers toppled,
> frost in the mortar has broken the gate,
> torn and worn and shorn by the storm,
> eaten through with age. The earth's grasp
> holds the builders, rotten, forgotten,
> the hard grip of the ground, until a hundred
> generations of men are gone. (3–9)

Though at one time, "bright were the buildings" (20),

> The seats of [slain swordsmen's] idols became empty wasteland,
> the city crumbled... So now these courts are empty,
> and the rich vaults of the vermilion roods
> shed their tiles. The ruins toppled to the ground,
> broken into rubble. (26–31)

The poem has physical or material waste (in the text itself) and visual waste (contrasted with a former gleaming city).[149] Confronted by former glory and present decay, this "ruined matter offers ways for interacting otherwise with the material world."[150] Like the it-narratives of the eighteenth century, Old English riddles insist on a perspective from an inanimate object's point of view. Here, the "rust-stained" wall "has seen one kingdom after another."[151]

Poetry acts as a precious gift of grace, restoring us in times of crisis.[152] After 9/11, I didn't know how to speak to my classes. What could I say? Do I just teach and pretend nothing had happened? I came to my Anglo-Saxon class on September 13, the Thursday after the event on Tuesday, and told my students I didn't know what to say. There was, however, an Anglo-Saxon poem I felt compelled to share with them, so I read aloud "The Ruin." Years later, a student shared that, whenever he thinks of 9/11, he thinks of that moment, requiring breathe, pace, and discipline—requiring waste poetry. These melancholic and riddling worlds suggest a way to read garbage as riddle, one to which we want an answer.[153] We humans cannot help ourselves—we want to define everything, even waste.[154] If Hamlet the elder in his prime, unburned Heorot, the glorious kingdom in "The Ruin," and the deliriously happy dreamer with his lord in "The Wanderer" can be read as riddles, their answer is, ultimately, the garbage that is history.[155]

Old English riddles provoke compassion for the mother of the alien cuckoo, the shield fighter, the tortured sheepskin/vellum/Gospel-book, and the crucifixion tree/cross. The Anglo-Saxon riddle jolts us out of our complacency through the pathetic fallacy/phallacy of the onion/penis.[156] We enter objects in terms of themselves, from their own points of view. Everything and everyone has a perspective, whether a key, a shield, or moon. In seeing how "inanimate things have a life of their own,"[157] Old English riddles allow us to experience the "kinship between people and things."[158] Tricking us to misread, riddles ultimately guide us to read ethically. False readings—penis for onion—in fact, open meaning up, making the interpreter vulnerable. Riddles and runic poems teach us the craft of reading. "Wulf and Eadwacer" functions metatextually, as a metaphor for our act of reading itself. Just as we get closer to meaning, "ungelic is us"—there is a difference between us. Opening meaning allows us to sense both the affinity and the distance between us and the text, between us and the Anglo-Saxons.

Another poem, another onion. Naomi Shihab Nye's "The Traveling Onion" journeyed far "just to enter my stew today." Its ancestry extends far back—to South Asia, Egypt, and Europe. Somehow, this "small and forgotten" vegetable possesses history. The poet "could kneel and praise / all small forgotten miracles" that allow this aromatic allium into her kitchen after its meandering pilgrimage. She focuses on the onion, so often ignored for the flashier "meat." No one comments

> on the translucence of onion,
> now limp, now divided,
> or its traditionally honorable career:
> For the sake of others,
> disappear.[159]

But the poet does. Nye's homage to the onion venerates its modesty, quietness, and decorum. This reverent poem salutes the humble vegetable's thing-power.

After a day of reading obfuscating object-oriented ontological texts—puzzling through challenging prose—I turned to Old English Riddle 3, "Wind."[160]

It blew me away.

"I shake / men's sanctuaries . . . I lace bright foam along [the breakers'] crests."[161]

Hit by a blast of cold air, the words summoned and spoke to me. I resisted on the rocky outcropping of my study chair, only to relent and let the poem ruffle my locks, chill my skin, and play with me as in a dance. I acquiesced to it in a submission thrillingly frightening—or frighteningly thrilling?—the acknowledgement of the Other—its affinity to me, with me, embracing me.

"Who will still it?"

This is why I came to literature in the first place, a place where I humble myself to the words that summon me to them.

I call back—

And the wind carries my words away.

CHAPTER 10

Trash Meditation: The Arts of Transience and Proximity

In the 1960s when I was a little girl, my family would spend the summer in Cape Cod. On the beach of Follins Pond in South Dennis, Alfred Short owned a piece of land, on which he had built—literally with his own hands—several cottages. Each summer he would construct an additional cottage, until by the early 1970s there were 12 cottages in all. In the early days, I remember how his teenage sons, Craig and Larry, would go down to the beach every morning and rake up the seaweed that had floated in on the tide over the previous 24 hours. Mr. Short would drive his old pickup truck to the beach. The boys would heave rakefuls of the sodden plant life over the wooden frame into the truck bed. The tires of the jeep would throw sand up into the air as it was maneuvered back onto the dirt drive, where seaweed was added in an ever-growing compost heap. Then, he and the boys would load up garbage cans from each cottage.

"Who wants to go to the dump?" Mr. Short would cry out. All of us kids wildly called out, eager to be the chosen ones. One day Maureen Duffy and I were so blessed. Six years old in our clammy bathing suits, we scrabbled into the cab—off we went to the dump, probably without our seatbelts on, if there even were seatbelts. Mr. Short would nonchalantly smoke his cigarette. Those were the days—safety never a concern. Just full-blown adventure. Once we got to the dump, he'd waggle the cans over the side and empty them, as the seagulls mewed their eerie cries. We would be allowed to stand on the truck bed, leaning against the cab as he finished his work. How proud we were to have gone to the rubbish dump, that mysterious treasure box of delights, smelling of moist life and salt air. As I make the journey in my memory, the excursion becomes a pilgrimage to the relics of summer pleasure and indolence, replete with clams shells, empty hours of dynamic idleness, and keen expectation.

I seem to have had an affinity to waste. The elementary school I attended grades three through six was brand new. Shaped like a doughnut, it was the latest thing in the 1960s. Sitting near the athletic fields where my brother had Little League practice, Woodland School had been constructed on a former landfill. I can recall running across the blacktop, tripping over pieces of rubber tire sticking out through the asphalt. The story is not just about me, a little girl, playing and stumbling. It is also the account of the ripped up tires and their histories as they wheeled a car down the highway. The chronicle of their origins in a rubber tree plant somewhere in Malaysia reverberates in the telling. Woven in this network of memories, stories, and histories, all of us—little girl, rubber tires, and liquid rubber dribbling from trees—act as nodes in positions of equivalence.

Memories of the past infiltrate my present. I compose a past of detritus into life-affirming fodder that feeds my emotional and spiritual self.

Lousy Limitlessness and Perfidious Perfectibility

L'environnement, c'est moi.

—With apologizes to an apochryphal phrase attributed to Gustave Flaubert[1]

Alexa Bingham articulates the conflicting "story-lines" in SD discourse. The dominant one prioritizes socioeconomic development while the alternative one prioritizes socioenvironmental sustainability.[2] SD suggests a closed system that has no built-in end time. Waste is produced within a teleological framework; something useful comes to lose its usefulness. Bingham elaborates on her two story lines: "Both prioritize social objectives but the developmentalist story-line claims that economic growth will achieve them, while the sustainability story-line argues that environmental priorities are the key." This sustainability story line tends to be marginalized or co-opted. SD discourse is increasingly being used to legitimize economic growth, but it is a vague term and "can be used to justify various" activities.[3] Bingham sees an inherent problem of *metaphor*. "The SD discourse works metaphorically reducing complex debates and disguising contradictory objectives . . . The crucial work of academics, NGOs and civil society organizations in highlighting the *incompatibility* of sustainability (in terms of social wellbeing and environmental vitality) and development (in terms of sustained economic growth) need to become widely accepted."[4] Now commonplace to see environmental and ecological concerns as critical to socioeconomic development,[5] we can safely assert that economic growth is not good for the environment.[6]

John Scanlan cautions us from overindulging in environmental positivism. Even environmentalism suffers under the hope of (ultimate, possible) perfectibility that Scanlan indicts as poisoning the groundwater of our existence: "Perfection is the idea that runs through Western culture; that emerges as the guiding principle of enlightenment reason, and that, in its creation of scientific technology, is ultimately the cause of more garbage."[7] Paradoxically, perfectionism worships the sanitary, all the while producing filth. The green critic and poet, Wendell Berry, has argued for the dire necessity of the "love of farming," an occupation that involves "getting dirty," hence modern man's aversion to it. As a culture, Berry argues, we have opted for a witches' brew: farming as an agri-industry, which has, in turn, created massive ecological devastation,

> destroyed rural communities and cultures; reduced our farming population almost to disappearance; yielded toxic food; and instilled an absolute dependence on a despised and exploited force of migrant workers . . . We have ahead of us a lot of hard work that we are not going to be able to do with clean hands. We had better try to love it.[8]

Separation from nature and the mastery of our environment are not only quixotic endeavors, but detrimental to ourselves and our world.[9] Berry warns that we "will keep on consuming, spending, wasting, and driving, as before, at any cost to anything and everybody but ourselves. This belief was always indefensible—the real names of global warming are Waste and Greed—and by now it is manifestly foolish."

This "fantasy" and "disease of limitlessness" can only destroy us and our world in the end, leading to "limitless violence, waste, war, and destruction."[10] Berry's limitlessness, cognate to Scanlan's perfectibility and Gay Hawkins's mastery, remains inherently impossible to achieve or reach. As Michael Pollan, whose writings have criticized contemporary agribusiness for intensifying the industrialization of the food chain, points out,

> Wendell Berry once wrote that when we took animals off farms and put them onto feedlots, we had, in effect, taken an old solution—the one where crops feed animals and animals' waste feeds crops—and neatly divided it into two new problems: a fertility problem on the farm, and a pollution problem on the feedlot.[11]

To combat material waste as pollution in our lives, we first need to combat our philosophical approach to our environment. The demonization, privatization, and societal repression of waste limit and harm our planet. Disposability constitutes waste as "*ethically* insignificant."[12] Only by making waste *ethically significant* can we hope to minimize disposability and waste itself.[13] Public policy can be supported by the change in consciousness that literature and the arts foster.

The Dangers of Linearity

> Þæs ofereode, þisses swa mæg. [This too may pass.]
>
> *Deor* in *The Exeter Book*

Waste, alas, does not pass. It is here to stay. And we had better acknowledge it. The dominant story line of waste in literature can be seen in the way traditional disciplines deal with aspects of waste. Interpretations of the world are grounded in and influenced by attitudes toward waste. Walter Benjamin's famous assessment in his "Theses on the Philosophy of History" forcefully and lucidly declares: "There is no document of civilization which is not at the same time a document of barbarism."[14] His sarcasm bubbles at the surface of his grim conclusion. As with Benjamin's Angel of History, we see "wreckage upon wreckage" piling up, the "pile of debris . . . grows skyward" before us. "This storm is what we call progress."[15] Even the vengeful God of the Pentateuch seems wed to this vision of history: one of wholesale destruction of the past to feed a pristine present that will likewise develop into a filthy future that will need to be cleansed.

Traces—waste—from earlier periods exist and transform, littering society and culture. There never is a clean break to signify "after." The poets express the precarious new state by evoking waste and destruction. The so-called digressions of *Beowulf* "—entrails, bits, [and] scraps" that constitute "garbage"[16]—remind the poet's listeners of tragic events in the past, events that haunt the present. In these digressions we learn that King Hreðel of the Geats took in his seven-year-old grandson, Beowulf, son of Ecgtheow, for whose feud with the Wylfings Hroðgar paid compensation. The tragic death of Herebeald at the hand of his brother, Hæðcyn, brings about their father's Hreðel's heartbreak and eventual death. The fratricide poses an impossible dilemma: as father of the slain man, Hreðel is obliged to exact revenge; as father of the killer, he must uphold kinship solidarity. Beowulf reminisces about the attack by the Swedes (2426–2509), which the messenger later recapitulates and recycles, predicting that this time the Swedes will destroy the Geatish kingdom. Another kingdom riddled with destruction and doom is that of Heremod, the stingy and murderous king of the Danes, who is eventually put to death by the Jutes (901–915, 1709b–23a). In the Lay of the Last Survivor (2247–2266), the speaker recalls the former glory of his now deceased people, expressing the sense of impending doom for once glorious kingdoms poignantly and succinctly (2262b–66).

Continually reminded of how any success in the present (the poem's present or our present) will inevitably be followed by decay, our social sympathy is at play when we hear of other doomed societies. Darwin in *The Descent of Man*, "attributes human morality to the capacity to compare present action with the past and the future and hence to assess the consequences of our acts in the light of our evolved feelings of social sympathy."[17] The "sampling" by the hall minstrel or *scop* suggests that stories emerge through the act of recycling. Digressions are detritus, rags and tatters, recycled moments from the past composted to be consumed by two audiences: the one present in the text and us. They exist simultaneously in the past—to the actors of the main "frame"—*and* in the present, a present that becomes our present too.

The politics of becoming and of seeing waste as integral to the self stands in opposition to the imagined control one has over waste as alterity. Levinas rejects Western philosophy's reduction of difference to the same. As he suggests, "the ethical exigency [is] to be responsible for the other . . . In ethics, the other's right to exist has primacy over my own."[18] Waste is both the other *and* of ourselves and forces us to confront our own natural instincts and temptations.

In the film *Wall-E*,[19] trash helps to save WALL-E, acronym for "Waste Allocation Load Lifter-Earth Class." His fire extinguisher knowledge saves him later when he is ejected from the escape pod about to self-destruct. A bricoleur, he collects junk to make beauty that he gives to his beloved, EVE. She, in turn, saves WALL-E with junk. In this anti-Eden, EVE is her own Mary and the source of redemption.[20] In wanting to return home, the captain cries out: "I don't want to survive, I want to live!" in a panged expression of "soliphilia . . . the love of and responsibility for a place, bioregion, planet and the unity of interrelated interests within it."[21] When a plant, against all odds, takes hold and grows, we bear witness to life emerging from junk. We the audience—along with WALL-E—will the plant to survive and thrive. Though the plant ends up in the trash chute, WALL-E saves it and we feel relief at its—and by implication *our communal*—survival.

Linear time orientates us to distinguish waste from non-waste.[22] While the "past may be repressed, [it] ultimately cannot be evaded."[23] Not only can it *not* be evaded, but also it *should not* be evaded. While in the Middle Ages, garbage was visible, with true progress and perfection only fulfilled in heaven, in modern life, garbage is made *invisible* so that time seems to be identical to progress. Modern life destroys the past; fashion builds into it its inherent outdatedness.[24] Providing many rich suggestions for understanding waste, Will Viney rewrites Douglas's dictum to suggest that waste is "matter out of time," distinguishing between "use-time and waste-time." "Use-time" projects into the future and "is a time of wearing, emptying, digesting, breaking or exhausting—it is a somewhat entropic time, a time of diminishing potential, a time oriented to and by an end." Once objects have become redundant, they have entered "waste-time,"[25] which "separates and divides," organizing our activities between "human use and non-use."[26] Waste interrupts use-time.

In *The Journey*, "Taking care of rubbish deepens one's relation to the stuff we use. That's why, if we are indeed thankful, we love rubbish; it doesn't matter that it's what we no longer need."[27] This is Leopold's point of view—we need to be mindful of rubbish. After emptying the rubbish barrels, "the rubbish still has to be spread out so that there is enough space for the next load as happens in any properly run dump." Even a dump has codification and order. "This place is not sad, it's a garden of freedom where everything that is dead is given over to itself... Such residue is free of all greed, humiliation done away with, all that is left is sun, wind, and rain, which are offered in peace to the useless treasures that the earth takes into its arms with an almost undetectable rustling."[28] The dump is an open site of ethical affinity where all are welcome.

Offering a sign of grace when Paul drinks wine with a family, whose grown daughter suggests a possible romance, the father says: "It was the wine of friendship that we shared. The journey will not separate us."[29] He continues:

> One must have a center, an unshakable quiet space that one clings to vigorously, even when one is in the middle of the journey, the unavoidable journey... an unmuddied sensibility free of rubbish, no left, no right, only the center, a constancy that does not change for the better or for the worse. I don't mean it strictly in a spatial sense. I mean it instead as a circuit, one that travels from hand to hand, from heart to heart and really exists.

Paul acknowledges that he understands what the man means:

> Despair is our fall from grace, here amid the rubble, our confiscated stolen property, our transmutable and disputable names, the heart's home, but not the heart—in short, everything that could be taken from us. The center, if I understand correctly, cannot be taken away from us. It travels with us and lifts us up from sin, from the rubble.

The man acknowledges that he and Paul understand one another: "When other people look into our faces, we have to try to reach out to them."[30] This face and this reaching out to the other is just what Levinas refers to in his analysis of *Crime and Punishment* and Sonia's look at Raskolnikov with "insatiable compassion." The adjective "insatiable" suggests consumption, eating, the body, surplus, and waste. We have an obligation to the Face;[31] the responsibility of each self to the Other's Face[32] in a motion of affinity not identity, since empathy just makes the self solipsistic.[33] The man says, "That is the grace amid which creation is woven and renewed." Paul agrees. "And grace is the journey."[34] He departs for the train station: "Perhaps the people understand why he's in such a hurry. He thinks he sees them waving, wishing him a safe journey, the rubbish and the rubble vanquished at last."[35]

Eric Leake evades the complacency inherent in "easy empathy" and the privileges that it implies, suggesting "a more difficult empathy that unsettles us."[36] "Difficult empathy challenges social positions, reader identities, and the otherwise prevailing relative comfort of an easier empathy."[37] Difficult empathy "must always be reading and never quite grasping... push[ing] the limits of our understanding in reaching out to those with whom we might not otherwise wish contact or association."[38] In easy empathy, we feel a character is likeable or deserving of empathy. Such empathy reaffirms power structures, totalizing and not challenging them, erasing differences between me and the object. Difficult empathy "blur[s] the boundaries"[39] between self and other, unsettling us.[40] Constantly at dis-ease, confronted by rubbish, by the Other, through the act of reading Adler's *The Journey*, we recognize our affinity to the Other, even the Nazis.[41] Transcending the victim/perpetrator dynamic, we walk through the camp and hear different voices. The journey is "on toward grace,"[42] where memory becomes a source of justice.

Teleological time and progressivist history makes us trash the past. A new conception of time might ameliorate our poisoned worldview where the idea of progress demands we jettison the past.[43] Elizabeth Povinelli argues that people are attracted to more spectacular and "sublime" events or deaths rather than cruddy decay. We need to look at less spectacular "events" of killing and dying. A story by Ursula LeGuin indicts the "ethics of empathy."[44] An indigenous woman argues with her alcoholic nephew: "No, that is not your body; that is my body. When you die, my body will suffer and die."[45] Acknowledge my entire body at every moment (including its less appropriate parts, like the genitals and excretory organs) *and* acknowledge the link between your body and mine. If your body suffers, so does mine.[46] Povinelli reads against present suffering as worthwhile for some future "redemption"; in other words, it is unethical to justify or tolerate someone's present misery for some future end. We should not see things in terms of a "redemptive" future (you suffer now but it will all be for the best). Seeing things synchronically, rather than diachronically, means my happiness exists at the same time as, and as a result of, your unhappiness. And does that make my happiness worth it?[47] "Some of us would-be humans have thought too that we should not be free at anybody else's expense."[48]

The emerging religious environmental movement (such as "Green Christianity") criticizes the resistance to action and lack of motivation to act concerning environmental degradation despite our scientific and technological ability to do so. Using the term "creation care," evangelicals see religion as a modality that can help make social change by changing the heart. Similarly, both literature and literary criticism can function as modalities of "restitution"[49] that can help motivate social change. Poetry, in Bataille's view, comes to be a pure kind of expenditure or sacrifice.[50] Poetry, as creative, ethical excess, "is not logic or / knowledge or philosophy; it is action and / action's pleasure."[51] Waste is a way to acknowledge the body; and, with this insight, comes an awareness of the interconnectedness of one's own body with those of others, enabling compassion for others.[52]

Waste Meditation

> We are the garbage, the waste, we make it and dump it, to be separated from it is a cancer causing delusion... We cannot separate ourselves, clean and perfect, from the trash we dump out back in the can. Clean is a vision of internal trash, not a mere separation.
>
> Gerald Vizenor, "Landfill Meditation"[53]

Native American writer Gerald Vizenor suggests we practice waste meditation, paying attention *to* waste, not as evil or negative, but as *itself*:

> Martin Bear Charme... teaches a seminar on refuse meditation, and circumscribes his unusual images and wise transformations in the material world on a refuse mount... Landfill meditation restores the tribal connections between refuse and the refusers.[54]

Martin Bear Charme nostalgically extols the past:

> Once upon a time taking out the garbage was an event in our lives, a state of being connected to action. We were part of the rituals connecting us to the earth, from the places where food grew, through the house and our bodies, and then back to the earth. Garbage was real, part of creation, not an objective invasion of can and cartons. Refuse meditation turns the mind back to the earth through the visions of real waste.

Pointing out how we have become repulsed by our trash, he recommends that we associate with it:

> "Now, rather than hating to clean up the kitchen, making it smell different, get right down there with the odors. Focus on the stench in the corners, take the odors in, you know, the same way we smell our underarms, because we are the bad smells we smell, separated from our own real kitchens in

> the mind... Go to a place in the waste to meditate," chanted the refuse mediator. "Come to our reservation on the landfill to focus on waste and transcend the ideal worlds, clean talk and terminal creeds, and the disunion between the mind and the earth. Come meditate on trash and swill odors and become the waste that connects us with the earth."[55]

While meditating on trash, scooping up the filth, and inhaling its odors like designer perfume may not appeal, Ian Bogost insists "our job is to get our hands dirty with grease, juice, gunpowder, and gypsum. Our job is to go where *everyone* has gone before, but where few have bothered to linger."[56] By lingering, even malingering, we pay attention, steadying our gaze. This rumination "means yoking: enacting or experiencing an intrinsic interconnectedness,"[57] finding not just sublimity in trash, but how we, too, are trash.[58]

After all, we are just bits of dust ourselves. We should consider what cosmologists like the Nobel Prize–winning George Smoot and his colleague Saul Perlmutter have determined: only 4 percent of the universe is made up of the material we formerly assumed it to consist of entirely. In this "ultimate Copernican revolution," we are not only *not* at the center of our solar system, but also not even comprised of the same matter as most of the universe. As Lawrence M. Krauss has provocatively posited, "We're just a bit of pollution... We're completely irrelevant."[59] If we, ourselves, are pollution, what can we produce but filth and waste?

Trash meditation takes multiple forms, provoked by written and visual works, from the terrifying *Hell* diorama (2000) by Dinos and Jake Chapman to the contemplative and sublime *Piss Christ* by Andres Serrano.[60] Dance, embodied movement, poignantly suggests the bionic fusion of human and waste. *The Trash Project*, a ballet of garbage trucks and their drivers, makes visible what we prefer not to acknowledge.[61] Allison Orr, the choreographer of *The Trash Project*, describes detritus gatherers in this way:

> These are just sweet, regular people that do creative things, that have bands, that make art, that write poetry, that have families, that love barbecue, that play football, that whatever. The stereotype that somehow somebody driving a trash truck ended up there because they couldn't think of anything else better to do is just not true. It's just not true in any way. You don't grow up saying you want to be a trashman. Who does? Well, actually, there are lots of 5-year-old kids who want to be a trashman, right?... Some of them, it *is* passed father to son; there are multiple generations of people who have worked with Solid Waste for many years.

As one critic has pointed out, "It's a job that's very choreographed."[62] The anthropomorphizing of the trucks suggests that the trash is our bodily excrescence, hence our filth. Yet the ballet-like quality of the trucks' movements renders the hulking iron and steel machines soft and fleshly, and invisible trash uncannily ephemeral and precious.

As we sat in the pouring rain, we found refuge in refuse: plastic trash bags made available to the audience. The workers emerged in florescent green jackets and then from far off we heard the low rumble of bees: 16 trucks emerged in the crepusculent light. One man gets on his recycling truck like he is flying. The symmetry in recycling cans placed in trucks and then dumped lends structure to this activity. The workers dance with their bins, a romance of cleaner and container that allows for their occupation to be fulfilled and fulfilling. This pair relies on each other—human and can.

The dump trucks, like ballerinas in syncopated rhythm, mimic unloading. The synchronized trucks pirouette, each claw carrying bins, like the cavalier carrying a beautiful dancer, lifted slowly up to haunting violin music. The lyrical moment is sustained as the bulky item pickup truck glides to a high beat. The driver lovingly swings his truck around, the metal ballerina with her iron legs spread whirling. The two metallic half-moons heave like bionic mammalian breasts in this machine and man *pas de claw*. One driver plays the harmonica. Joined by other drivers who emerge and put on glasses, gloves, and hats, they do a dance with their trash bags. The brushy pickup dance is fast paced as sticks are thrown into bags. The small dead animal pickup weaves a pretty wavy pattern: lyrical and serpentine, man and truck realize the sacred task of taking away the dead.

My drenched and miserable son asks, "Mom, when is this going to be over?"

"It'll never be over. There's always trash with us."

PART III

Recycling and Composting: Form as Restitution

CHAPTER 11

Waste Aesthetics: Puns, Litter-ature,* and Intertextuality

"Second Harvest" is a concept well-known to archeologists through their analysis of coprolites. Coprophagy, or the eating of feces, could occur in times of famine. There is evidence that certain foodstuffs were processed in this way. For example, pitahaya dulce pears, consumed by the Guaicura in early summer, contain thousands of tiny seeds that would pass undigested through the intestine. An eighteenth-century German Jesuit among the California Indians claims to have witnessed the reuse by Indians collecting pitahaya seeds from excrement, which they then processed by roasting and grinding before (re-)eating.[1] Cabeza de Vaca, while amongst the Mariames comments, "They eat dirt and wood and whatever they can get, as well as deer excrement and other things I will not talk about."[2] W. W. Newcomb speculates in his majesterial work on Texas Indians that what de Vaca refuses to talk about is this reuse of seeds from human excrement.[3] While some argue these seeds gleaned from excrement would have been vital in times of famine for sustaining a group, others say their use was not restricted to hard times.

Writers likewise reuse, pilfering from earlier texts in a second harvest for a new poetic banquet. Taking a tip (pun intended!) from Ruth Evans, we could consider waste literature along the lines of Lacan's pun "poubellication," which translator Bruce Fink says "is a condensation of *poubelle*, garbage can (or dust-bin), and *publication*, publication." Evans suggests the alternate translation of "litter-ature."[4]

Linguistic Excess: Provoking Puns and Wild Words

> *[P]oo-belle*... a pun that points to the trope of writing as the transformation of shit into beauty.
>
> Ruth Evans[5]

When I first told my students about my research into medieval excrement, one shy student raised her hand and asked how I found enough material to work on. Enough? I had too much and had to cut the book by a third. In fact, I continued to explain, I'm thinking of writing a second book on excrement. The same bashful student raised her hand again and meekly said, "You could call it Number Two."

The pleasure in this story is manifold: first, the timid student's unexpected boldness in voicing this pun, "Number Two"—referring to my putative second book on the subject—with the euphemism for shit.[6] Excrement demands punsters. A woman protesting her picky neighborhood association handed out 17 toilet bowls to neighbors to decorate and put on their front lawns in the so-called Bowel Movement. As her husband says, "This neighborhood's gone to pot."[7] Puns, non-reified language work, inevitably produced by those confronted with excrement, force us beyond our everyday, often restricted, thinking into unexpected avenues of thought and feeling.

The graveyard scene in *Hamlet* illustrates a fundamental aspect of death, that it is the "'garbaging' of the body." Hamlet senses, as Scanlan articulates, that "the things we destroy (things that did have a value) are also made void; and the point is that all things, all objects, begin from nothing (garbage) and eventually return to nothing (garbage again)."[8] As he holds up Yorick's skull, Hamlet reflects upon how he had loved the living man: "Here hung those lips that I have kissed I know not how / oft" (V.i.178–179). The disappearance of this flesh and the decay of the body to bone signifies how we all return, ultimately, to dust. When Hamlet reflects on Yorick's skull, now tongueless and eyeless, and on the fate we all face, he points out how we end in an end: "To what base uses we may return, Horatio! / Why may not imagination trace the noble dust of / Alexander till 'a find it stopping a bung-hole?...Alexander died, Alexander was buried, Alexander / returneth to dust, the dust is earth, of earth we make / loam, and why of that loam whereto he was converted / might they not stop a beer-barrel?" (V.i.192–201). This blasphemous parody of the Eucharistic Proclamation ("Christ has come. Christ has risen. Christ will come again") nihilistically reduces man to matter without form, flesh without soul. "Bung-hole," meaning "the hole in a cask, which is closed with the bung [or cork]," comes to have the transferred meaning of "the anus" as early as 1571. Once Hamlet points out how we all end in an end—surely a pun Hamlet would not be averse to—in the anus that suggests our affinity, does he, fully embodied, act.

Puns are a form of waste and excess.[9] Puns gesture toward additional meanings that were layered in the nexus of words all along, textured into the memory or "waste" of the linguistic matter. Falstaff wastes words on his bodily waist and wasteful habits. When the Lord Chief Justice chastises Falstaff, "Your means are very slender, and your waste is great," Falstaff wryly comments, "I would it were otherwise; I would my means were greater and my waist [slenderer]" (*Henry IV, Part II*, I.ii.142–143). Protesting his thrift, Falstaff yokes homonyms with quips: "Indeed I am in the waist two yards about; but I am now about no waste" (*The Merry Wives of Windsor*, I.iii.41–43).[10] Although puns can become domesticated, although they become the said, puns interrupt, provoking new lines of thought. Low language for high-falutin' subjects percolates into humiliation and mockery. The peasant Marcolf punctures Solomon's grandiose wisdom in a medieval Latin dialogue translated into various vernacular languages. Typically, Marcolf uses an established order that he then parodies, mirroring back the closed knowledge system of Solomon with life-affirming and bodily driven folk wisdom. For example, when Solomon says "A black boss is ideally suited to a white shield," Marcolf replies, "A black asshole is ideally situated between white cheeks."[11] "Very fitting is a fine sword next to my flank," comments Solomon, to which Marcolf responds, "Very fitting is a turd next to my hedge."[12] Marcolf parrots the sentence structure of the *sententia* of his better, all the while poking fun. "Four evangelists hold up the world," solemnly utters the king. "Four posts hold up the latrine, lest he fall in who sits over it," responds the practical peasant. The target of his rejoinders "is the seemingly unassailable authority of Solomon's discourse."[13] Not only is he clever and funny, but also he brings us repeatedly back to an everyday rural realm, in which the "fertilizing scatology"[14] of Marcolf's speech functions integrally with growth and fecundity. In a very *Till Eulenspiegel*-like misunderstanding, Solomon asks for a pot of fresh milk "covered." Marcolf's mother covers it with a flan that Marcolf promptly eats. He then covers the milk pot with a cow turd. Outraged, Solomon says that wasn't what he had "stipulated" and Marcolf responds, "That is how I understood it."[15] Reprimanded for spitting on a bald courtier's head, Marcolf defends himself:

> I didn't befoul, I fertilized. Manure is placed on barren ground so that the crops may be multiplied there in abundance . . . I saw his forehead denuded of hair and, believing it to be bare ground, I spat on it. He certainly should not be angry about this since I did it for his own benefit. If his forehead were frequently watered like this, the hairs would regenerate.[16]

Humiliating scatology may rejuvenate. Punning is *ethical* in so far as it can open up to something novel and unexpected.

Punning suggests transgression, a deviance that demands comeuppance. "No wonder punsters must be punished."[17] Our moral responsibility and duty is to use language well. Flattery, as excessive language, is appropriately punished with another surplus—shit—in Circle 8 of Dante's *Inferno*.

In *Gulliver's Travels*, the protagonist's equine Master cannot understand a lie, the thing that is not—a waste of language, in other words. Lawyers squander their talent since they lie.[18] Many of the early sixteenth-century German tales of the rascal and trickster Till Eulenspiegel involve waste, typically actual turds.[19] While the misunderstandings of Amelia Bedelia, that dauntless babysitter in children's picture books who understands only literal language, inevitably end in gentle humor and acceptance, Till uses excrement to discipline others by doing what people tell him to do *literally*, which is not what they *intend* him to do.[20] Excessive—metaphoric—language is punished by another excessive other—shit itself. For example, when Till becomes a sacristan, the priest farts as Till is arranging his vestments. Till reprimands the priest, who angrily replies that he even has the "right to shit in the middle of this church." Till bets him a barrel of beer. The priest proceeds to evacuate his bowels and demands the barrel of beer. But Till points out that the priest said he would shit in the *middle* of the church; Till measures carefully, discovering that the priest had taken a dump far from the middle of the church. So Till wins.[21] The priest actually meant that he would simply shit *in* the church, but used a metaphor of the "middle" that Till cleverly takes literally. While Till may be offensive, his insistence on the literal use of language purifies his filthy deeds and linguistic machinations.[22] Till's insistence on the literal use of language, a form of satire, mocks those who speak metaphorically and pokes holes in people's arguments.

Gary Snyder's reading of language in "Language Goes Two Ways" asks us to reread the category of linguistic waste. Wordplay seems like symbolic excess when language is viewed as order. Snyder argues, however, that language is "wild" as expressed in wordplay and the vernacular.[23] In the film *Ball of Fire* (1941), the garbage man comes and wants help with a "quizzola." This slang term allows the stiff professor Bertram Potts to realize he needs to get in touch with present-day reality. Slang, the linguistic garbage used by the lower classes as embodied by a literal garbageman, catapults the plot, leading to love, death threats, and the inevitable happy ending.

Litter-ary Larks

In Rabelais's *Gargantua*, Gargamelle, about to give birth to Gargantua, eats a huge amount of tripe, itself shitty stuff. Once born, Gargantua "beshit himself every hour; for, to speak truly of him, he was wonderly phlegmatic in his posteriors."[24] The ideal opposite of the follower of *civilitas* as described by Erasmus, Gargantua plays in filth:

> Still he wallowed and rolled himself up and down in the mire and dirt; he blurred and sullied his nose with filth; he blotted and smutched his face with

> any kind of scurvy stuff... He pissed in his shoes, shit in his shirt, and wiped his nose on his sleeve; he did let his snot and snivel fall into his pottage, and dabbled, paddled and slobbered every where.

The excessive euphemisms for Gargantua's penis when he is five reveals the foolishness of euphemisms, which set up a binary between vulgar or crude language and flowery and excess sigifiers: "Her pillicock, her fiddle-diddle, her staff of love, her tickle-gizzard, her gentle-titler. Another, her sugar-plum, her kingo, her old rowley, her touch-trap, her flap dowdle."[25] And it goes on. Gargantua talks about what he likes to use to wipe his ass, telling his father when he is six that "in all the country there was not to be found a cleanlier boy than he... I have... by a long and curious experience, found out a means to wipe my bum, the most lordly, the most excellent, the most convenient that ever was seen."[26] He describes women's velvet masks, silk that was "voluptuous and pleasant to my fundament." He doesn't like spangles that scrape his skin, but there are many other items he tries, including using a scratching cat when he "[dungs] behind a bush." He wipes his posteriors with "straw, with thatch-rushes, with flax, with wool, with paper," as well as pillows, kerchiefs, panniers, a shaggy hat, and, best of all, a goose's neck![27] This psychologist of ass wiping identifies the process of finding stuff to swab himself. Nowadays we don't want to question who squeezed the Charmin—let alone borrow it from Gargantua. Packed adjectives, nouns, multiple synonyms, and excessive signifiers for one signified in a cornucopia of meanings prolierate in *Gargantua*, as opposed to *Till Eulenspiegel*, a work that breaks euphemisms down. Style works as homeopathy. The therapeutic effect of laughter aids poor constipated Silvia after the visiting Ratclife has her figure out a riddle: "A noted person's passing by, / Who's known the city round: / The moment that you guess his name, / Your certain cure is found." After many failed attempts, she finally, piteously, says, "I'm tir'd upon my word, man; / Nor have I any hopes of ease, / Except it be TOM Turd-man." Told that she is correct, Silvia laughs "so heartily, / That her imposthume broke." [*The Grub-street Journal*, August 22, 1734).[28] Through explosive hilarity, excess is expended and cures socially and medically.

That which is not useful or utilitarian may be waste in terms of *form*. In genre fiction, plot is key. The verbiage exists to propel the story line along, whether in a romance, crime fiction, or a Western. Formula fiction narrative, such as mass market romances or murder mysteries, is utterly utilitarian—linear and plot-focused with little waste. Yet this very absence of waste renders the work into a potboiling piece of low literature.[29] So-called quality literature is often messier than the more generally popular "trashy" literature—a consumable of closure that gives the illusion of clarity in a cluttered world. Literature filled with waste suggests a porous aesthetics of promise and becoming. In "good" literature, "style [functions] as 'excess.'"[30] That something is "useless" or "waste" should not automatically compel us to condemn it.

Stories need to select; authors need to excise.[31] Cleanliness in narrative structure can be seen in the *Eyrbyggja Saga*, about the establishment of law and Christianity from late ninth- to early eleventh-century Iceland. The narrator periodically tells us what adheres to the story and what does not. "There is no need to record here the settlements of people who do not come into our story... Now Thorarin is out of the story... The next part of the story tells how Gizur the White and his son-in-law Hjalti came to Iceland to preach the faith."[32]

But what happens when an author seemingly never selects, simply junking everything onto the garbage heap of his literary dumps? Sometimes the literary excretions are literal. The margins of medieval manuscripts, replete with pooping apes and pissing hybrid monstrosities, contain evidence for the practice of amendment. There one "can come along and alter, add, erase, and recreate what is there."[33] The book has been seen as static and closed only since the nineteenth century—think of the abundance in Rabelais, Cervantes, Sterne. Some works, such as James Joyce's *Ulysses*, reflect the bloat of existence with their joyful exhumation of details and particulars. Such literature contains beloved nuggets amid the junk.[34]

Tristram Shandy is a novel seemingly without an end, nothing is everything, where the plot is mitigated by exegesis. Shandy refers to "this Fragment of Life,"[35] addressing the reader countless times. "I Know there are readers in the world, as well as many other good people in it, who are no readers at all,—who find themselves ill at ease, unless they are let into the whole secret from first to last, of every thing which concerns you."[36] "—Shut the door.—"[37] is a prelude to the section on his conception that only some readers may want to read. Signaled on the page with lines, in words ("Shut the door"), he allows the reader the freedom to skip this section. He was born at an ill time: "I wish I had been born in the Moon, or in any of the planets... for it could not well have fared worse with me in any of them (tho' I will not answer for *Venus*) than it has in this vile, dirty planet of ours,—which o' my conscience, with reverence be spoken, I take to be made up of the shreds and clippings of the rest... for which cause I affirm it over again to be one of the vilest worlds that ever was made."[38] The planet is bricolaged from scraps just like his life/book.

Telling the reader how and what to read, Sterne's narrator puffs the story. "I insist upon it that you read the two following chapters, which contain such a sketch of [Yorick's] life and conversation, as will carry its moral along with it.—When this is done, if nothing stops us in our way, we will go on with the midwife."[39] There exist two black pages after Yorick dies and the quote from Hamlet is set off in a box.[40] "But I must here, once for all, inform you, that all this will be more exactly delineated and explain'd in a map, now in the hands of the engraver, which, with many other pieces and developments to this work, will be added to the end of the twentieth volume,—not to swell the work,—I detest the thought of such a thing."[41] A swollen text contains the vibrancy of the narrator's life force.

Tristram asks if it is possible to tell a story in a straight line in the way a "muleteer drives on his mule." Carrying on the analogy, Sterne writes, "[The historiographer or the muleteer] might venture to foretell you to an hour when he should get to his journey's end;—but the thing is, morally speaking, impossible: For, if he is a man of the least spirit, he will have fifty deviations from a straight line to make with this or that party as he goes along, which he can no ways avoid."[42] Like the ragpickers we saw in chapter 8, narrative is peripatetic. The more he narrates, the more he has to narrate, once the narration has becomes part of his life:

> To sum up all; there are archives at every stage to be look'd into, and rolls, records, documents, and endless genealogies, which justice ever and anon calls him back to stay the reading of;—In short, there is no end of it;—for my own part, I declare I have been at it these six weeks, making all the speed I possibly could,—and am not yet born:...you see the thing is yet far from being accomplished.

Tristram therefore plans "to go on leisurely, writing and publishing two volumes of my life every year;—which, if I am suffered to go on quietly, and can make a tolerable bargain with my bookseller, I shall continue to do as long as I live."[43] There is no prospect of closure. As with Funes, Tristram can never hope to contain all his memories.

In Volume IV, Tristram points out that

> I am this month one whole year older than I was this time twelve-month; and having got, as you perceive, almost into the middle of my fourth volume—and no farther than to my first day's life—'tis demonstrative that I have three hundred and sixty-four days more life to write just now, than when I first set out; so that instead of advancing, as a common writer, in my work with what I have been doing at it—on the contrary, I am just thrown so many volumes back—was every day of my life to be as busy a day as this—And why not?...[A]t this rate I should just live 364 times faster than I should write—It must follow, an' please your worships, that the more I write, the more I shall have to write—and consequently, the more your worships read, the more your worships will have to read...I shall never overtake myself.[44]

This ceaseless accumulation is like the prospect of a never to-be-topped trash heap. While *Tristram Shandy* has moments of "unreadability," the work purposely does not adhere to "the ingest-and-excrete model of 'text in/text out,' nor the production of 'slag,' but the production of a 'fruitful remainder,' the remainder of desire: something that cannot be signified but that continues to provoke new meanings."[45] You could read *Tristram Shandy* as a satire of autobiography; or, rather, a satire of readers and their expectations.[46] The digressive details are not just part of a wastefully digressive joke; he shows what happens when you try to explain "all." The black pages mockingly deny readers details of Yorick's ghost; the afterlife of ghosts cannot be explained, so has to be filled in as a placeholder. He will annotate his life as if he has not already gone into enough digression.[47] Wasteful or irrelevant digressions elaborate and show variation, the witness to life.

Tristram's narrative is no tragedy; at least, from the reader's perspective. If we read the text as a body, then digressions and excess are the excrement, that which gives life. Resisting closure, digressions work against a closed text. Even footnotes can be read as privies, full of potential fertilizer. Shandy knows he will never finish; the autobiography is waste. Calvino concurs that personal history can be read as junk, concluding his own paean to garbage, with a series of discrete phrases junked together, including "autobiography as refuse."[48] Meaning lies in the very meaningless of the wasteful digression.[49]

The irreconcilable waste of the literary text forces us to face our own ethics, ethical position, and subjectivity. In this way, the form of waste literature contributes to restitution, a kind of compensation or amends. Waste literature both forgives us for our actions that have soiled the world and urges us to rectify those actions.

Intertextuality

Whan that Aprill with his shoures soote.

Geoffrey Chaucer, *General Prologue*, *The Canterbury Tales* (I.1)

April is the cruelest month.

T. S. Eliot, *The Waste Land*

The metaphors of cultural recycling—in concepts such as intertextuality, *bricolage*, and appropriation—become theoretical ways of justifying and explaining literary detritus for a generation discovering that originality is virtually impossible.[50] Intertextuality can be used to both comic and tragic ends. The waste of the past pervades T. S. Eliot's *The Waste Land* (1922), giving voice to the inevitability of historic decay and destruction. Even the title prepares the reader for the fragmentary detritus of early twentieth-century culture. The opening parody of Chaucer argues that fertility rouses past memory. The dead land and its "forgetful snow" was almost a kindness. "Winter kept us warm, covering / Earth in forgetful snow, feeding / A little life with dried tubers." Just as we saw in "The Wanderer," it is cruel to have *Sehnsucht* aroused. To be conscious is to be unhappy. "What are the roots that clutch, what branches grow / Out of this stony rubbish? Son of man, You cannot say, or guess, for you know only / A heap of broken images." This evil plurality of images rots before us. The sun brings shadows—memory, past, history—and with them the reality of death. April, the advent of spring and fertility, is paradoxically cruel. "Your shadow at morning striding behind you / Or your shadow at evening rising to meet you; / I will show you fear in a handful of dust." The light exposes the disguise each one of us wears to evade consciousness. Innocence and sunshine are exposed as mere masks. Dehumanized people have no hope of God or salvation in this "Unreal City, / Under the brown fog of a winter dawn." Not even detritus remains to indicate that civilization, however filthy, endures. "The river bears no empty bottles, sandwich papers, / Silk handkerchiefs, cardboard boxes, cigarette ends / Or other testimony of summer nights. The nymphs are departed." Their abandonment of this world simultaneously signals an abandonment of pure nature, which never existed in the first place.

Medieval detritus girds *The Waste Land*. The reference to the Grail legend underpins the entire poem. The choir of children sings at the ceremony of foot washing that precedes the restoration of the wounded Anfortas (Fisher King) by the Knight Parzival and lifting of the curse from a wasted land, in a reference to the Maundy Thursday foot-washing ritual. The washing of feet in soda water parodically debases a glorious and legendary past. Yet the Unreal Cities of "Falling Towers," like those decayed societies of "The Wanderer," "The Ruin," and "Ozymandias," all civilizations that thrived and died, still have the ability to communicate.[51] Unreal London will be remembered. As people crawl through the scum of city life, memories keep order. "And upside down in air were towers / Tolling reminiscent bells, that kept the hours / And voices singing out of empty cisterns and exhausted wells." Bells should recollect sacred life to us in a ritual of memorialisation. "In this decayed hole among the mountains / In the faint moonlight, the grass is signing / Over the tumbled graves, about the chapel / There is the empty chapel, only the wind's home." Nature sings through empty church. While there is little hope that Christianity can be saved or save, its ruins endure.

Rather than seeing waste as that which is not useful, as Francesco Orlando's monumental work with its taxonomy of ruin suggests,[52] Viney argues that concepts of waste "represent a specific relation to or augmentation of the time we make through things." Rather than "use" and "waste" being polarized terms, we should "seek to understand the transient co-dependence and cross-contamination of these terms." For example, Viney reads Eliot's poetry as "contemplat[ing] the leftovers of literature," constituting a "poetics" of "discards" and "residues."[53] Pound, who saw the poem as "fascinatingly excremental," cleansed the text of some dirt, including the adjective in "dirty camisoles" in "The Fire Sermon" and also excised "the more cloachal passages in Joyce's *Ulysses*."[54] Only more recently has there come a fuller understanding of how the poem, according to Tim Armstrong,

> accumulates detritus... Eliot cannot fully incorporate all his materials; his fragments remain undigested. *The Waste Land* thus bespeaks a simultaneous fascination with, and revulsion from, waste. The poem seems to revel in excess... The process of waste-production is knitted into its cultural moment: it cannot (and Pound cannot) "edit out" all the waste, because it *is* waste material... There can be no production without waste.[55]

This creates a conundrum for Eliot: while waste must be "eliminated from the poem," it remains "central to its production."[56] Eliot shovels cultural waste—from Wagner to Shakespeare—alongside heaps of sexual trash sealed with "one final patronising kiss." The cracking of civilization with its "Falling towers" portends the "tumbled graves" of physical death and spiritual despair. "Shantih shantih shantih."

Intertextuality grounds parody. Academic language and bureaucratic bloat haunt the elevated, even convoluted, discourse of Gerald Vizenor's character named Shaman Truth Lies. In the story "Crossblood Coffee," this trickster comes up with a scheme by which his reservation "would corner and control international coffee markets." He announces this at an

> academic assessorization conference... The college provosts and deans at the conference were neither surprised nor displeased with this unusual proclamation. Tricksters were well known in higher education circles for exploiting familiar forums of instruction, revealing unnatural visions, and exposing banal dreams.[57]

He tells how Indians tricked white men and how coffee properly belongs to tribal peoples, who almost lost coffee to "enterprising white men."[58]

> The administrators at the conference on academic assessorization were so pleased with the entertaining benediction that they voted to name the trickster an honorary provost of the college of his choice. Truth Lies became provost of Kresge College at the University of California, Santa Cruz. The trickster returned the gesture; he named the provosts and deans honorary tricksters, members of an urban reservation of their choice. They became landfill meditators on their own academic trash.[59]

This parody of scholastic drivel reveals the utter ridiculousness of academic speech through the excessive adaptation of the (supposedly) rational discourse by a speaker in a dubious position (trickster).

The intertextual background for *The Journey* includes that most German of genres: the fairy tale. A world replete with actual demons sporting swastikas makes the unreal aura of fairy tales cruelly apt. Rather than an edifying and uncanny lesson for youngsters, Adler uses the *Märchen* atmosphere to heighten the creepiness and pathos of Zerlina's fate. He depicts the path to her murder and fiery death in the crematorium through the fractured lens of fairy tales, making her plight all the more poignant. Zerlina, "my pale Snow White,"[60] caught in the web of perverse fairy tale detritus, cannot even get her father's death to be recognized as murder. The camp is like a bewitched place: "Have you not often said yourself that in Ruhenthal we were under a spell and really just sleeping?"[61] Who speaks to her? Frau Ilsebill from Brothers Grimm or other women gluing boxes? Zerlina regrets having told the women fairy tales of eventual freedom; their talk is

> disturbing and unsettling, a delusion that bedecks all of the boxes with a poisonous dust. Yet the delusion contains nothing, it is only a halfhearted murky shadow that dissolves into an untraceable odor. It smells of the night, of bare trees, vermin, rubbish, the thin layer of dust left by the fog, and all of it falls into the pits . . . The witnesses to this destitute journey waste away as well.[62]

Zerlina hopes to escape and live in woods like a rabbit. In the wake of people's consumption, Zerlina "wants to take the broom and sweep the room. The vermin in the cracks will be pleased. Cleanliness is half of healthiness . . . Disturbed fleas leap up in shock like the seven dwarves."[63] Adler jumbles time and place, disorienting us just as the characters themselves confront the baffling, inexorable onslaught. Fairy tales give shape to "shapeless and unknowable" events,[64] though there is no doubt that Zerlina is killed.

In the ultimate of intertextual partners, *Endgame*, a metaphysical play about the ultimate play, the one directed and written by God, echoes *Genesis* and *The Tempest*. Beckett acts like "a real rag-and-bone writer who collects and recycles a large number of biblical, literary, and philosophical references."[65] Hamm, prismed through his avatars—the storyteller, director, God, Prospero—counts his sufferings highest beyond those of his creatures—father, mother, or dog. His first words—"Me—(*he yawns*)–to play"[66]—signals the theatricality of the content, the drama enacting the wastage to which we are all heir. Nagg tells a funny story, that of the tailor who needs more time to make a pair of striped trousers for an Englishman. The tailor repeatedly tells his customer to return the following week, until the Englishman, in utter desperation and being fed up, bursts out:

> God damn you to hell, Sir, no, it's indecent, there are limits! In six days, do you hear me, six days, God made the world. Yes Sir, no less Sir, the WORLD!

> And you are not bloody well capable of making me a pair of trousers in three months! (*Tailor's voice, scandalized.*) But my dear Sir, my dear Sir, look—(*disdainful gesture, disgustedly*)—at the world—(*pause*) and look—(*loving gesture, proudly*)—at my TROUSERS![67]

Endgame dramatizes the abundance of waste in God's creation. Despite this, no one wants it—creation, life, the play—to end. Hamm hesitates "to . . . to end. Yes, there it is, it's time it ended and yet I hesitate to—(*he yawns*)—to end." And yet, the play mitigates utter metaphysical despair through humor. Clov and Hamm listen to the alarm clock. Clov: "The end is terrific!"[68] Like the play itself, Clov offers hope to the audience that something will ring true, but Hamm claims, "I prefer the middle." In fact, all of life IS the middle; we merely impose a narrative with a beginning and an end onto it. Hamm: "It's finished, we're finished . . . Nearly finished . . . There'll be no more speech."[69] Yet the play isn't over, so there is more speech, creating a circle in which we the audience participate in mutual affinity.

Beckett interweaves biblical history, voiding an easy allegory of redemption to his work. We learn that Hamm was "a father to [Clov],"[70] who appeared when he was a child. A man had crawled there seeking food for his child, leaving him "deep in sleep," "three whole days."[71] How can we not think of Christ, dead for three days in hell, before his resurrection? Hamm assures Nagg that he'll "soon have finished with this story. (*Pause.*) Unless I bring in other characters. (*Pause.*) But where would I find them? (*Pause.*) Where would I look for them?"[72] The creator/author/storyteller must always ask himself for inspiration as to where he would get more characters. Convinced Hamm will "make up another," Clov nonetheless fears the story will end soon: "Keep going, can't you, keep going!" Hamm: "I don't know. (*Pause.*) I feel rather drained. (*Pause.*) The prolonged creative effort."[73] The play draws inevitably to a close, as does God's creation; our lives likewise near the end. Pointing to the theatricality of the misery and waste of our temporary antics, Hamm responds angrily, "An aside, ape! Did you never hear an aside before?"[74] Hamm worries there are more complications. "Not an underplot, I trust"[75]—that would only prolong the agony of the play, existence, life, creation. Clov announces toward the end, "This is what we call making an exit." Hamm acknowledges that they were "obliged to each other."[76] Man and God need each other, character and playwright need each other in a mutually anguished dance of blighted existence. Clov says, "I use the words you taught me. If they don't mean anything any more, teach me others. Or let me be silent."[77] Like Caliban, he is frustrated by the limits the language imposed on him by his director, Hamm/Prospero. Hamm: "Gone from me you'd be dead." Clov: "And visa versa."[78] The playwright needs his characters just as they need him to come into being. But, Clov asks, "Then how can it end? Hamm: You want it to end?"[79] Hamm: "We've come to the end. I don't need you any more." The playwright never needs the characters after the end of the play, nor God his creatures. Hamm is left with "A few words . . . to ponder . . . in my heart,"[80] like the Virgin Mary, whose Son was kindled by the creator Himself. Hamm evokes Prospero: "Our revels now are ended."[81] Hamm as Prospero, creator, writer, sees only the end, an end he fears yet is desperate to see. Hamm impatiently inquires after the funny story about the tailor, "Have you not finished? Will you never finish? (*With sudden fury.*) Will this never finish?"[82]

Clov asks, "What is there to keep me here? Hamm: "The dialogue. (*Pause.*) I've got on with my story. (*Pause.*) I've got on with it well. (*Pause. Irritably.*) Ask me where I've got to."[83] We are kept by history, by story, we keep going in our own plot, story, and play.

We only exist in conjunction with another—in dialogue with another and with other texts.

CHAPTER 12

Gleaning Aesthetics: Poetry as Communal Salvage

In a passage from the writings of his followers, St. Francis's love of the lark is noted, the bird gaining Francis's affection by the lowliness of its food and the spontaneity of its food collection. "Sister lark . . . is a humble bird who goes cheerfully along the road to find herself some corn, and even if she finds it among the dung of beasts, she takes it out and eats it."[1] Gleaning nutrition from what most might disdain as mere waste, her humility is admirable.

A basis of Judeo-Christian culture stipulates that waste in the form of gleanings should be left for the poor. In *Leviticus* (19: 9–10; 23: 22), gleanings are to be left after harvesting. There is profound virtue in gleanings, in both those who leave and those who gather. Waste *should* be left for the poor and alien to gather. Gleaners are those people cut off from the community, their low status a result of codification. Both actions reinforce the hierarchy and distinction between those who leave and those who gather; those who can afford to waste and those who can only afford to survive on charitably donated waste. Those who leave gleanings for those who are cut off from community are acting *for* the community and the common good. In this way, gleaners are made gleanings for society, and as such constitute an integral part of the community.

The Hebrew Bible emphasizes the plight of the foreigner. *Leviticus* suggests how to deal with wasted people, including ostracism and separation. But in several passages the Jews are instructed to treat the "stranger" like themselves, suggesting an affinity between each group (*Leviticus* 16: 29; 17: 8–16; 24: 10–15). The punishment for disobeying remains becoming aliens themselves. "And they that are left of you shall pine away in their iniquity in your enemies' lands; and also in the iniquities of their fathers shall they pine away with them" (*Lev.* 26: 39). The exile is a reject—as an immigrant, he is human waste, but a powerful symbol, laden with the potential to become one of the chosen. *Leviticus* speaks to the protection of aliens and how one should protect strangers:

> And if a stranger sojourn with thee in your land, ye shall not vex him. But the stranger that dwelleth with you shall be unto you as one born among you, and thou shalt love him as thyself; for ye were strangers in the land of Egypt: I am the LORD your God. (*Lev.* 19: 33–34)
>
> And if thy brother be waxen poor, and fallen in decay with thee; then thou shalt relieve him: yea, though he be a stranger, or a sojourner; that he may live with thee. (*Lev.* 25: 35)

God promises to lay waste to cities that disobey his edicts (26: 31). After all, as the Lord tells the Israelites, "The land shall not be sold forever: for the land is mine; for ye are strangers and sojourners with me" (*Lev.* 25: 23). All are alien to God.

The Ruth Effect

The Book of Ruth most movingly presents the "Ruth effect"—the expression of and acting upon utter ethical affinity. One woman stays with another (daughter-in-law with mother-in-law after the husband/son dies). A foreigner, worse, a vilified Moabite, Ruth elects to stay with Naomi. "But Ruth cleaved unto her" (*Ruth* 1: 14). The model for *hesed* (steadfast love and faithfulness) for Israel is embodied in a Moabite.[2] "For whither thou goest, I will go; and where thou lodgest, I will lodge; thy people shall be my people, and thy God my God; / where thou diest, will I die, and there will I be buried; the Lord do so to me, and more also, if aught but death part thee and me" (*Ruth* 1: 16–17).[3]

Ruth proposes to Naomi that they glean among the ears of corn in the fields of Boaz in Bethlehem. Boaz hears of her gleaning and tells Ruth to glean there by his

> "maidens. Let thine eyes be on the field that they do reap, and go thou after them; have I not charged the young men that they shall not touch thee? and when thou art athirst, go unto the vessels, and drink of that which the young men have drawn." Then she fell on her face, and bowed down to the ground, and said unto him: "Why have I found favour in thy sight, that thou shouldest take cognizance of me, seeing I am a foreigner?" (*Ruth* 2: 8–10)

She eats with his maidens and he even helps her gleaning: "And when she was risen up to glean, Boaz commanded his young men, saying: 'Let her glean even among the sheaves, and put her not to shame. / And also pull out some for her of purpose from the bundles, and leave it, and let her glean, and rebuke her not'" (*Ruth* 2: 15–16). Naomi tells Ruth to wash herself and go to Boaz and lie down at his feet after she has uncovered them. Boaz says to her, "Thou has shown more kindness in the end than at the beginning, inasmuch as thou didst not follow the young men, whether poor or rich. And now, my daughter, fear not; I will do to thee all that thou sayest; for all the men in the gate of my people do know that thou art a virtuous woman" (*Ruth* 3: 10–11). He refrains from having sex with her, sending her back to Naomi with six measures of barley. Boaz then buys up all that belonged to Naomi's husband and sons. "Moreover Ruth the Moabitess, the wife of Mahlon, have I acquired to be my wife, to raise up the name of the dead upon his inheritance, that the name of the dead be not cut off from among his brethren, and from the gate of his place; ye are witnesses this day" (*Ruth* 4: 10). Ruth conceives Obed, grandfather of King David. Naomi's own redemption comes through Ruth's unwavering loyalty and Boaz's pure intentions. Through the "total relinquishing of her self,"[4] Ruth appropriates "Naomi's existence and fate [to the] disappropriation of her own,"[5] thereby acting as a perfect example of Levinasian responsibility to the other.

Regard for those supporting gleaning persists throughout the Middle Ages. In the *Prologue* to *Piers Plowman*, the angel speaks to the Commons: "*Si ius nudatur, nudo de iure metatur; / Si seritur pietas, de pietate metas*'" ["If you strip law bare, bare law will be your share. / If you sow pity, / you'll be sitting pretty"] (*Prologue* 137–138). No gleanings result in an absence of salvation for the harvester. Grace can come only if it is earned through selfless generosity. Holy Church exhorts the rich to have pity on those who are less fortunate:

> Forthi I rede yow riche, haveth ruthe of the povere,
> Though ye be myghty to mote, beeth meke in youre werkes,
> For the same mesure that ye mete, amys outher ellis,
> Ye shulle ben weyen therwith whan ye wenden hennes.
>
> [Therefore I implore rich persons to have pity on the poor: / Though you're mighty men at law, be meek in your deeds, / For the same measure you mete out, amiss or otherwise, / You shall be weighed with it when you go hence] (I.175–178).

Faith without works is "feblere than nought" ["worse than nothing"] (I.186). While some churchmen think being clean in body is enough, they lack charity. "Manye chapeleyns arn chaste, ac charite is aweye" ["Many chaplains are chaste, but their charity is missing"] (1.190). Holy Church cites Luke 6: 38: "Give and it shall be given unto you": "And that is the lok of love that leteth out my grace, / To conforten the carefulle acombred with synne" ["That is the love-latch that lets my grace out / To comfort the careworn overcome with sin"] (I.202–203). Those supporting gleaners and gleaning glean goodness and grace themselves.

The gleaning tradition extends into the modern period. According to Sir William Blackstone's Commentaries on the Laws of England of 1768: "The poor are allowed to enter and glean upon another's ground after the harvest, without being guilty of trespass."[6] Gleaning has regained popularity in economically hard times.[7] Allan Stoekl has argued for the generous "ethics (and aesthetics) of filth, of orgiastic recycling," where the salvaged object carries a sacred dimension leading to community, as in Agnès Varda's film *Les glaneurs et la glaneuse* (*The Gleaners and I*) (2000). Gleaning from "postsustainable generosity" fosters community.[8] While recycling is not always utilitarian or economically most sensible, it has been argued that it is nonetheless important to do ethical work.[9] The focus should be on the "social rather than environmental impacts of recycling,"[10] through the spirit of "corporeal generosity."[11]

> Rather than a minor change in habits that confers virtue on the dutiful recycler, waste practices based on corporeal generosity would involve a body that was open to the difference of waste and the nonhuman world; a body that was aware of how its ways of living depended on the gifts of others. It would mean a body that was not indifferent to waste's alterity but aware, instead, of the intersubjective links that always connect us to what we discard.[12]

The act of recycling and gleaning can lead to a kind of societal redemption. As Calvino points out about putting out the trash:

> The reform that seems to be most pressing and important will be that of separating the rubbish according to its qualities and different destinies, incineration or recycling, so that at least part of what we have wrung from the treasures of the world may not be lost forever but discover the path of regeneration and re-use, the eternal return of the ephemeral.[13]

This "regeneration" springs not just from dynamic trash, but the lively act of recycling restores the human actors in this process. We can see this in the character played by Dana Andrews in *The Best Years of Our Lives* (1946), Fred Derry, who has bombed German cities, including Düsselfdorf. Fred, suffering from some form of posttraumatic stress disorder, visits a sea of junked planes in Boone City. Sitting in one, he sweats out bad memories. Thereafter he gets a job helping junked planes turn into prefab houses, rebuilding, in a sense, the houses he had bombed earlier. His bricolage and recycling contribute to community through providing shelter, perhaps even for Fred and his promised beloved.[14]

In *Seedfolks*, a vacant trash lot in Cleveland becomes a community garden tended to by immigrants and people of color. The urban garden becomes a locus of community, thanks to the tenacious efforts of its multiethnic caretakers. Nora, the British nurse, comments that when she goes to the observation tower in Cleveland to find the community garden, "which loomed so large to its tenders, [it] was hopelessly hidden from view by buildings. I looked at all the tourists, who'd no notion it existed, who thought they were seeing all of Cleveland, and restrained myself from pointing and shouting out, 'The Gibb Street garden is there!'"[15]

Each short chapter tells a story in a different voice, so that we glimpse briefly into one person's life, a garden of different shoots and growths hinting at the life within. The lot at first is called "that awful jungle of junk,"[16] becoming a kind of incubator, with the seeds laid into the ground "as gently as sleeping babies."[17] A Vietnamese girl, Kim, initiates the project by carrying dried lima beans and a spoon, a utensil to dig with, foreshadowing what she will use to eat with later. She decides to garden so that her dead father's spirit, one that "had no memories of me,"[18] could see her. Voyeurs decide to help Kim in her endeavor, a dream that accretes more and more participants. Wendell, a school janitor, decides to use his cleaning up skills on the lot. "There's plenty about my life I can't change... But a patch of ground in this trashy lot—I *can* change that... I dragged the garbage off to the side and tossed out the biggest pieces of broken glass."[19] Gonzalo tells how he is made by his parents to babysit his Tío Juan from Guatemala where he had been a farmer. Here in the United States, Tío Juan cannot work or even speak Spanish, let alone English. "Here he became a little baby."[20] The garden becomes his incubator; before long "he'd changed from a baby back into a man."[21] Virgil says that his dad's "lettuce was like having a new baby in the family... It was like a baby always crying for its milk."[22] Maricella, a pregnant 16-year old, keeps wishing her fetus would die. As part of a program for pregnant teens, she resents participating in planting seeds in the garden. During a thunderstorm, all the electricity in the neighborhood goes out. Leona, an African American woman who understands Maricella's resentment toward her baby, tells the girl:

> "Whole city shuts down, but the garden just keeps going"... She talked on, how plants don't run on electricity or clock time, how none of nature did. How nature ran on sunlight and rain and the seasons, and how I was part of that system. The words sort of put me into a daze. My body was part of nature. I was related to bears, to dinosaurs, to plants, to things that were a million years old. It hit me that this system was much older and stronger than the other. She said how it wasn't some disgrace to be part of it. She said it was an honor... And for just that minute I stopped wishing my baby would die.[23]

The one white, non-immigrant character, retired from working "for different groups, promoting world government, setting up conferences on pacifism, raising money, stuffing envelopes,"[24] points out, how he used "to try to patch up the whole world... Sewing up the rips in the neighborhood."[25] He is an urban bricoleur, patching up diversive groups by participating in this garden project. Bricolage involves the "creative reuse with whatever is at hand... [and] the human capacity to imagine and create new uses and meanings for things... it involves an active reappropriation of things in different contexts."[26] The bricoleur engages in "dialogue."[27]

This bricoleur in *Seedfolks* intensifies the communal spirit by initiating a contest for children to figure out how they can get water more easily into the garden and not have to carry it "like coolies."[28] The British nurse, Nora, pushes her charge Mr. Myles in his wheelchair through "a miniature city. Some plots sported brick pathways and flower borders, while others looked haphazard. One had a gate that was in fact a car door. Within, beans climbed a propped-up set of bedsprings."[29] One owner fashions a gate and climbing stick out of junk. Nora points out that, even though different groups grew different types of fruits and vegetables and that there were language blocks, "we were all subject to the same weather and pests, the same neighborhood, and the same parental emotions toward our plants. If we happened to miss two or three days, people stopped by on our return to ask about Mr. Myles' health. We, like our seeds, were now planted in the garden."[30] As the Indian Amir points out, "the garden's greatest benefit, I feel, was not relief to the eyes, but to make the eyes see our neighbors."[31] An old Italian lady had once called Amir "a dirty foreigner." After they become friendly in the garden, he reminds her of that time and she is horrified to remember. "Back then, I didn't know it was *you*."[32] And in the spring following a hard, frost-laden winter, Florence is happy to see the regulars at the garden.

"We waved and waved to each other."[33]

PART IV

Source Reduction and Reuse: Compassion through Generous Metaphor

CHAPTER 13

Compost Aesthetics: The Poet[h]ics of Metaphor

In "A Load of Shit," John Berger comes to see the importance of not-naming:

> Not far from where I dug the hole, a lilac tree is coming into flower. The wind must have changed to the south, for this time I can smell the lilac through the shit. It smells of mint mixed with a lot of honey.
>
> This perfume takes me back to my very early childhood, to the first garden I ever knew, and suddenly from that time long ago I remember both smells, from long before either lilac or shit had a name.[1]

This prelapsarian moment allows the narrator to cherish this mingling of the sweet and the foul.

What happens when we *do* name? How can we name ethically? Metaphor opens us to ethical understanding, what Nussbaum has called "metaphorical imagination."[2] Metaphors and similes create new characteristics for objects, feelings, people, nature, and animals. In Walt Whitman's "I Sing the Body Electric," in which "the bowels [are] sweet and clean," we encounter a poem that is, as Nussbaum points out, against shame and disgust, where "all bodies are equally worthy of respect."[3] Not only all bodies, but also all *parts* of bodies. Imagination as excess is generous waste.[4] Berger sees beauty in "the smell of decay . . . of putrefaction, of corruption . . . It has nothing to do—as puritanism with its loathing for the body has consistently taught—with shame or sin or evil. Its colours are burnished gold, dark brown, black: the colours of Rembrandt's painting of Alexander the Great in his helmet."[5] Shit and Rembrandt—material filth and priceless painting—yoked together through the affinity of color.

Compassionate Materiality

Disgust "pose[s] a particular threat to compassion."[6] For Nussbaum, the "capacity to imagine [another's] experience," key to undermining the politics of disgust,[7] allows us to "see the other as a center of perception, emotion, and reason, rather than an inert object."[8] Key to this process is literature; the arts can change hearts.[9] If you empathize with characters, particularly those you would never be likely to meet in your life, you simulate emotions of empathy that imprint within you the feeling of sympathy, thus fostering a literary imagination that may extend into real-life encounters with others. Narrative allows us to participate in what Keats identified as Shakespeare's great capacity for "negative capability," a kind of erasure of one's self in the quintessence of literary imagination.[10] Negative capability allows us to remain in doubt; by not being cemented in our own perspective, we can be open to those of others. Or perhaps, like Walt Whitman, we enact *positive* capability—rather than the *erasure* of oneself, the *expansion* of oneself to enter the perspective of every other creature—human and nonhuman. "All goes onward and outward, nothing collapses" (*Song of Myself, Song* 6).

In *Philoctetes*, Neoptolemos ultimately resists Odysseus's treachery, even asking, "Would you like me to hold you—to hold you close?"[11] Neoptolemos hands Philoctetes back his bow in a moving moment of concilation and Levinasian ethical acknowledgment of Philoctetes's affinity to himself and the Greeks: "Take this weapon of yours from my right hand."[12] Returning the bow both sacred and vital to Philoctetes's survival, Neoptolemos humbles himself by privileging Philoctetes over himself and the Greeks. These moments of deferring to the Other stand as ethical pools of hope in a tradition that witnesses the all too frequent privileging of self. Socially positive results can come out of acknowledging a common humanity rooted in waste.[13] Ethics are "embodied," "emerg[ing] out of experiential networks of obligation."[14] Building on Eileen Joy's "ethics of interdependence," we need to become "*more*, and not less, human,"[15] by recognizing we *all* are wasted humans, humans with waste, humans who produce waste.

By compassionately recognizing anxieties about one's own body and society, one could come to self-love and then wield compassion for others.[16] By shaping forms of life "aesthetically," Ottmar Ette reminds us that "fictionality creates a space of experimentation in which readers, in serious playfulness, can test out different life situations, with which they can engage to collect experiences that they could not have in 'real life.'" Like muscle memory, *emotional memory* may kick in when faced with an experience or situation structurally similar to one encountered in a narrative. Having practiced compassion, as it were, that simulacrum of pity, empathy, and sympathy could be fostered in real-life scenarios. "Unlike philosophy, literature can translate life knowledge into experiential knowledge."[17]

The real-life situation has affinity to the narrative moments cherished by devoted readers.

As Whitman writes in *Song of Myself*, "Whoever degrades another degrades me" (*Song* 24).[18] Like John Donne who enjoins us, "Do not ask for whom the bell tolls," Whitman insists on the similarities among humans, our connections, our affinities, hence his assurance to the runaway slave (*Song* 10). Through the introduction of an unexpected comparison and linked image, a previously neutral or unmarked entity becomes transformed into a marked or non-neutral one. This bond highlights difference and similarity; for example, in this passage:

> A child said, *What is the grass*? fetching it to me with full hands;
> How could I answer the child? I do not know what it is any more than he.
> I guess it must be the flag of my disposition, out of hopeful green stuff woven.
> Or I guess it is the handkerchief of the Lord,
> A scented gift and remembrancer designedly dropped,
> Bearing the owner's name someway in the corners, that we may see and remark, and say *Whose*?
> Or I guess the grass is itself a child, the produced babe of the vegetation...
> And now it seems to me the beautiful uncut hair of graves.[19]

This plethora of metaphors associates grass with a flag, God's handkerchief, the child himself, and the "hair of graves." This entire verse epic constitutes an embrace of everything: "Not an inch nor a particle of an inch is vile" (*Song* 3). He weaves a song of himself out of everything and everyone—"husband" and "prostitute," "opium-eater" and "drover," "quadroon girl," and the "malform'd" (*Song* 15).

How could waste literature help compassion?[20] As Ette comments, "the question arises of how to translate the literary experiment into one's own ways of living, perhaps even into certain groups' practices of everyday life."[21] In a sense, he is asking, can literature make us better people? And where is the role of waste literature in this endeavor? Waste literature is not a genre; rather it is a literature that takes our blinders off to take in the layers of our world. While Mary Douglas's seminal work on waste where dirt is "matter out of place" typically emphasizes what has been excluded or is incapable of being categorized, Rob Smith, the Chief Flusher of Thames Water, makes the significant point that the sewer is "an equalizer." "It's a great leveller" down there.[22] Waste allows us to see the fundamental similarity among us all, just as metaphor or simile allows us to see the affinity between two things or states of being not previously perceived. In this way, waste is inherently metaphorical. It acquires the patina of positive or negative characteristics depending on the literary and linguistic meaning imbued in it. Like simile and metaphor, waste produces a perception of affinity and connection while simultaneously disrupting through difference. In Whitman's poem, such affinity is overt:

> Or I guess [grass] is a uniform hieroglyphic,
> And it means, Sprouting alike in broad zones and narrow zones,
> Growing among black folks as among white.

The physical metaphors in poems dedicated to describing decay and waste can be read as verbal ballets acknowledging affinity among all living creatures, one normally denied. "The Excrement Poem" by Maxine Kumin (1978) begins by recognizing,

> It is done by us all, as God disposes...
> We eat, we evacuate, survivors that we are.[23]

Flesh as the common basis of us all allows for carnal understanding. Kumin honors shit not just for its transience,[24] but also for how it gestures to the past, present, and future of all creatures.

Caroline Bird's "Joy is Like a Hungry Pig"[25] invokes fecal matter with the enthusiastic wonder of children, whose parents "have to hose them down in the garden." Joy ensues from revelling in waste, defying civilization's supposed proper norms. Transgression transforms the transgressor into a pig, until "you start to oink," a reminder of our inherent animal nature. Lifeless drudges who wield "knives and folks" with propriety snub the threatened "pandemonium" that could ensue from "HUNGRY PIG[S]" lathering up "mad crazy joy." The men and women with knives and forks are like the cultivated betters of Norbert Elias's civilizing process, restricting and disapproving in their infertile and orderly "bowling alleys." The joy of waste and dirt threatens a burst of delighted energy, and you, the "only one clean," the one imposed on by the civilizing process, remain alone, without community, without a fertile field springing up with life.

Indeed, the inevitable state of us all—to produce and ultimately become waste—emerges from waste literature. Rather than separating us into positions of dominance and submission, this affinity links us in similarity. Such a comparison "renders visible ways" in which we are alike as opposed to different.[26] The point is "to re-cognize comparison."[27] By rethinking and reconceptualizing comparison, we see ourselves not *in opposition* to but *in addition* to others.[28] Our horror at recognizing our own inevitable decay may vaccinate us to become more ethical and more humane in our dealings with others, particularly those "wasted" beings denigrated by society.

Whitman combs the meaning of grass in *Song of Myself*:

> I bequeath myself to the dirt to grow from the grass I love,
> If you want me again look for me under your boot-soles.
> You will hardly know who I am or what I mean,
> But I shall be good health to you nevertheless,
> And filter and fibre your blood.
> . . .
> I stop somewhere waiting for you.

The belief in the future compels us to act for the distant and future Other.

Poetry and literary criticism can function as modalities of "restitution"[29] to help motivate social change. Waste is my *Ansatzpunkt* or "point of departure," as Erich Auerbach would say. In his assessment of *Weltliteratur*, Auerbach points out the burdens of literary history considering how "man not only lives on earth, but that he is in the world and in the universe." Nonetheless, he urges us to remember that "our philological home is the earth: it can no longer be the nation." In this time when ecological collapse threatens, a philological waste project can, I hope, offer a beneficial path "for one who wishes to earn a proper love for the world."[30]

Ottmar Ette proposes that

> knowledge for living together [*Zusammenlebenswissen*] is knowledge of the conditions, possibilities, and limits of living together as the literatures of the world have shaped it aesthetically and have tested it experimentally from radically different perspectives. The concept has not only social, political, and economic dimensions but also cultural and ethical ones.[31]

The study of waste—how it reflects the human capacity and potential for creativity, evil, and magical grace—suggests "how people might live together in difference."[32] It is up, then, to the reader, to take the next step. Acting on knowledge—even of something as repugnant as waste and the whereabouts of its disposal—prevents alienation, ultimately leading, as Aristotle would argue, to pleasure and happiness.[33]

Poetry is creative, ethical excess, what Ammons calls an "enchantment," charming listeners into silence.[34] "Anything, / anything, anything is poetry . . . it is a plentiful waste and / waste of plenty . . . my / poetry is strawbags full of fleas the dogs won't / sleep on or rats rummage."[35] Despite—no—*because of*—his low view of us, writes Ammons, "that is why / I love us or try to move to love us."[36]

Perhaps that is the point of poetry: to help us to love.

CHAPTER 14

Poetry as Homeopathy: The Poet as Ragpicker

> I become a responsible or ethical "I" to the extent that I agree to depose or dethrone myself—to abdicate my position of centrality—in favour of the vulnerable other.
>
> Emmanuel Levinas[1]

What if that "vulnerable other" is a mephitidae?

In the 1940s my parents went on a camping trip in the Boundary Waters Wilderness Area between Minnesota and Canada. Carrying backpacks, they canoed on lakes and rivers for a month. One night, after falling asleep in their sleeping bags in the tent, they heard a noise.

"Maybe it's a bear," said my mother, a little scared.

"Oh, no, I'm sure it's just the wind," said my dad who was very tired and turned over to go back to sleep.

"There it is again!"

Daddy pretended to sleep.

"Maybe you should just check."

He got up and peeked out of the tent flaps. The moon was almost full. He looked across the tiny bay to a grassy meadow.

"Is it a bear?"

"No, it's a skunk."

"A skunk? What's it doing?"

"It has a tin can on its head."

"A tin can? How could it have a tin can on its head?"

"Someone who camped here must have brought a can along, used it, and just left it on the ground. The skunk's head is stuck inside the can."

My mom looked out the tent flaps. "Oh, well, you must save it."

"Save it? What can I do?"

"Pull the can off!"

"But what if it sprays me?"

"What if the skunk has little babies at home, waiting for it to return with food?" My mother's eyes filled with tears.

Daddy thought for a long time. They had no tomato juice with them. So he took off his clothes until he was naked. Then he left the tent, climbed into the canoe, and paddled across the narrow inlet. He landed, quietly getting out.

He neared the skunk, bent over, and pulled the tin can from the skunk's head.

The story goes that the skunk started to run away, then stopped, and looked my father straight in the eye. The skunk tipped his hat, bowed, and said, "Thank you, brave sir, for your kind deed."

Most of that story is true. Maybe all of it.

Waste in the form of a tin can provoked a rapprochement between man and animal. Or rather, the understanding between two animals, equal subjects in a piney forest.[2]

New Materialism: Body as Becoming

Be not afraid of my body.

Walt Whitman, *Children of Adam*[3]

Myra J. Hird proposes how "new materialism" can "reorient our approaches to bodies and sexual difference."[4] Nonlinear biology and new materialism present "the body as a positive event rather than a negated origin," where matter can be seen as "active, mobile and subject to constant, dynamic change."[5] To dehierarchize the body is to break down the binaries that vilify the lower bodily stratum. We need to see and sense our bodies already with us. The body is not simply a bounded, enclosed object out of which disgusting fluids and solids are ejected. The matter within us touches an exterior. Our bodies, then, embrace a world beyond the envelope of our skin. You can attempt to transcend your open, porous body or accept and revel in it.

In the Anglo-Saxon *Wonders of the East*, on an island in the Red Sea, the Donestre live. They not only "know all human languages" but also are "shaped like soothsayers from the head down to the navel, and the other part is similar to a man's body."

> When they see a man of foreign race, they call him and his fellows with the names of known men, and with lying words they deceive him and seize him. And then, after that, they devour all of him, except the head, and then they sit and weep over the head.[6]

This passage allegorizes the rhetoric of othering: the lies we tell about others that enable us to "devour" them through othering, and the subsequent profound melancholy we create in ourselves by denying their humanity in difference.

As Michael Uebel has asked, what if you changed your relation to the Other?[7] An openness to the "Other" or "Face" is a sign of the ethical. Such openings provoke moral agency. As Levinas writes, "Ethics is, therefore, *against nature* because it forbids the murderousness of my natural will to put my own existence first."[8] Waste, in all its manifestations, including excrement, provides us with a reason for acknowledging affinity among all people, one that we normally deny. The display of Grendel's arm in the hall, like a hunting trophy, is the visible sign of culture's triumph over the monstrous (833b–6b). As Benjamin points out, "Whoever has emerged victorious participates to this day in the triumphal procession in which the present rulers step over those who are lying prostrate . . . The spoils are carried along in the procession [or, in the case of Grendel's arm, hung on the wall of Heorot]. They are called cultural treasures."[9] The gesture of putting Grendel's arm on the wall makes waste into art(ifact). Yet, as Benjamin cautions, we need to contemplate such trophies with horror. Only by destroying Grendel's bodily integrity can the humans reassure themselves of their own.[10] The monster is partial and disturbing, making "us" whole and "normal." The hanging of Grendel's arm as a trophy serves to reinforce the fictive wholeness of the human body.

Sianne Ngai schematizes a "possible grammar of disgust," pointing out that "[a] form of sociability or mutual attunement [is] facilitated by turning away. Paradoxically, the other is included in the subject's response of disgust via his act of exclusion, his shunning of the object."[11] As she develops her argument, Ngai points out that "this act of exclusion, of turning away from the object, creates an absence or void *that becomes the space for the other to step in*. The outwardness of the subject's gestures of exclusion, in other words, enables the mutual attunement between him and the other."[12] While "disgust is intrinsically exclusionary,"[13] it could allow for affinity. Grendel, constructed as a monstrous other, as waste, by the humans, is oddly similar to those who seek him out for destruction, as the horrific discovery of Æschere's head exposes.[14] We are all simultaneously whole and fragmented; we all contain our inevitable potential to become—inevitably—trash. "The restoration of the subjective life of one's own body serves to place the self in the world with other subjective bodies."[15] Why do people hate "wasted humans"? Because we see ourselves in them.[16] Anyone could become an exile; anyone could become "monstrous"[17] and fragmented. "The line separating a seat of power from a rubbish heap" is "thin and frail."[18] Though Grendel is viewed as waste, as garbage, and dispensable, he, like his counterparts, the human colonizers who invade, kill, raid, and conquer, exists as part of a cycle of destruction and mastery within which human achievement is, in its turn, doomed to failure. Any triumph is temporary, fleeting, and caught up with barbarous acts, implicating any present victory as unethical brutality. The poem enjoins us to remain thoroughly mindful of our own inevitable decay. Not unlike Baudelaire's "hypocrite lecteur," Grendel is "mon semblable, mon frère."[19]

Myra Seaman has characterized the human, as "presumed by traditional Enlightenment and post-Enlightenment humanism,"[20] to be a rational male being at the center of the world with independent agency. A fully embodied human—even to the extent that she/he defecates—is repulsive and disturbing. Those who handle filth, literally or figuratively, become tainted by it morally and socially. But as Seaman points out, "without vulnerability, a society of extremely rational beings experiencing no loss to decay or disease would find little need to express their experiences, their individual selfhood through art . . . Identity does not seem possible without stories."[21] My "waste human"—who has more in common with a mama's boy monster living in a dank mere than with a triumphant, "unified, cohesive 'human'"—adheres to the posthumanist emphasis on "mutation, variation, and *becoming*."[22]

In *Beowulf* the monsters emulate human behavior. Grendel's mother is only doing what the humans do—avenging a killed loved one with another killing. When imitative behavior is practiced by those perceived as outsiders (as other or waste) to the dominant culture—in this case, for example, monsters imitating humans—the emotional response of the dominant culture is not empathy but hatred. While, as Hrothgar explains, the people at Hrothgar's court had heard tell of a woman associated with Grendel (1345–1351), no one had expected her to attack; no one had empathized with her, imagining *her* grief, anger, and desire for revenge for her son. Grendel's mother, in choosing Æschere to behead, on the other hand, clearly could empathize with Hrothgar; she deliberately chose his most beloved companion knowing the pain it would induce in him. Though Grendel's mother resembles a woman ("idese onlīcnæs," 1351), the narrator nudges us to view her as other, if only in her environs. Like the water of the mere, her position is murky. This both/and, both like a woman yet not a woman, this hybridity, produces the "uncanny," and renders her powerful, vulnerable, eerie, and even—possibly—sympathetic.

In Mary Shelley's *Frankenstein*, the excluded monster is actually a sign of us, for us. Lennard Davis analyzes the novel in this light and asks, much as we might of Grendel, "[The creature's] body is a zone of repulsion; the reaction he evokes is fear and loathing. The question one wants to ask is why does a physical difference produce such a profound response?"[23] Frankenstein's monster, patchworked from the waste of dead skin, that which we are all encased in, is more easily able to empathize with others than his own creator is able to with him. The creature, patched together with dead material, exposes how vital and vibrant matter not controlled by humans can be.[24] Like the creature, waste is "man's creation that keeps escaping its creator... Humans can no longer destroy it but only be destroyed by it."[25] Embodying "the force of things,"[26] the creature's lack of passivity—suggested by the patched dead matter of which he is comprised—disturbs.

The violent response to the creature's body emanates from the fear of touch, as though his wasted composition would infect with an insidious decay.[27] With his body, a creation of many parts, constructed out of bits and pieces from the cemetery, prison, and slaughterhouse, the monster feels connected to the whole in a way inaccessible to the father/creator who singularly birthed him. Shelley portrays the monster as acting out of respect for his composite body by living his brief life as a vegan, eating nothing to which he is so intimately related.[28] We all are the monster, fashioned out of our skin.[29]

This skin, this flesh, is our common basis, allowing communal and carnal metaphors for a totality of understanding. Cleaning toilets is generally seen as "unsexy" or humiliating. But how profound a gesture is it to fellow human beings. The Chinese dissident artist Ai Weiwei tells how his 60-year-old father, after running into trouble with Communist authorities, was exiled in the late fifties and had to "clean out the village's public toilets" for 16 years. As his son explains,

> He had never done physical work in his life and he had to start doing it. Every night, he comes home very, very dirty. But he says, "For 60 years, I don't know who cleans my toilets. So now I do something for them." That's something I learned from him. He became very powerful in terms of his thinking. He made the toilet so clean, he would see it as a work of art—like a museum, like MoMA.[30]

Filth management has composted into art, a compassionate gesture of selfless and generous expenditure.

Affectionate Affinities

> Trash is what people is dat puts dirt on de head er dey fren's en makes 'em ashamed.
>
> Jim in Mark Twain's *The Adventures of Huckleberry Finn*

Devastated by Huck's malicious and thoughtless trick of pretending that Jim just dreamt that they were separated, Jim realizes the truth upon spying the "leaves and rubbish on the raft"—as though the "smashed oar," whispered in Jim's ear that Huck was lying.[31] We could do worse than follow

Erasmus's compassionate enjoinder to "be lenient towards the offences of others. This is the chief virtue of *civilitas*, of courtesy... If one of your comrades unknowingly gives offence... tell him so alone and say it kindly. That is civility."[32] Fecopoetics—the poet's use of scatological discourse—emerges from the embodied mind and acknowledges the porous boundaries of the contingent body. A critical humanism, which "takes the human to be an open-ended and mutable process,"[33] could help us to see those humans traditionally viewed as "waste" as akin to ourselves.

Hildegard von Bingen's songs include *Antiphons* short religious chants to the Virgin Mary, whom she praises as "that lucid [clear] matter / Through whom the Word [God] breathed forth everything of value."[34] For Hildegard, the Virgin Mary exemplifies the best of womanliness. "Your womb contained joy / Just as the grass was infused with / Greenness, when the dew sank into it; / Therefore Mother everything joyful has been Created through you."[35] Hildegard redeems it and, thus, women's bodies.[36] Taking the lower bodily order of Bakhtin, Hildegard transforms the womb into the center of goodness. "But, O Dawn, from your womb / a new sun came forth, / who wiped away all the sins of Eve."[37] Hildegard associates the "matter," from which God created all creatures, with the Latin word *mater* for "mother" or "maternal." Indeed, the Indo-European root for "matter" is "māter," which birthed words such as mother, maternal, material, matter, and even De*meter*, the Roman goddess of crops.[38] If Mary Douglas's dictum is "dirt is matter out of place," what happens if we return to the etymology of material, māter? Does dirt become *mother* out of place?[39] One becomes noticed when one is not where one "should" be. A woman is dirt when she is present in a man's world. Only in her own domain is she not perceived as dirty. Rather than suggesting that matter, long connected to women, as the source of all evil, Hildegard redeems it and, thus, women's bodies. Indeed, Hildegard von Bingen writes a song in praise of the Virgin and reclaims women's bodies as containing the possibility of virtue and goodness. "A bright Maiden overcame [death], / and so the highest blessing / in all of creation / lies in the form of a woman."[40] Hildegard's construct allows matter—and women—to transcend binaries labelling matter or dirt as filth and bad.

As an act of "oppositional consciousness," we coalesce not through *identity*, which leads only to fragmentation, but through *affinity*.[41] Waste is the great leveller, linking us all through effective—and elective—affinities.[42] In *Cleanness*, Daniel describes to Balthazar the madness and descent into beastliness of Nebuchadnezzar, who ate the food of beasts and went on all fours, his beard growing down to the earth. This lasted for seven summers,

> Til he wyst ful wel who wroght alle myghtes,
> And cowthe uche kyndam tokerve and kever when hym lyked.
> Thenne he wayned hym his wyt that hade wo soffered,

That he com to knawlach and kenned hymselven.
Thenne he loved that Lorde, and leved in trawthe
Hit was non other then he that hade al in honde.

[Until he readily recognized Who created all powers, / And could carve up every kingdom and restore them at will. / Then He restored his senses, who had suffered such sorrow, / So that he regained his reason and remembered himself. / Then he lauded the lord and believed in truth / That He alone held all in hand] (1699–1704).

When he understood, he gained reason. Nebuchadnezzar must be like an animal, have affinity with the animals, to have natural reason return and know himself.

The Christian tradition could see "the body as evil, but as humanizing, as a source of sustenance more than a locus of sin and corruption."[43] In *Cleanness*, the passage citing Matthew 22: 1–14 and Luke 14: 16–24, the parable of the wedding feast, describes a lord inviting many people to a feast. The lord specifically enjoins his servants to scour the land for hidden people:

What kyn folk so ther fare, feches hem hider.
Be thay fers, be thay feble, forlotes none,
Be thay hol, be thay halt, by thay on-yyed,
And thagh thay ben bothe blynde and balterande cruppeles,
That my hous may holly by halkes by fylled.

[All those who abide there, escort them here. / Whether they be fierce or feeble, forgo no one, / Whether healthy or halting or having but one eye, / Although they be both blind and bumbling cripples, / So that my court may be crammed to the corners] (100–104).

Compare this open invitation to the distaste present in *Leviticus* to those who are not whole in body. Yeats's poem, "Crazy Jane Talks With the Bishop," likewise suggests that Christ (Love) celebrated all aspects of the body, even the lowliest. "A woman can be proud and stiff / When on love intent; / But Love has pitched his mansion in / the place of excrement; / For nothing can be sole or whole / That has not been rent."[44] God informs Abraham that he intends to destroy Sodom and awakens Abraham's concern: "Sir, with yor leve, / Schal synful and sakles suffer al on payne?" ["Sir, by Your leave, / Shall the sinful and sinless suffer the same pain?"] (715–716). If 50 worthy people were in the town, would they have to suffer, too? God responds:

"Nay, for fyfty," quoth the Fader, "and thy fayre speche,
And thay be founden in that folk of her fylthe clene,

I schal forgyve alle the gylt thurgh my grace one,
And let hem smolt al unsmyten smothely at ones."

["No, for fifty," the Father said, "and your fair speech, / If they can be found there, free of their filth, / Through grace alone I shall forgive the guilt of them all, / Allow them to escape unscathed altogether"] (729–732).

Abraham is delighted and asks permission for him to continue this conversation, he "that mul am and askes" ["although I am but ashes and dust"] (736). If 5 of those 50 were wanting, what would God do? He would save them all. And if 40 were free from sin and those 10 were not? God says he would "voyde away my vengaunce, thagh me vyl thynk" ["set aside vengeance, though to do so seems vile"] (744). Abraham continues this line of argument, continually reducing the number of worthy souls to ten. "'I graunt,' quoth the grete God. 'Graunt mercy,' that other" ["'I agree,' said the great God, 'I give thanks,' that other"] (765). God departs from Abraham's gate, "by those grene wayes" ["along those green paths"] (767), suggesting that Abraham's discourse has provoked affinity in God for some souls, like that of Lot.

Adam's kin are told there is only one law: "Ther was no law to hem layd bot loke to kynde, / And kepe to hit and alle hit cors clanly fulfylle" ["No law was laid down to them but look to nature, / And keep it and cleanly conform to its ways"] (263–264). Of course, the humans disobey God, who becomes angry and inflicts the flood on them, saving only Noah and his family. Mothers try to save their children by taking them to higher ground, all in vain. Bittersweet farewells, inciting pity in the reader, ensue in the wake of the flood:

Uuche burde with her barne the byggyng thay leves...
Then uche a segge sey wel that synk hym byhoved;
Frendes fellen in fere and fathmed togeder,
To dryy her delful destyné and dyyen alle samen;
Luf lokes to luf, and his leve takes,
For to ende alle at ones and for ever twynne.

[Each woman with her little ones runs away from her house... Then all saw for certain they must sink in the end. / Friend embraced friend, falling to each other, / To endure their dreadful destiny and die together. / Lover looks to beloved and takes leave at last, / To end all at once, evermore to be parted] (378, 398–402).

In the face of death, they see their affinity in one another. Though God's pity has been withdrawn, the dire situation awakens theirs.

Proposals to our current ecological disaster—the sixth big die-off looms—include minding the "material and corporeal dimension of 'what it means to be human.'"[45] Emphasizing the "embodied nature of cognition,"[46] Emilia Ferraro and Louise Reid pay tribute to seeing humans as simply one organism among many others in a web or mesh. Craft needs to be integrated into discussions about sustainability,[47] including that of the poet as *homo faber*—the shaper, crafty crafter, of words.[48]

The *Beowulf* poet points out how, when cultural codes fail and war ensues, colonization occurs. These colonizing forces are as terrifying as the monstrous forces men try to subdue. There are numerous moments in the poem alluding to the conquering of various tribes. Even the initial moment of Scyld's arrival suggests an unspoken and silenced conquest and oppression. In the opening (3–6), Scyld Scefing and his campaigns are mentioned and how those conquered must pay tribute ("gomban ġyldan," 11a). We are told about how Beowulf lands on Finland, Ecgtheow goes to the South-Danes, and Breca goes to the Brondings. Colonizers recur naturally, that is, inevitably, within the economy of the poem. The most successful adaptive culture is that of predatory colonizers, such as the Swedes, who will take over the Geats, as forecast at the conclusion of the poem. Yet the figures we empathize with are the victims of colonization and restrictive cultural codes, including Hildeburh in the *Fight at Finnsburg* passage and the father voicing his anguish in the Father's Lament, who are impossibly caught between death and revenge. In the final lines of *Beowulf*, the Geatish woman, with her hair bound up, utters a lament, full of the knowledge of her people's inevitable destruction. Her lament is only alluded to; her actual words are not related. The visionary wisdom of this wordless lament has been rendered literally invisible, in that the lines are not even extant because of the deterioration—the decay—of the physical manuscript itself. This creates an aesthetic impression that echoes within the reader/listener: all humans—victims and victimizers—end up as waste. In the final lines of poem (3156–3182), Beowulf's people construct a monument, into which they place rings, jewels, and ornaments from the hoard. This serves as a valediction for the ephemeral nature of the poem itself.[49] Even the manuscript containing *Beowulf* barely survived a fire; pages were scorched and are crumbling. The poem, oscillating between bleak and grim determination and religious injunctions, remains, in the end, our only solace.

Ann Murphy criticizes the "ahistorical rendering of responsibility." Reason is put as highest good, not heart: "In this context, shame and guilt become problematic insofar as they hamper political rationality."[50] She points out that Derrida and Levinas have privileged "the stranger, the foreigner, the radically other," and that this tradition worries "that there is something in the experience of [guilt and shame] that is perhaps narcissistic, indulgent, and even patronizing." For Levinas, "Responsibility *does not limit but founds my freedom*."[51] The only way to redeem wasted and forgotten people is to embrace them, a gesture of both protection and yet integrity. As Murphy so eloquently and urgently argues, "the problem is not that we cannot be responsible for history, but rather that we cannot be responsible enough."[52] The most powerful moments in literature illustrate this. At the conclusion of Chapter Fifteen in *The Adventures of Huckleberry Finn*, Huck regrets his gesture of "*trash*," in mocking Jim, saying, "It was fifteen minutes before I would work myself up to go and humble myself to a nigger; but I done it, and I warn't ever sorry for it afterwards, neither."[53]

Dirty Homeopathy: Poetry as Social Cure

Grooming does for / baboons most of what words do for us.

A. R. Ammons, *Garbage*[54]

The ability for that which is unclean to purify has its basis in homeopathy.[55] Dirt is good for you biologically.[56] Anxiety about cleanliness both within and about oneself, physically and spiritually, can be seen in the modern manifestation of censoring filth—inappropriate "waste" in terms of physical filth or sexual transgression. Geoffrey Chaucer offers us a case study for how a poet might be perceived of as "dirty." As we have seen, metaphor lies at the base of human communication; embodied metaphorical thought is rooted deep in our Indo-European heritage. We need to examine both metaphor and the recent history of censorship to assess whether and how Chaucer may be viewed as "dirty" and how this helps us understand the effects of waste literature on its audience.

"Dirty," as applied to a story or joke, first appears, the OED tells us, in 1599. "Dirty word" has its own entry, meaning vulgar or "smutty" word, starting in 1842, the same year, coincidentally, as the earliest entry for the word "pornography."[57] In an English case from the mid-nineteenth century, Benjamin Hicklin (whose name has adhered to this view) ruled in favor of an "obscene" work "on the basis that [the author's] purpose was good, that he had no intention of corrupting public morals." While this case was subsequently reversed,[58] it nevertheless established one paradigm for deciding as to whether something was obscene or not, whether it was "dirty": you should not read a passage out of context.

In 1873, Anthony Comstock became a "special agent" who could inspect mail he suspected might be obscene in any US post office.[59] The Act for the Suppression of Trade in, and Circulation of, Obscene Literature and Articles of Immoral Use soon became known as the Comstock Act,[60] which made it a crime to "send information and advertisements about obscene publications, contraception, or abortion through the mails."[61] Various cases stemming from this Act allude to Chaucer. For example, in 1917, a censorship case involving Théophile Gautier's *Mademoiselle de Maupin* was ultimately decided for the defendant accused of having sold it. Judge [William S.] Andrews uses the "whole book" concept: "No work may be judged from a selection of [vulgar and indecent] paragraphs alone. Printed by themselves they might, as a matter of law, come within the prohibition of the statute. So might a similar selection from Aristophanes or Chaucer or Boccaccio or even from the Bible. The book, however, must be considered broadly as a whole."[62] In the 1920s, an attorney in Baltimore ordered French classics, such as *Candide*, to be seized. Postal officials eventually admitted that the ruling "was made by a man who 'did not make a practice of reading much' and could not answer with certainty whether Geoffrey Chaucer was a living or a dead, let alone an obscene, poet."[63]

Did reading a work "in its entirety" have "the effect of promoting lust"? This question dominates the case concerning Joyce's *Ulysses*. The "Hand" opinion, so named for the two judges ruling in favor of *Ulysses*, Learned Hand and Augustus Hand, narrowed the definition of what might be determined to be obscene: "only if, *taken as a whole*, it tended to arouse lust." However, if only *some* passages were "thought 'filthy,' or sexually 'immoral' or likely to 'corrupt' or to have an aphrodisiacal quality," then it would not be sufficient grounds for it to be condemned.[64] Chaucer would have his own censorship woes soon enough.[65] In May 1953, San Antonio Mayor A. C. (Jack) White contended, "I believe we should stamp all books (in the library) written by known Communists." The 118 authors in question produced in total 574 books that appeared on a list by the House Committee on Un-American Activities. These books include editions of Melville and Chaucer, with illustrations by Rockwell Kent, who even had his passport revoked.[66]

In 1986, "The Miller's Tale" and *Lysistrata* were of concern in an elective humanities class for Columbia County High School students in Lake City, Florida. The textbook had been used in the state for ten years. In 1985, the daughter of a fundamentalist minister was in class and objected, even though these two selections were not assigned, simply referred to with sections read aloud by the teacher.[67] After the complaint, an advisory textbook committee at the high school read the offending pieces. This advisory committee recommended that the textbook be retained in the curriculum, but that the passages under consideration not be assigned "as required reading." The school board rejected this and voted to confiscate all copies and lock them in the book room. To avoid the appearance of censorship, they also voted to keep one copy in the library where it was placed on the "mature" shelf.[68]

Then, in 1988, the American Civil Liberties Union submitted a brief against the school board,[69] arguing that the board suppressed the free thought and speech of students.[70] The Court's determination was that the school board decided to remove this material for a number of reasons, including "that the sexuality of the selections was violative of the socially and philosophically conservative mores, principles and values of most of the Columbia County populace."[71] The minister was concerned over words such as "ass" and "fart" in "The Miller's Tale," as well as the levity of tone in referring to adultery.[72] Included in court documents is the assessment that "the specific objections of the protesting parents included: 'bad language, sexually explicit story, promotion of women's lib.' Superintendent Pittman commented: 'that any literature in which God's name is used in vain is not appropriate for use in the classroom.'"[73]

The US Court of Appeals for the 11th Circuit[74] pointed out that the book was available in the library. Nevertheless, the Court affirmed two points in favor of the plaintiffs: "It has long been clear that public school students do not 'shed their constitutional rights to freedom of speech or expression at the schoolhouse gate.'"[75] And "it is of course true, as plaintiffs so forcefully point out, that *Lysistrata* and 'The Miller's Tale' are widely acclaimed masterpieces of Western literature."[76] The Court found it "difficult to apprehend the harm that could conceivably be caused to a group of eleventh- and twelfth-grade students by exposure to Aristophanes and Chaucer. Indeed, authorities on Western literature are virtually unanimous in their high praise for the works of these authors."

Although optional, since the readings were included in a school-approved textbook, "such materials would obviously carry the imprimatur of school approval."[77] The "Court assumes the limited role of determining whether sexuality or vulgarity are at all present in the removed materials, and if so, determining whether the measure taken to remove the sexuality and vulgarity was at all reasonable."[78] While the Court did "not endorse the Board's decision,"[79] it nonetheless determined that "sexuality and vulgarity [were] both unquestionably present in the contested materials."[80] Basing its decision on various precedents, the US Court of Appeals for the 11th Circuit[81] concluded no constitutional violation had occurred.[82]

How might we read *The Canterbury Tales*, given Hicklin's "whole book" test for obscenity? Fragment VI of Chaucer's *The Canterbury Tales* suggests how poetry can be curative. "The Physician's Tale" presents an inviolate female body, with no honorable option for being porous. We are witness to an Old Testament version of the perfect sacrifice as stipulated in *Leviticus*. Virginia's perfection is summed up in her name—ontologically she must be a virgin to exist. Like the animals to be sacrificed in *Leviticus*, she must be without blemish and clean; only the innocent creature can be killed. Once defiled, she can no longer be clean and whole. The Old Testament references, particularly to Jephtha (*Judges* 11: 1), indicate the tale's roots in Old Testament divisions between clean and dirty. The Roman setting is key—it is pre-Christian. Virginia is denied the options offered her post-Christianity sisters. The early third-century Perpetua, for example, not a virgin but a chaste mother, guides the sword of the Roman soldier to her throat, much as Virginia allows her father to kill her. Saint Agatha, who is threatened with defilement, proclaims that even rape would not sully her: intent and will determine the soul's purity. These role models for women are not available to Virginia, who must die to retain her pure being. This Physician's "medicine" is that of surgery, a treatment increasingly suspect after the Fourth Lateran Council. Virginia feigns sickness to avoid parties. She asks for a remedy—the only one being surgery. The father is then expelled, like a leper. It is not justice that is sick, but the code of honor that sanctions cruel sacrifice. Indeed, she is referred to that "Gemme of chastitee" (223), possibly as used by the *Pearl*-poet to symbolize inviolate purity, like that embodied pearl, the Virgin Mary. The price of cleanliness is death.[83]

As Calvert Watkins has shown in his work on Indo-European roots, the root of shit, *skei*, devolves into *scit* means to *cut or divide*, to *separate*.[84] The body while shitting divides itself from its digested matter, cuts this matter from its body, projecting it to another space. This disembodied material then becomes designated as "dirty," thereby rendering the one who defecated as "clean"—at least until the next time. Similarly *turd* comes from *der*, to *tear or split*—clearly a turd is torn or split from the body. In all these etymological cases, there is division made between two states. To cut is to make clean, hence Virginius cuts Virginia so she will not become dirty.

The trope of "dirty literature" or "dirty Chaucer" ultimately goes back to the inherently metaphorical way humans think. The host claims that the story of the Physician was so sad he almost got a heart attack; the story was like bad medicine:

> But wel I woot thou doost myn herte to erme,
> That I almoost have caught a cardynacle.
> By corpus bones! but I have triacle,
> Or elles a draughte of moyste and corny ale,
> Or but I heere anon a myrie tale,
> Myn herte is lost for pitee of this mayde. (VI.312–317)

The Host can be cured by a tale, with the medicine from that unlikely source, the Pardoner. But can language work a cure, particularly when the tale is uttered by the oh-so-slimy and untrustworthy con man, the Pardoner?

The Pardoner invokes the medieval controversy about what happens to the Eucharist after it enters the human body. "O wombe! O bely! O stynkyng cod, / Fulfilled of donge and of corrupcioun! / At either ende of thee foul is the soun. / How greet labour and cost is thee to fynde! / Thise cookes, how they stampe, and streyne, and grynde, / And turnen substaunce into accident / To fulfille al thy likerous talent!" (VI.534–540). The solution for medieval theologians was to proffer the accident (the material bread of the host and material wine) and the substance (the actual body and blood of Christ). The substance of "The Pardoner's Tale" is the moral story, while the accident is the immoral frame, indeed, his immoral self. Thomas Aquinas dealt with a cognate issue in his *Summa Theologica*.[85] He asks whether or not a wicked priest can consecrate the Eucharist. While

> It seems that a wicked priest cannot consecrate the Eucharist... *I answer that*... the priest consecrates this sacrament not by his own power, but as the minister of Christ, in Whose person he consecrates this sacrament. Hence it is evident that priests, even though they be not godly, but sinners, can consecrate the Eucharist.

Analogously, an immoral man *can* tell a moral tale.

Based in homeopathy, dirt purifies. Christ cures a deaf-mute (Mark 7: 33–35) with his own spit in a Scriptural version of this long-established folk remedy. *Cleanness* tells us how Christ healed filth: "So clene was his hondelyng uche ordure hit schonied" ["All filth fled, so fair was His touch"] (1101). While the excreta of animals and even humans was accepted in prescriptions to remedy ailments, less known is the implication that divine waste was highly regarded for its curative properties. There was even a belief that Christ's own excrement could function like a healing relic.[86] Is Christ a ragpicker, a garbageman, a detritus gatherer? Perhaps the trash collector is our present-day saint, cleaning up literal filth. Chaucer's own "drysty rymyng" (VII.930) [crappy verse] purges the filth of the Prioress's anti-Semitic privy, expunging the insidious filth from the previous tale by his own poetry.

In his graphic description of the seven deadly sins, Chaucer's Parson drags us through the "dong and in the mire" (X.418), just like the trains of the superfluous clothes worn by the proud exhibitionists he excoriates. The material "trailyng is verraily as in effect wasted... and roten with donge" (X.418). Worse, the excess fabric denies the poor a few pitiful clothes by jacking up the price of material for everyone. While the pungent and polluted sermon unsparingly details the filth of sin and consequent punishments, the effect of such "dirty" speech drives positive action. The potty-mouth Parson succeeds in inspiring Chaucer the pilgrim to confess his filthy sins to become clean in the *Retraction*. Readers can take heart and learn from the proffered cures for offenses. Might we not be acting in the best interests of ourselves and ecology (ourselves as part of ecology) through "humylitee, or mekenesse. / That is a vertu thurgh which a man hath verray knoweleche of hymself, and holdeth of hymself no pris ne deyntee, as in regard of his desertes, considerynge evere his freletee" (X.475–476). The Parson argues for humility as a remedy for pride. This traditional medieval injunction could be equally applied as a rejoinder against anthropocentrism, where humans privilege themselves at the cost of all others, human and nonhuman.

Metaphoric Medicine

Just calling certain texts "dirty" puts us in the realm of the metaphoric. In the hierarchy of solid waste management, reduction is an ultimate goal, suggestive for literature. Might not metaphor be a version of slow composting and aerobic digestion? After all, it is a conflation of concepts into one image, thus reducing verbiage but replete with flavor.[87] Cultural ecology, like metaphor, "thinks together *connectivity* and *diversity*, *relationality* and *difference*."[88]

Metaphor works by noting similarities, what is proper to it. Building on Aristotle, Derrida points out, "One may speak properly or improperly of what is not proper to the thing, its accident, for example... The space of language, the field of its divisions, is opened precisely by the difference between... the proper, and the accident." Metaphor brings us "closer to the thing's... proper truth."[89] In Christian theological terms, the proper is the substance, the improper is the accident. The accident drives the force of metaphor.

In his analysis of metaphor, Graham Harman builds on José Ortega y Gasset's neglected "An Essay in Esthetics by Way of a Preface."[90] Ortega, Harman tells us, distinguishes between observing and being: "To look at a thing is not the same thing as to stand in its place and undergo its fate, even if what we are observing is our own psychic lives."[91] As Ortega puts it, "Just as there is an I-John Doe, there is also an I-red, an I-water, and an I-star... Everything, from a point of view within itself, is an 'I.'"[92] While we can never fully grasp the "inwardness of things," aesthetics (art) has the potential ability to *seem* to "confront the impossible depth of objects." One tool for art's "magic power" is metaphor.[93] Inessential qualities make effective metaphors work because of the "mesmerizing implications for the thinghood of things."[94]

Using Ortega's example, we can see that saying "the cypress is a conifer" fails metaphorically because the names are too closely identical with one another, whereas "the cypress is a flame" works because the cypress and flame, while similar in shape (the basis for the metaphor) resist being identified with one another. The metaphorical gesture makes the new object "a vaporous hybrid of both: one that cannot even be described in terms of definite tangible properties."[95] This new thing touches us emotionally, allowing us to enter "the magic world of a cypress-flame-feeling-thing."[96] For Ortega, style is "nothing other than a specific mode of de-creating images and recreating them as feeling-things."[97]

The metaphors that come from our embodied state hold true for much of our Indo-European language roots. In Indo-European, *substance* means *to stand*, while *accident* means *to fall. Healing* means *to make whole. Substance* is what is *clean*, the *dirt* is the *accident*, and what falls away is *cut* or *split* from the body.[98] The healing is in the making whole of these—the uniting, the healing by filthing, the standing by falling. Poetry creates new and unexpected metaphors, enabling profound insight. One in a long line of dirty poets, Chaucer, by confronting us with our inevitable decay, would act as spiritual homeopath.

Speaking to the Dead, Caring for the Living

> "Show me the manner in which a nation cares for its dead, and I will measure with mathematical exactness the tender mercies of its people," William Gladstone, the British statesman, is said to have observed . . . People who have learned how to care tenderly for the bodies of the dead are almost surely people who also know how to show mercy to the bodies of the living.
>
> Thomas G. Long, "Chronicle of a Death We Can't Accept"[99]

An undertaker in Japan hurried to care for the corpses who littered the coastal town of Kamaishi after the earthquake and tsunami of March 2011. As the author of a work chronicling Atsushi Chiba's care for the dead, Kota Ishii says that "this story is ultimately about how small acts of kindness can bring a little humanity, even in a tragedy that defies all imagination." Mr. Chiba massaged the limbs of the dead so they would not appear so disturbingly awkward to their relatives. He wiped their faces clean. Mr. Chiba's own words describe why he found it necessary to gently work with the corpses so that they became less contorted. "I thought that if the bodies were left this way, the families who came to claim them wouldn't be able to bear it . . . Yes, they are dead. But in Japan, we treat the dead with respect, as if they are still alive. It's a way to comfort the living." He even spoke to the dead themselves: "You must be so cold and lonely, but your family is going to come for you soon so you'd better think of what you're going to say to them when they arrive."[100]

The dramatic rituals consecrated to the deaths concluding *Beowulf* and *Hamlet* function as ways to recuperate loss and integrate that waste positively into the culture. The solution to the wasted kingdoms at the close of those works is found in ritual as structure, as expressed through poetry. Poetry itself is rooted in the body. Oral poetry, embodied language, replicates with a meter the beat of the human heart. Language emerges from "embodied minds . . . The bodily gesture is . . . a celebration of the world, poetry. The body is a sensing sensed."[101] In *Seedfolks*, the Korean dry cleaner Sae Young tells her story in broken English about how she was attacked and robbed and thereafter became afraid. Her story is written in staccato colorless words until she encounters the garden. Then she speaks in similes: "But just be near people, nice people, feel good, like next to fire in winter."

For her, the garden is "almost like family."[102] Poetry reflects and enables community.

The "pharmacy of language" poisons, yet it also has the antidote: it allows belief in eternity.[103] Bauman cites George Steiner: "It is because 'if' sentences . . . can, spoken at will, deny, reconstruct, alter past, present and future, mapping otherwise the determinants of pragmatic reality, that existence continues to be worth experiencing. Hope is grammar." The ethics of self-limitation can succeed if we believe in eternity.[104] This kind of ethics requires us to think long term. Keats expresses the thingness of the urn. Projected into the future, the urn is eternal, unlike us. "Heard melodies are sweet, but those unheard / Are sweeter." Heard melodies are commodities, the unheard never become obsolete. Anticipatory, always pre-consumption, always in present tense, unheard melodies gesture toward eternity.

Walt Whitman's "This Compost" addresses spring and herbs, feeling disgust from the uncanny sense that the dead lie buried beneath the beauty:

> Are they not continually putting distemper'd corpses within you?
> Is not every continent work'd over and over with sour dead?
> Where have you disposed of their carcasses?
> Those drunkards and gluttons of so many generations?
> Where have you drawn off all the foul liquid and meat?
> I do not see any of it upon you to-day, or perhaps I am deceiv'd,
> I will run a furrow with my plough, I will press my spade through
> the sod and turn it up underneath,
> I am sure I shall expose some of the foul meat.

He speaks then to the reader:

> Behold this compost! behold it well!
> Perhaps every mite has once form'd part of a sick person—yet behold!
> The grass of spring covers the prairies,
> The bean bursts noiselessly through the mould in the garden,
> The delicate spear of the onion pierces upward,
> The apple-buds cluster together on the apple-branches,
> The resurrection of the wheat appears with pale visage out of its graves . . .
> The summer growth is innocent and disdainful above all those strata of
> sour dead.

Then the speaker marvels at the "chemistry" that allows for the mixture of death and life, while remaining terrified of this perverse way of nature, how it

> grows such sweet things out of such corruptions . . .
> It renews with such unwitting looks its prodigal, annual, sumptuous crops,

It gives such divine materials to men, and accepts such leavings
from them at last.

The Earth "distills" and "renews," just like Whitman's verse. "Has any one supposed it lucky to be born? / I hasten to inform him or her it is just as lucky to die, and I know it" (*Song of Myself, Song* 7).

Poet as Ragpicker: Spiritual Homeopathy

I can find nothing lowly in the universe:
. . .
moss, beggar, weed, tick, pine, self, magnificent
with being!

A. R. Ammons, "Still"[105]

There is a long tradition of seeing poetry as a kind of alchemy.[106] Waste embodies "ambivalence," yet, as Zygmunt Bauman argues, it also carries a "divine" and "sublime" "power, equivalent to that of the alchemists' philosopher's stone." Waste is essential to the "creative process."[107] Waste poets, those poets who confront us with our inevitable decay, would act as spiritual homeopaths, curing the alienation and disgust we all too often feel toward our own and others' waste and decay. Yeats writes in "The Circus Animals' Desertion," "Those masterful images because complete / Grew in pure mind, but out of what began? / A mound of refuse or the sweepings of a street, / Old kettles, old bottles, and a broken can, / Old iron, old bones, old rags, that raving slut / Who keeps the till. Now that my ladder's gone, / I must lie down where all the ladders start / In the foul rag-and-bone shop of the heart." Here, in this junked emotional furnace, poetry is forged. As Baudelaire writes,[108] the poet is analogous to a ragpicker or *chiffonier*: both are revolutionaries:

In the heart of some old suburb, muddy labyrinth,
Where humanity crawls in a seething ferment,
One sees a rag-picker go by, shaking his head,
Stumbling, bumping against the walls like a poet.

The task, the duty, the ethical obligation, of the poet is to record and recognize us all in what Baudelaire calls this "commingled vomit" that is humanity.[109] Only by cobbling together the pieces of a life and shards of memory can the bricoleur-poet hope to survive. "These fragments I have shored against my ruins."[110] As Czeslaw Milosz observes, "Man constructs poetry out of the remnants found in ruins."[111] We can read literary waste, the scop's scavenged gleanings, as a form of bricolage, not in an attempt for humans to establish identity in opposition to materiality, but rather "as evidence of what Latour calls the 'networked quality of things,'" an aspect of *Dingpolitik*.[112]

Ammons dedicates *Garbage* "to the bacteria, tumblebugs, scavengers, / wordsmiths—the transfigurers, restorers."[113] Rather than extinction being imminent, perhaps we can detect *increasing* life on earth due to microbes and bacteria proliferating to deal with our waste in landfills.[114] Bacteria are us. What is more, "we [humans] are, above all, *dependent*"[115] on the earth, on the invisible microbes that sustain and dissolve us into soil and gas. We rely on the earth and on each other. As Ammons articulates in praise of detritus: "garbage has to be the poem of our time because / garbage is spiritual."[116] Garbage is piled high like an offering to the gods. "There is a mound, / too, in the poet's mind dead language is hauled / off to and burned down on, the energy held and / shaped into new turns and clusters, the mind / strengthened by what it strengthens: for / where but in the very asshole of comedown is / redemption."[117] As Timothy Morton points out, "A poem is not simply a representation, but rather a nonhuman agent,"[118] capable of upsetting the "digestive system" of "the capitalist order of things."[119] Ammons points to the performative nature of verse: "We are nearly / alone in words: but the words do for us what other / languages do for others—they warn, inform, / reassure, compare, present: we may be alone in / words but we are not singular in language . . . words are a way of fending in the / world."[120] The process of revision is likened to garbage that needs to be composted.[121] Poetry itself

> reaches down into the dead pit / and cool oil of stale recognition and words and / brings up hauls of stringy gook . . . there / is some untransformed material, namely the poem / itself; the minute its transmutations end, it / becomes a relic sometimes only generations or / sets of countrywide generations can degrade.[122]

The ragpicker-poet, *sans domicile*, is, like the chiffonier, peripatetic. On writing a poem, Ammons suggests, "you keep your mind / open and on the move and eventually there is a / trace of feeling."[123] Poetry, constructed out of ruins, found in garbage, *is* garbage, healing homeopathically.

The garbage collector, like the poet, sees the truth and has a deep insight into human nature. When the garbage truck driver gets to the landfill and gets out of his truck to look off "into the rosy-fine / rising of the day," he senses "the gateway to beginning, here the portal / of renewing change, the birdshit, even, melding / enrichingly in with debris, a loam for the roots / of placenta."[124] The "garbage spreader gets off his bulldozer and / approaches the fire: he stares into it as into / eternity . . . he stands in the presence / of the momentarily everlasting, the air, about / him sacrosanct, purged."[125] As Calvino points out, garbage men are seen

> not just as emissaries of the chthonic world, gravediggers of the inanimate, Charons of a beyond of greasy paper and rusty tin, but as angels too, as indispensable mediators between ourselves and the heaven of ideas in which we undeservedly soar...heralds of a possible salvation beyond the destruction inherent in all production and consumption, liberators from the weight of time's detritus, ponderous dark angels of lightness and clarity.[126]

Underestimating "the sacred nature of the operations involved in refuse removal,"[127] Calvino points out that even the newspaper has a "moment of resurrection"[128] when used to line the *poubelle*. For Calvino's poetic twin Ammons, garbage terns hover like angels. Reflecting on the bottle and the yellowjackets buzzing around the specks of stale wine that will inevitably dry up, "the bulldozer man thinks about that / and concludes that everything is marvelous...all is one, one all: / hallelujah."[129] Evoking Eliot's *Four Quartets*—"in your / end is my beginning"—the chapel has become a dump and vice versa.[130] Poetry acknowledges the poignancy of materiality; it functions homeopathically for the alienation and disgust we all too often feel toward our own and others' bodily waste and decay. Poetry, salvage and sacrifice, leads to the creation of a community of affinity.

The humanities can play a role in the movement to "degrow" and decenter, to what Serres calls *tenancy*, as opposed to *ownership*, of the world.[131] Literature, as Ette writes, as

> a mutable and dynamic storehouse of knowledge for living...devises and aesthetically shapes blueprints for how to live.[132]

Literature both reflects the ways humans perceive waste and, yet, at the same time, can reshape reality—it is a space where possible other worlds open up, both good and bad. And through our sharing in these other moments of potential, we can reshape our relationships to and how we see other people, the world, and ourselves.

I have, *pace* Keats's nightingale, "been half in love with" vital trash, a kind of sickness of *trashitis*, an illness of trash obsession.[133] It has not meant loving the landfill with its potential to leach harmful chemicals into the earth and poison the water table in the surrounding land and disrupt scenic beauty; it has not meant admiring the hatred that the privileged exercise toward the weak and disempowered; nor has it meant wishing to emulate those who obsessively accumulate stuff. But all these less-than-admirable traits are human. Nothing human is alien to us. We need to see in these actions and attitudes their affinity to us. This connection poetry can help us see, not always a welcome revelation, such as the human proclivity of "wasting" other humans—of rendering metaphorically and sometimes literally fellow beings into waste. We need to take Whitman's notion of "hospitality" and extend it to waste.[134] By seeing ourselves in our depravity, literature can offer us ways to emerge from this baseness to accept others and ourselves in a generous and ethical way.

Calvert Watkins has shown in his magisterial study how Indo-European poetics are woven in a web of connected tissue. In sum, he shows the trope of the poet as healer in the Indo-European poetic tradition. In the ancient formula, "verbal magic is the homeopathic, analogic magic of a charm or incantation,"[135] with specific focus on the dragon/serpent/worm destroyed by the spoken word. The charm, Watkins argues, is the domain of the poet-healer, "the professional of the spoken word."[136] Specifically, poetry can create unexpected metaphors, triggering epiphanic insight, even when imagining desert landscapes of hopeless abandonment. Poetry can help startle us into recognizing our kinship[137] with others: "The young fellow drives the express-wagon (I love him, though I do not know him)" (*Song of Myself, Song* 14). The poet makes "holy whatever I touch or am touch'd from" (Whitman, *Song* 24), even the "cotton-field drudge or cleaner of privies" (*Song* 40).

In the Middle Ages, human matter was ejected into barrel latrines that contained both hay and sedge for wiping and tailor's woolen scraps deeply stained with women's blood, the rejected bits of material used as rags to collect menstrual fluid. The poet is no less than a gongfermor, a farmer of cesspits, hired to collect the sewage only to sell it as fertilizer, profiting doubly from dung. Is that what poetry is, the barrel latrine, the cesspit of our imaginings, where all the detritus and filth is ejected only to be recycled for the common good, the health of crops that feed a population of hungry people? A midden heap is a junkyard of shells and bones—or a treasure-house of information, of lives long ago. Writings are the rubbish heap or composted waste of the mind. The poet mosaics together shards, recycles litter, and salvages fragments. The poet is a garbage collector, a detritus gatherer, a waste gourmet. Perhaps that is the beauty in poetry: it lets us feel viscerally the terror of waste, yet revel in the beauty of artifacts of loveliness—the word, the meter, the sound, the hush of decay. A gleaner, the poet delicately cradles each morsel hidden in the rubbish tip like a gleaming jewel.

Notes

Introduction The Waste-ern Literary Canon in the Waste-ern Tradition

1. Zygmunt Bauman, *Wasted Lives: Modernity and Its Outcasts* (Cambridge: Polity Press, 2004), 26.
2. Mary Douglas, *Purity and Danger: An Analysis of the Concepts of Pollution and Taboo* (London: Routledge, 1966/2002), 2, 44.
3. Susan Signe Morrison, *Excrement in the Late Middle Ages: Sacred Filth and Chaucer's Fecopoetics* (New York: Palgrave Macmillan, 2008), 153–158. The book enacts what Dana Phillips labels "excremental ecocriticism." "Excremental Ecocriticism and the Global Sanitation Crisis," in *Material Ecocriticism*, ed. Serenella Iovino and Serpil Oppermann (Bloomington: Indiana University Press, 2014), 184.
4. Morrison, *Excrement*, 123.
5. Dana Phillips and Heather I. Sullivan, "Material Ecocriticism: Dirt, Waste, Bodies, Food, and Other Matter," *Interdisciplinary Studies in Literature and Environment* 19.3 (Summer 2012): 447. "Our trash is not 'away' in landfills but generating lively streams of chemicals and volatile winds of methane as we speak." Jane Bennett, *Vibrant Matter: A Political Ecology of Things* (Durham: Duke University Press, 2010), vii.
6. Bennett, *Vibrant Matter*, viii.
7. Ibid., vii.
8. See Figures 1 and 2 in Vincent B. Leitch, *Literary Criticism in the 21st Century: Theory Renaissance* (London: Bloomsbury, 2014).
9. Pippa Marland and John Parham, "Remaindering: The Material Ecology of Junk and Composting," *Green Letters: Studies in Ecocriticism* 18.1 (2014): 1.
10. Scott Slovic, "Editor's Note," *Interdisciplinary Studies in Literature and Environment* 20.3 (2013): 456.
11. Mark Edmundson, "Against Readings," *The Chronicle of Higher Education*, April 24, 2009: B7–B10.
12. Véronique Bragard, "Introduction: Languages of Waste: Matter and Form in Our Garb-age," *Interdisciplinary Studies in Literature and Environment* 20.3 (2013): 459. The special issues were *Critical Inquiry* 30 (2004) and *PMLA* 125 (2010).

13. James Joyce, "The Grandeur That Was Rome," *Ulysses* (New York, 1961), 131. Cited by Richard Neudecker, *Die Pracht der Latrine. Zum Wandel öffentlicher Bedürfnisanstalten in der kaiserzeitlichen Stadt.* Studien zur antiken Stadt, vol. 1 (Munich: Verlag Dr. Friedrich Pfeil, 1994), 7.
14. Anonymous reader of a proposed article on excrement in the Middle Ages, which ultimately became a chapter of Morrison, *Excrement.*
15. Heinrich Böll, "In Defense of 'Rubble Literature,'" in *Stories, Political Writings, and Autobiographical Works*, ed. Martin Black, trans. Leila Vennewitz (New York: Continuum, 2006), 274.
16. Ibid., 272–273.
17. Patricia Yaeger, "Editor's Column: The Death of Nature and the Apotheosis of Trash; or, Rubbish Ecology," *PMLA* 123 (2008): 338.
18. Ibid., 325.
19. Sarah Lyall, "Whimsical Works of Art, Found Sticking to the Sidewalk," *The New York Times*, June 14, 2011, accessed November 22, 2014, http://www.nytimes.com/2011/06/14/world/europe/14muswell.html?_r=1&ref=global-home.
20. Rubbish ecology witnesses "the act of saving and savoring debris." Yaeger, "Editor's Column," 329. Also Morrison, *Excrement*, Chapter 6.
21. See Morrison, *Excrement,* for a fuller discussion of waste studies with a focus on *Piers Plowman*, *Wynnere and Wastoure*, and Chaucer, especially 129–151; see also Eleanor Johnson, "The Poetics of Waste: Medieval English Ecocriticism," *PMLA* 127.3 (2012): 460–476; and Vin Nardizzi on an assessment of "Medieval Ecocriticism," *postmedieval: a journal of medieval cultural studies* 4.1 (2013): 112–123.
22. Steven Johnson, "Tool for Thought," *The New York Times Book Review*, January 30, 2005, accessed October 18, 2014, http://www.nytimes.com/2005/01/30/books/review/30JOHNSON.html.
23. W. J. T. Mitchell, "Medium Theory: Preface to the 2003 *Critical Inquiry* Symposium," *Critical Inquiry* 30 (2004): 328.
24. Elizabeth V. Spelman, "Combing Through Trash: Philosophy Goes Rummaging," *The Massachusetts Review* 52.2 (2011): 324. Will Viney contemplates the academic compulsion to examine waste in *Waste: A Philosophy of Things* (London: Bloomsbury, 2014), 181.
25. Susan Stanford Friedman, "Why Not Compare?," *PMLA* 126.3 (2011): 753–762. She uses I. A. Richards's foundational articulation about how metaphor works, whereby metaphor uses the *vehicle* (an image) "to explain something else (the *tenor*)." The tenor remains primary, the vehicle secondary, though, as Friedman points out, "in much poetic language, the vehicle steals the show, however much it appears to serve the tenor... In a world structured in dominance, why should one thing exist to explain another instead of being seen as a thing in itself?" (754).
26. Ibid., 755. In fact, she concludes, "in the end it is worse not to compare than to compare" (756).
27. Thomas Claviez, "Done and Over With—Finally? Otherness, Metonymy, and the Ethics of Comparison," *PMLA* 128.3 (2013): 609, 613. Since the other "is always contingent on the self and thus never stands in a metaphoric,

comparative relation to the self... this relation has to be reconceived as a metonymic one" (613).

28. Timothy Morton, *Ecology without Nature: Rethinking Environmental Aesthetics* (Cambridge, MA and London: Harvard University Press, 2007), 175.
29. Friedman, "Why Not Compare?," 758. She cites Djelal Kadir, "Comparative Literature Hinternational," in *Comparative Literature: States of the Art*, ed. Djelal Kadir, Special issue of *World Literature Today* 69.2 (1995): 245–303, here 246; and Kenneth Reinhard, "Kant with Sade, Lacan with Levinas," *MLN* 110.4 (1995): 785–808.
30. Friedman, "Why Not Compare?," 759.
31. Lois Parkinson Zamora, "Eccentric Periodization: Comparative Perspectives on the Enlightenment and the Baroque," *PMLA* 128.3 (2013): 690–697. See Tim Edenson, "Waste Matter: The Debris of Industrial Ruins and the Disordering of the Material World," *Journal of Material Culture* 10.3 (2005): 312.
32. Nirvana Tanoukhi, "The Movement of Specificity," *PMLA* 128.3 (2013): 669. See Chimamanda Ngozi Adichie, "The Danger of a Single Story," *TED Talks*, October 2009, accessed November 27, 2014, http://www.ted.com/speakers /chimamanda_ngozi_adichie.
33. Vilashini Cooppan, "Net Work: Area Studies, Comparison, and Connectivity," *PMLA* 128.3 (2013): 616. I am aware of the inherent dangers of articulating literary waste relations in this way. See Robert T. Tally, Jr., "World Literature and Its Discontents," *English Language and Literature* 60.3 (2014): 401–402.
34. Kate Soper, "Response Essay: The Postmedieval Project: Promise and Paradox," *postmedieval: a journal of medieval cultural studies* 1 (2010): 259–260; also Kate Soper, "Waste Matters," *Capitalism, Nature, Socialism* 14.2 (2003): 129. Gee contends that waste is "always *made*, not found." Sophie Gee, *Making Waste: Leftovers and the Eighteenth-Century Imagination* (Princeton, NJ: Princeton University Press, 2010), 9.
35. "About Discard Studies," accessed November 22, 2014, http://discardstudies .com/about/.
36. Ibid., and http://discardstudies.com/2014/09/04/why-discard-studies-why-not -waste-studies-2/. In her column "Why 'Discard Studies'? Why not 'Waste Studies?'" from September 4, 2013, Liboiron continues, "While discard studies is absolutely about the material and especially the materiality of what is thrown away, it is more about the mass of social, political, cultural, technical and economic systems around the object that premises or supports its status as waste or wasted."
37. Waste specifically signifying useless or worthless refuse or trash dates from c. 1450 according to the *Middle English Dictionary*. See also Steven Connor, "Exhaust: On Aerial Rejectamenta," in *Aesthetic Fatigue: Modernity and the Language of Waste*, ed. John Scanlan and John F. M. Clark (Newcastle upon Tyne: Cambridge Scholars, 2013), 72.
38. For more on the history of the word, see Viney, *Waste*, 18.
39. Valerie Allen, "Commentary on Susan Signe Morrison's Waste Studies—A New Paradigm for Literary Analysis: Something Is Rotten in the Denmark of *Beowulf* and *Hamlet*," The 2009 Compass Interdisciplinary Virtual

Conference, accessed October 18, 2014, http://compassconference.files.wordpress.com/2009/10/civc-commentary-valerie-allen-john-jay-college-cuny-on-waste-studies-a-new-paradigm-for-literary-analysis-susan-signe-morrison.pdf.

40. Additionally, as Suzanne Raitt argues, "To call something 'waste' [as opposed to 'rubbish, 'garbage,' and 'litter'] is to invoke its history." "Psychic Waste: Freud, Fechner, and the Principle of Constancy," in *Culture and Waste: The Creation and Destruction of Value*, ed. Gay Hawkins and Stephen Muecke (Lanham, MD: Rowman & Littlefield, 2003), 73. For Gee, Raitt's point shows how "a narrative of its production is what distinguishes waste from all other types of leftover." Gee, *Making Waste*, 9.
41. "Metaphorically capacious," waste encompasses "categories of garbage, shit, sexual excess, economic surplus, unproductive labor, idleness, and aesthetic imbalance." Christopher Schmidt, *The Poetics of Waste: Queer Excess in Stein, Ashbery, Schuyler, and Goldsmith* (New York: Palgrave Macmillan, 2014), 16.
42. Gee, *Making Waste*, 5.
43. Serenella Iovino and Serpil Oppermann, "Material Ecocriticsm: Materiality, Agency, and Models of Narrativity," *Ecozon@* 3.1 (2012): 83.
44. Tadeusz Sławek, "The Vase and Broken Pieces: Productivity and the Margin of Waste," in *Rubbish, Waste and Litter: Culture and Its Refuse/als*, ed. Tadeusz Rachwał (Warsaw: Wydawnictwo SWPS Academica, 2008), 19.
45. Bragard, "Languages of Waste," 460.
46. Levi R. Bryant, "Substantial Powers, Active Affects: The Intentionality of Objects," *Deleuze Studies* 6.4 (2012): 542.
47. Zsuzsa Gille, *From the Cult of Waste to the Trash Heap of History: The Politics of Waste in Socialist and Postsocialist Hungary* (Bloomington: Indiana University Press, 2007), 18.
48. Schmidt, *The Poetics of Waste*, 16.
49. Gille, *From the Cult of Waste*, 19–20.
50. "Surely when a word can mean everything it risks meaning nothing. But the interesting thing is that in spite of differences in the scope of our definitions we all meet everyday certain statements that everyone recognizes as metaphor and class by that name." Wayne C. Booth, "Metaphor as Rhetoric: The Problem of Evaluation," in *On Metaphor*, ed. Sheldon Sacks (Chicago: University of Chicago Press, 1979), 48.
51. Graham Harman, "Gold," in *Prismatic Ecology: Ecotheory beyond Green*, ed. Jeffrey Jerome Cohen (Minneapolis: University of Minnesota Press, 2013), 106.
52. Slavoj Žižek, *Living in the End Times* (London: Verso, 2010), 35.
53. Steve Mentz, "Brown," in *Prismatic Ecology: Ecotheory beyond Green*, ed. Jeffrey Jerome Cohen (Minneapolis: University of Minnesota Press, 2013), 195, quoting Timothy Morton, *The Ecological Thought* (Harvard: Harvard University Press, 2012).
54. Christine Temko, "Regulation and Refuse Matter in Don DeLillo's *Underworld* and Eugene Marten's *Waste*," *Interdisciplinary Studies in Literature and Environment* 20.3 (2013): 505.

55. Gary Snyder, *The Real Work: Interviews and Talks: 1964–1979* (New York: New Directions, 1980), 174, quoted in Marland and Parham, "Remaindering," 3.
56. Robert Eaglestone, *Ethical Criticism: Reading after Levinas* (Edinburgh: Edinburgh University Press, 1997), 167.
57. Derek Attridge's contention. "What Does It Make You Feel? Responding Affectively to Literature," Literary and Critical Theory Seminar, Institute of English Studies, School of Advanced Study, London, February 14, 2011.
58. William Rathje and Cullen Murphy, *Rubbish! The Archaeology of Garbage* (New York: HarperCollins, 1992), 33.
59. "Non-Hazardous Waste Management Hierarchy," accessed November 22, 2014, http://www.epa.gov/waste/nonhaz/municipal/hierarchy.htm.
60. Simin Davoudi, "Planning for Waste Management: Changing Discourses and Institutional Relationships," *Progress in Planning* 53 (2000): 171.
61. Slovic, "Editor's Note," 454.
62. "There is no 'away' to throw our waste or pollution and thus makes every 'there' a 'here.'" Samantha Clark, "Strange Strangers and Uncanny Hammers: Morton's *The Ecological Thought* and the Phenomenological Tradition," *Green Letters: Studies in Ecocriticism* 17.2 (2013): 99.
63. Bruno Latour, "*Realpolitik* to *Dingpolitik*—Or How to Make Things Public," 1–31, accessed January 29, 2013, http://www.bruno-latour.fr/sites/default/files /96-DINGPOLITIK-GB.pdf). Bill Brown, "All Thumbs," *Critical Inquiry* 30 (2004): 456.
64. Schmidt, *The Poetics of Waste*, 157.
65. Ottmar Ette, "Literature as Knowledge for Living, Literary Studies as Science for Living," trans. Vera M. Kutzinski, *PMLA* 125 (2010): 983.
66. Ibid.
67. All scholars want to make a difference with our academic work. The danger with "making a difference" is if we stay, as Roger Griffin has articulated, "splitters," specializing only in our home fields. If we bridge disciplines as "lumpers," we run the risk of antagonizing those whose field we trespass on. Yet we can possibly effect more of a "difference" by making forays into other disciplinary "cultures." Roger Griffin, "The Rainbow Bridge," The 2009 Compass Interdisciplinary Virtual Conference, accessed June 4, 2010, http:// compassconference.wordpress.com/2009/10/19/griffin/.
68. Personal communication.
69. Jane Bennett, "Systems and Things: A Response to Graham Harman and Timothy Morton," *New Literary Theory* 43 (2012): 225.
70. Schmidt, *The Poetics of Waste*.
71. Ibid., 5.
72. Ibid., xii.
73. Ibid., 5.
74. Yaeger, "Editor's Column," 327. Walter Moser has pointed out, "Garbage, long considered alien and impure, something to be excluded from cultural production, has in recent decades made a progressive entry into the systems of art and culture in many and diverse ways." "Garbage and Recycling: From Literary Theme to Mode of Production," *Other Voices* 3.1 (2007): accessed April 18, 2012, http://www.othervoices.org/3.1/wmoser/index.php.

75. Viney, *Waste*, 35. Italo Calvino notes the silence of plastic garbage lids in "La Poubelle Agréée," in *The Road to San Giovanni*, trans. Tim Parks (New York: Pantheon, 1993), 95, 97.
76. Elizabeth Shove, Matthew Watson, Martin Hand, and Jack Ingram (eds.), *The Design of Everyday Life* (Oxford: Berg 2007), 105–106. See Graham Harman, *Tool-Being: Heidegger and the Metaphysics of Objects* (Peru, IL: Open Court, 2002), especially 20–35; and Harman's article, "The Well-Wrought Broken Hammer: Object-Oriented Literary Criticism," *New Literary History* 43.2 (2012): 183–203. Timothy Morton amusingly plays with Harmans's take on Heidegger's hammer. Timothy Morton, "An Object-Oriented Defense of Poetry," *New Literary History* 43 (2012): 215.
77. Slovic, "Editor's Note," 454.

1 Codification: The Anxiety of Ambiguity

1. As Bill Brown points out citing Marcel Mauss's anthropological work, "however material stable objects may seem, they are, let us say, different things in different scenes." Bill Brown, "Thing Theory," *Critical Inquiry* 28.1 (2001): 9.
2. Mary Douglas, *Purity and Danger: An Analysis of the Concepts of Pollution and Taboo* (London: Routledge, 1966/2002), 85.
3. Nigel Barley points out that "problems of pollution are logical problems, problems of classification, and therefore physical hygiene presupposes mental hygiene." There is a "desire to categorise and assign boundaries" and "a concern for the symbolic load of bodily secretions and processes." "The Letter to Brother Edward," *Neuphilologische Mitteilungen* 79 (1978): 23–24.
4. *Babylonian Talmud: Tractate Berakoth.* Thanks to Teresa Ann Ellis for this fascinating material, accessed October 23, 2014, http://www.come-and-hear.com/berakoth/berakoth_23.html.
5. "Q. When are we supposed to recite the blessing 'over the washing of hands?' A. The blessing 'over the washing of hands' is recited when one washes his hands upon rising from bed. It is also proper to recite this blessing when one comes out of a privy, and washes one's hands in order to recite the Minha services." Irving A. Agus, *Rabbi Meir of Rothenburg: His Life and His Works as Sources for the Religious, Legal, and Social History of the Jews of Germany in the Thirteenth Century* (New York: Ktav, 1970), 179.
6. Ibid., 183.
7. Thanks to Don Singer for this valuable information. Rabbi Richard Sarason, *Asher Yatsar [Our Bodily Needs]*, Mishkan T'filah, 194–195. June 26, 2008, Week 241, Day 4, 23 Sivan 5768, accessed October 23, 2014, http://tmt.urj.net/archives/4jewishethics/062608.htm.
8. "Dissolving the barrier between ingestion and excretion" produces anxiety, due to the subsequent intermingling of "oppositions such as inside/outside, top/bottom, before/after, form/formlessness, positive/negative." Barley, "The Letter to Brother Edward," 23–24.
9. John T. McNeill and Helena M. Gamer (eds. and trans.). *Medieval Handbooks of Penance: A Translation of the Principle Libri Poenitentiales* (New York: Columbia University Press, 1938/1990), 191.

10. Ibid., 185.
11. Ibid., 186.
12. Ibid., 197.
13. Ibid., 199.
14. Ibid., 191.
15. "The production of 'official fear' is the key to the power's effectiveness," due to "human vulnerability and uncertainty." Zygmunt Bauman, *Wasted Lives: Modernity and Its Outcasts* (Cambridge: Polity Press, 2004), 50.
16. All references from Medieval Sourcebook: Twelfth Ecumenical Council: Lateran IV 1215, accessed February 15, 2010, http://www.fordham.edu/halsall/basis/lateran4.html.
17. McNeill and Gamer, 188.
18. *Eyrbyggja Saga*, trans. Hermann Pálsson and Paul Edwards (London: Penguin, 1989), 28.
19. Ibid., 29.
20. Ibid., 30.
21. Ibid., 34.
22. Ibid., 37.
23. Kevin J. Wanner, "Purity and Danger in Earliest Iceland: Excrement, Blood, Sacred Space, and Society in *Eyrbyggja saga*," *Viking and Medieval Scandinavia* 5.5 (2009): 214.
24. Pálsson and Edwards, *Eyrbyggja Saga*, 68–69.
25. Ibid., 69.
26. Ibid., 70.
27. Ibid., 79.
28. Rex Ferguson, "Garbage and Gatsby," in *Aesthetic Fatigue: Modernity and the Language of Waste*, ed. John Scanlan and John F. M. Clark (Newcastle upon Tyne: Cambridge Scholars, 2013), 206–219.
29. F. Scott Fitzgerald, *The Great Gatsby* (New York: Scribner, 1925/2004), 5.
30. Ibid., 10.
31. Ibid., 107.
32. Ibid., 176.
33. Ibid., 3.
34. Ibid., 180.
35. How Lord Voldemort is referred to. By naming him, Harry Potter undermines the power of his enemy.
36. George Bataille, *The Accursed Share: An Essay on General Economy, Vol. II: The History of Eroticism, Vol. III: Sovereignty*, trans. Robert Hurley (New York: Zone Books, 1993), 56.
37. "Shit is what is left behind undifferentiated." John Berger, *Keeping a Rendezvous* (New York: Vintage, 1991), 40.
38. Ibid., 37.
39. Ibid., 38.
40. Bauman, *Wasted Lives*, 101.
41. Don DeLillo, *Underworld* (London: Picador, 1997), 77, quoted in John Scanlan, *On Garbage* (London: Reaktion Books, 2005), 179.
42. Pálsson and Edwards, *Eyrbyggja Saga*, 62.

43. Christopher Schmidt, *The Poetics of Waste: Queer Excess in Stein, Ashbery, Schuyler, and Goldsmith* (New York: Palgrave Macmillan, 2014), 41.
44. Julia Kristeva, *Powers of Horror: An Essay on Abjection*, trans. Leon S. Roudiez (New York: Columbia University Press, 1987), 103.
45. Douglas, *Purity and Danger*, 66.
46. Gerald Vizenor, *Landfill Meditation: Crossblood Stories* (Hanover, NH: Wesleyan University Press/University Press of New England, 1991), 58.
47. Ibid., 64.
48. Fitzgerald, *The Great Gatsby*, 12–13.
49. Ferguson, "Garbage," 209.
50. Fitzgerald, *The Great Gatsby*, 13.
51. Ibid., 130.
52. All references to William Shakespeare, *The Arden Edition of the Works of William Shakespeare: Hamlet*, ed. Ann Thompson and Neil Taylor (London: Thomson, 2006).
53. Miles Wilson, "Everything," in *Line of Fall* (Iowa City: University of Iowa Press, 1989), 18.
54. Michel Foucault, *The Order of Things: An Archaeology of the Human Sciences* (New York: Vintage Books, 1973), xv.
55. "The monstrous quality that runs through Borges's enumeration consists, on the contrary, in the fact that the common ground on which such meetings are possible has itself been destroyed." Foucault, *The Order of Things*, xvi. However, Steve Mentz, "Brown," in *Prismatic Ecology: Ecotheory beyond Green*, ed. Jeffrey Jerome Cohen (Minneapolis: University of Minnesota Press, 2013), 209, argues that the list is not as random as it appears.
56. Foucault, *The Order of Things*, xviii.
57. Donald Barthelme, "Nothing: A Preliminary Account," in *Sixty Stories* (New York: G. P. Putnam's Sons, 1981): 245, 246.
58. Ibid., 247.
59. Mentz, "Brown," 208, citing the work of Ian Bogost.

2 The Fragmented and Corruptible Body: Gendered Waste

1. A. R. Ammons, *Garbage* (New York: W. W. Norton, 2002), 105.
2. Michael Thompson, *Rubbish Theory: The Creation and Destruction of Value* (Oxford: Oxford University Press, 1979), 10.
3. Theodor W. Adorno, "Trying to Understand *Endgame*," trans. Michael T. Jones, *New German Critique, Critical Theory and Modernity* 26 (Spring–Summer, 1982): 133.
4. Alice Sebold, "Living with the Dead," *The New York Times*, September 11, 2005, accessed October 16, 2014, http://www.nytimes.com/2005/09/11/opinion/11sebold.html?_r=0.
5. James Schmidt, ed., *What is Enlightenment? Eighteenth-Century Answers and Twentieth-Century Questions* (Berkeley: University of California Press, 1996), 78–79.
6. Julia Kristeva, *Powers of Horror: An Essay on Abjection*, trans. Leon S. Roudiez (New York: Columbia University Press, 1987), 70.

7. Linda Holler, *Erotic Morality: The Role of Touch in Moral Agency* (New Brunswick, NJ: Rutgers University Press, 2002), 90. In its most pernicious extreme, mind-body duality can cause societal dysfunction and has been shown to be prevalent in those with mental illness. "The Cartesian certainty, 'I think, therefore I am,' appears similar in its solipsism to forms of psychosis." This is a legacy we have inherited. Holler, *Erotic Morality*, 62.
8. Emmanuel Levinas, *On Escape. De l'évasion*, trans. Bettina Bergo, Introduction by Jacques Rolland (Stanford, CA: Stanford University Press, 2003), 64–65.
9. Heinrich Böll, "In Defense of Washtubs," in *Stories, Political Writings, and Autobiographical Works*, ed. Martin Black, trans. Leila Vennewitz (New York: Continuum, 2006), 274–275.
10. Susan Strasser, *Waste and Want: A Social History of Trash* (New York: Henry Holt, 2000), 16.
11. Italo Calvino, "La Poubelle Agréée," in *The Road to San Giovanni*, trans. Tim Parks (New York: Pantheon, 1993), 103.
12. Ibid., 104.
13. Ibid., 103.
14. Ibid., 93, 102.
15. Ibid., 104.
16. Kristeva, *Powers of Horror*, 3–4. As she writes, "The body's inside, in that case, shows up in order to compensate for the collapse of the border between inside and outside... Urine, blood, sperm, excrement then show up in order to reassure a subject that is lacking its 'own and clean self'" (53).
17. "If waste is one of our most immediate others, and establishing our difference and separation from it the condition of possibility for a self, then its persistence, its refusal to go, is a primordial threat to the drive for wholeness." Gay Hawkins, *The Ethics of Waste: How We Relate to Rubbish* (Lanham, MD: Rowman & Littlefield, 2006), 35.
18. "The abnormal body suggests that one's coherent identity is actually a dreamlike construction that merely conceals the fundamentally fragmentary nature of identity." Vincent B. Leitch on Lennard J. Davis, "Enforcing Normalcy: Disability, Deafness, and the Body," in *Visualizing the Disabled Body: The Classical Nude and the Fragmented Torso*, reprinted in *The Norton Anthology of Theory and Criticism*, ed. Vincent B. Leitch (New York: Norton, 2001), 2399.
19. Davis, "Enforcing Normalcy," 2403.
20. Compare to the homo sacer of Roman culture as discussed by Giorgio Agamben, *Homo Sacer: Sovereign Power and Bare Life*, trans. Daniel Heller-Roazen (Stanford, CA: Stanford University Press, 1998).
21. Sophocles, *Philoctetes*, trans. Carl Phillips (Oxford: Oxford University Press, 2003), 70, lines 926, 943.
22. Ibid., 40, lines 179–181.
23. Ibid., 42, line 223.
24. Ibid., 53, line 496.
25. Ibid., 35, line 45.
26. Ibid., 59, lines 646–647.
27. Ibid., 78, lines 1142–1147.

28. Ibid., 85, line 1341.
29. See Jacqueline Vanhoutte, "Denmark's Rotting Reconsidered," *Philological Quarterly* 91.3 (2012): 397. Hayden White comments that "perceived differences between men had less significance for Greeks and Romans than they had for Hebrews and Christians. For the former, differentness was perceived as physical and cultural; for the latter, as moral and metaphysical." "The Forms of Wildness: Archaeology of an Idea," *Tropics of Discourse: Essays in Cultural Criticism* (Baltimore, MD: Johns Hopkins University Press, 1978), 156. Quoted in Michael Uebel, *Ecstatic Transformation: On the Uses of Alterity in the Middle Ages* (New York: Palgrave Macmillan, 2005), 18.
30. Lothario Dei Segni (Pope Innocent III), *On the Misery of the Human Condition: De miseria humane conditionis*, ed. Donald R. Howard, trans. Margaret Mary Dietz (Indianapolis: Bobbs-Merrill, 1969), 71.
31. Davis, "Enforcing Normalcy," 2410.
32. Ibid., 2405. While evoking Bakhtin's concept of the grotesque, Lennard Davis points out that, nonetheless, "what the term [grotesque] has failed to liberate is the notion of actual bodies" (2418).
33. "Gigantic features, scatological references, inverse political power were all hallmarks of the grotesque—an aesthetic that ultimately was displaced by humanistic notions of order, regularity, and of course power during the Renaissance." Ibid., 2418.
34. Jean Baudrillard, *Simulacra and Simulation*, trans. Sheila Faria Glaser (Ann Arbor: University of Michigan Press, 1994), 105. Baudrillard continues, "The body can be nothing but the waste product of its own residue, the fallout of its own fallout." *Simulacra and Simulation*, p. 148. Mikhail Bakhtin, *Rabelais and His World*, trans. Hélène Iswolsky (Bloomington: Indiana University Press, 1968/1984), 320. See also Michael Mayerfeld Bell, "Deep Fecology: Mikhail Bakhtin and the Call of Nature," *Capitalism, Nature, Socialism* 5.4 (1994): especially 73–75. "The grotesque body is a popular-festive body that, outgrowing itself, threatens forms of established order for the sake of its own immediate self-celebration and for the long-term goal of promoting social cohesion and purposiveness from below." For Bakhtin, then, the body is of interest as an "instrument of political critique." Gail Kern Paster, *The Body Embarrassed: Drama and the Disciplines of Shame in Early Modern England* (Ithaca, NY: Cornell University Press, 1993), 14–15. See Bakhtin, *Rabelais and His World*, 147.
35. Bakhtin, *Rabelais and His World*, 321.
36. Thompson, *Rubbish Theory*, 10; see also, famously, Mary Douglas, *Purity and Danger: An Analysis of the Concepts of Pollution and Taboo* (London: Routledge, 1966/2002).
37. Caroline Walker Bynum, *Fragmentation and Redemption: Essays on Gender and the Human Body in Medieval Religion* (New York: Zone Books, 1991), 285; Margaret E. Owens, *Stages of Dismemberment: The Fragmented Body in Late Medieval and Early Modern Drama* (Newark: University of Delaware Press, 2005), 53. "If in Christ and in the martyrs part is whole, perhaps we too shall be—even are—whole in every fragment no matter how threatened by consumption, death, and decay." Caroline Walker Bynum, The *Resurrection of the*

Body in Western Christianity, 200–1336 (New York: Columbia University Press, 1995), 316. It would be appropriate to explore the relic according to Harman's take on Levinas. Graham Harman, in referencing *Totality and Infinity*, 160, suggests that things "must always be there, always supporting the thing as a whole," not just when humans regard them. Graham Harman, "Levinas and the Triple Critique of Heidegger," *Philosophy Today* 53 (2009): 412.

38. Hermann Pálsson and Paul Edwards, trans., *Eyrbyggja Saga* (London: Penguin, 1989), 75.
39. Charles Scribner III, "Celestial Eyes—From Metamorphosis to Masterpiece," *Princeton Library Chronicle* 53.2 (1992): 141–142.
40. F. Scott Fitzgerald, *The Great Gatsby* (New York: Scribner, 1925/2004), 13.
41. Ibid., 23.
42. Ibid., 63.
43. Ibid., 69.
44. We later hear of the hair in nostrils again. Ibid., 171.
45. Ibid., 85.
46. Ibid., 105.
47. H. G. Adler, *The Journey: A Novel*, trans. Peter Filkins (New York: Random House, 2008), 82.
48. Ibid., 233.
49. Ibid., 99.
50. Kristeva, *Powers of Horror*, 109. Georges Bataille equates "defecation . . . death and the cult of cadavers (above all, insofar as it involves the stinking decomposition of bodies) . . . shit, gods, and cadavers." Georges Bataille, "The Notion of Expenditure," in *Visions of Excess: Selected Writings, 1927–1939*, ed. Allan Stoekl, trans. Allan Stoekl, et al., Theory and History of Literature, vol. 14 (Minneapolis, MN: University of Minnesota Press, 1985), 94. As Dominique Laporte puts it, "Corpses are no more and no less than waste that one buries." Dominique Laporte, *History of Shit*, trans. Nadia Benabid and Rodolphe el-Khoury, Introduction by Rodolph el-Khoury (Cambridge, MA: MIT Press, 1993), 60. See also how Peter Ackroyd equates the dead with rubbish in *London: A Biography* (New York: Doubleday, 2000), 337. "To rise above the body is to equate the body with excrement . . . T]he peculiar human fascination with excrement is the peculiar human fascination with death . . . Excrement is the dead life of the body." Norman O. Brown, *Life Against Death: The Psychoanalytical Meaning of History* (Middletown, CT: Wesleyan University Press, 1959), 295.
51. Thomas G. Long, "Chronicle of a Death We Can't Accept," *The New York Times*, October 31, 2009, accessed October 18, 2014, http://www.nytimes.com/2009/11/01/opinion/01long.html.
52. Joyce Carol Oates, "Landfill," *The New Yorker*, October 9, 2006, accessed April 20, 2010, http://www.newyorker.com/magazine/2006/10/09/landfill.
53. From Jenkins in William Shakespeare, *The Arden Edition of the Works of William Shakespeare: Hamlet*, ed. Harold Jenkins (London: Methuen, 1982), 551.
54. All quotes from Oxford, Bodleian Library, MS Ashmole 1434, 292–294. The manuscript consists of three sections or works sewn together; the final one

dating from, apparently, 1586. See Will Stockton, *Playing Dirty: Sexuality and Waste in Early Modern Comedy* (Minneapolis, MN: University of Minnesota Press, 2011), xv, for more on excrements and hair; and Will Fisher, "The Renaissance Beard: Masculinity in Early Modern Europe," *Renaissance Quarterly* 54 (2001): 168, 174, 177 [155–187] on the beard as excrement.

55. Levinas, *On Escape*, 53.
56. Ibid., 66.
57. Samuel Beckett, *Endgame* (New York: Grove Press, 1958), 23.
58. Ibid., 49.
59. This "portmanteau" word, coined by Teresa Mack, combines "monstrous" and "menstrual."
60. Martha C. Nussbaum, *From Disgust to Humanity: Sexual Orientation and Constitutional Law* (Oxford/New York: Oxford University Press, 2010), 7.
61. Martha C. Nussbaum, *Upheavals of Thought: The Intelligence of Emotions* (Cambridge: Cambridge University Press, 2001), 205.
62. Kristeva, *Powers of Horror*, 65.
63. Elizabeth V. Spelman, "Woman as Body: Ancient and Contemporary Views," *Feminist Studies* 8.1 (1982): 109–131.
64. Emily Martin, "The Egg and the Sperm: How Science Has Constructed a Romance Based on Stereotypical Male-Female Roles," in *Feminist Theory and the Body: A Reader*, ed. Janet Price and Margrit Shildrick (New York: Routledge, 1999), 180. In discussing various bodily fluids, Cyrille Harpet comments, "Le sang garde sa valeur 'excrémentielle' ou 'de perte' seulement dans le cas des menstrues du fait de sa libération spontanée et 'naturelle' par un orifice dit 'inférieur'." *De Déchet: Philosophie des Immondices Corps, Ville, Industrie* (Paris: L'Harmattan, 1998), 124.
65. Gary Waller, *The Virgin Mary in Late Medieval and Early Modern English Literature and Popular Culture* (Cambridge: Cambridge University Press, 2011), 40.
66. David Darst, "Witchcraft in Spain: The Testimony of Martín de Castañega's Treatise on Superstition and Witchcraft (1529)," *Proceedings of the American Philosophical Society* 123 (1979): 309. "These three problems of madness, illegitimacy, and anomaly came together in the figure of woman, who within the medico-moral legacy of Galen and Aristotle was typically regarded as an anomalous and monstrous creature, or a secondary creation." Bryan S. Turner, "The Body in Western Society: Social Theory and Its Perspectives," in *Religion and the Body*, ed. Sarah Coakley (Cambridge: Cambridge University Press, 1997), 24. As Georges Bataille has commented, "Menstrual blood seems to have condensed the abhorrence and the fear." Georges Bataille, *The Accursed Share: An Essay on General Economy, Vol. II: The History of Eroticism, Vol. III: Sovereignty*, trans. Robert Hurley (New York: Zone Books, 1993), 65.
67. Nussbaum, *From Disgust to Humanity*, xvii.
68. Nussbaum, *Upheavals of Thought*, 302.
69. It has been argued that liquid was linked to goddesses in old matrilineal tribes. "Why is Beowulf so hostile to the sea that he must swim with his sword in his hand? Perhaps because the water was especially thought the domain of the

chthonic female deity." Frank Battaglia, "The Germanic Earth Goddess in *Beowulf*?" *The Mankind Quarterly* 31 (1997): 427 [415–446].

70. Editions consulted were *Klaeber's Beowulf and the Fight at Finnsburg*, ed. R. D. Fulk, Bobert E. Bjork, and John D. Niles, 4th ed. (Toronto: University of Toronto Press, 2008) and *Beowulf: An Edition*, ed. Bruce Mitchell and Fred C. Robinson (Oxford: Blackwell, 1998).
71. Alcuin Blamires, ed., *Woman Defamed and Woman Defended: An Anthology of Medieval Texts* (Oxford: Clarendon Press, 1992), 44; Isidore of Seville, *Isidori Hispalensis Episcopi, Etymologiarum sive Originum*, ed. W. M. Lindsay (Oxford: Clarendon Press, 1962), XI. i. 140–141.
72. Bx 8558 fol. 152r; B77.04.06.
73. Dei Segni, *On the Misery of the Human Condition*, 9.
74. Bede, *The Ecclesiastical History of the English People*, trans. Leo Sherley-Price (Harmondsworth: Penguin, 1990), 84.
75. Dana M. Oswald, "'Wigge under Wætere': Beowulf's Revision of the Fight with Grendel's Mother," *Exemplaria* 21 (2009): 64.
76. Hawkins, *The Ethics of Waste*, 46.
77. See Jacqueline Murray, "Men's Bodies, Men's Minds: Seminal Emissions and Sexual Anxiety in the Middle Ages," *Annual Review of Sex Research* 8 (1997): 1–26.
78. Susan Signe Morrison, *Excrement in the Late Middle Ages: Sacred Filth and Chaucer's Fecopoetics* (New York: Palgrave Macmillan, 2008), 52.
79. Hildegard of Bingen, *Scivias*, trans. Mother Columba Hart and Jane Bishop (New York: Paulist Press, 1990), 493, quoted by Madeline H. Caviness, *Visualizing Women in the Middle Ages: Sight, Spectacle, and Scopic Economy* (Philadelphia: University of Pennsylvania Press, 2001), 172.
80. Hildegard, *Scivias*, 508. Italics in original.
81. Michael Swanton (trans. and ed.), *Anglo-Saxon Prose* (London: Everyman, 1993), 181. See a parallel Gertrude Stein's poem as discussed by Christopher Schmidt, *The Poetics of Waste: Queer Excess in Stein, Ashbery, Schuyler, and Goldsmith* (New York: Palgrave Macmillan, 2014), 43.
82. Schmidt, *The Poetics of Waste*, 43. A stank is a pond or ditch filled with water.
83. Peter Suchsland and Erika Weber, *Deutsche Volksbücher in Drei Bänden, Zweiter Band* (Berlin/Weimar: Aufbau Verlag, 1968), 150; Paul Oppenheimer, trans., *Till Eulenspiegel: His Adventures* (New York: Routledge, 2001), 184.
84. Waller, *The Virgin Mary*, 34.
85. Ibid., 14. Chapter One, describing a number of these savage attacks, perceptively analyzes the psychological anger behind them. Later chapters eloquently describe the tragic effect this had on many of the Catholic faithful. Sacred items such as altars or holy water fonts were turned to profane use such as tables or chamber pots (15).
86. For the full argument, see Susan Signe Morrison, "Waste Space: Pilgrim Badges, Ophelia, and Walsingham Remembered," in *Walsingham in Literature and Culture from the Middle Ages to Modernity*, ed. Gary Waller and Dominic Janes (Farnham, England: Ashgate, 2010), 49–66.

87. Peter Seng, "Ophelia's Songs in *Hamlet*," *Durham University Journal* 25 (1964): 77–85.
88. "Yet [Ophelia] 'speaks' through snatches of ordinary, albeit disjointed, discourse: popular ballads, traditional legends, routine pieties, even familiar expressions of greeting and farewell. The important work her psyche attempts, to pervert convention in order to find a vehicle through which she can protest her lifetime of repression, is ultimately undermined by the very *conventionality* of her discourse." Her conversation is "irrational and disconnected." Gabrielle Dane, "Reading Ophelia's Madness," *Exemplaria* 10 (1998): 419.
89. The act of processing waste even becomes associated with women when, in 1903 in Woolwich, England, a new refuse destructor had engines baptized "'Flavia', 'Wilhelmina', 'Gertrude', and 'Muriel' after the names of the wives of the engineer and the Electricity Committee." J. F. M. Clark, "'The incineration of refuse is beautiful': Torquay and the introduction of municipal refuse destructors." *Urban History* 34 (2007): 273.
90. Fitzgerald, *The Great Gatsby*, 24.
91. Ibid., 79.
92. Dorothy Allison, *Trash* (New York: Plume, 1988/2002), xvi.
93. Ibid., 11.
94. Ibid., 12.
95. Ibid., 20.
96. Ibid., 26.
97. Ibid., 23.
98. Ibid., 25.
99. Ibid., 30.
100. Ibid., 44.
101. Elfriede Jelinek, *Services*, in *Gate Biennale*, trans. Nick Grindell (London: Methuen, 1996), 82.
102. Ibid., 96.
103. Ibid., 128.
104. Michel Serres, *Malfeasance: Appropriation through Pollution?*, trans. Anne-Marie Feenberg-Dibon (Stanford: Stanford University Press, 2011), 28–29.
105. Ibid., 31.

3 The Civilizing Process: Divisive Divisions

1. Roald Dahl, *The BFG* (New York: Scholastic, 1982), 67.
2. Georges Bataille, *The Accursed Share: An Essay on General Economy, Vol. II: The History of Eroticism, Vol. III: Sovereignty*, trans. Robert Hurley (New York: Zone Books, 1993), 53.
3. See Charlotte Allen, "A Dark Age for Medievalists: At Their Annual Congress in Kalamazoo, It's no Longer Your Grandfather's Middle Ages," *The Weekly Standard* 13.36 (June 2, 2008), accessed October 17, 2014, http://www.weeklystandard.com/Content/Public/Articles/000/000/015/146etleh.asp and the blogosphere response.
4. Paul Oppenheimer (trans.), *Till Eulenspiegel: His Adventures* (New York: Routledge, 2001), 31.

5. Peter Suchsland and Erika Weber, *Deutsche Volksbücher in Drei Bänden. Zweiter Band* (Berlin/Weimar: Aufbau Verlag, 1968), 135; Oppenheimer, *Till Eulenspiegel*, 168.
6. Bataille, *The Accursed Share*, 63.
7. Ibid., 62–63.
8. Ibid., 63.
9. Gay Hawkins, *The Ethics of Waste: How We Relate to Rubbish* (Lanham, MD: Rowman & Littlefield, 2006), 1.
10. Bataille, *The Accursed Share*, 66.
11. Villain as a person of low repute or dubious intentions and actions comes from the related word in Anglo-Norman and Old French: *vilein*, *villain*, or *villain*.
12. "Dirt indicates not only a lack of civilization but the nature of peasant toil, productive, useful, but unpleasant and so degrading as to remove the unfortunate victim from full consideration as human." Paul Freedman, *Images of the Medieval Peasant* (Stanford, CA: Stanford University Press, 1999), 143; see also 150.
13. Oppenheimer, *Till Eulenspiegel*, 178–180.
14. David Rollo, *Kiss My Relics: Hermaphroditic Fictions of the Middle Ages* (Chicago, IL: University of Chicago Press, 2011), 126.
15. "In short, French was the language of honour, of chivalry, and even of justice, while the far more manly and expressive Anglo-Saxon was abandoned to the use of rustics and hinds, who knew no other." Sir Walter Scott, *Ivanhoe*, Chapter One. "It has been said that we have a synonym at each level—popular [English], literary [French], and learned [Latin]," Albert C. Baugh and Thomas Cable, *A History of the English Language*, 6th ed. (Upper Saddle River, NJ: Prentice Hall, 2013), 182. See also Caroline Walker Bynum, "Why All the Fuss about the Body? A Medievalist's Perspective," *Critical Inquiry* 22 (1995): 7; Valerie Allen, *On Farting: Language and Laughter in the Middle Ages* (New York: Palgrave Macmillan, 2007), 11; Thomas W. Ross, "Taboo-Words in Fifteenth-Century English," *in Fifteenth-Century Studies: Recent Essays*, ed. Robert F. Yeager (Hamden, CT: Archon, 1984), 140; Steven Pinker, *The Stuff of Thought: Language as a Window into Human Nature* (New York: Viking, 2007), 319–320, 344–346.
16. Morrison, *Excrement in the Late Middle Ages: Sacred Filth and Chaucer's Fecopoetics* (New York: Palgrave Macmillan, 2008), 16–17.
17. See Morrison, *Excrement*, Chapter 5; William Ian Miller, *The Anatomy of Disgust* (Cambridge, MA: Harvard University Press, 1997), 154; Caroline Holmes, *The Not so Little! Book of Dung* (Thrupp: Sutton, 2006), 38; Karma Lochrie, *Covert Operations: The Medieval Uses of Secrecy* (Philadelphia: University of Pennsylvania Press, 1999), 164.
18. Sarah Lyall, "Why Can't the English Just Give Up that Class Folderol?" *The New York Times*, April 26, 2007, accessed October 18, 2014, http://www.nytimes.com/2007/04/26/world/europe/26britain.html?pagewanted=print.
19. Cited in Dolly Jørgensen, "The Metamorphosis of Ajax," *Early English Studies* 3 (2010): 1–31; from John Harington, *A New Discourse of a Stale Subject, Called the Metamorphosis of Ajax*, ed. Elizabeth Story Donno (New York: Columbia University Press, 1962), 186. See also J. F. M. Clark, "'The Incineration

of Refuse is Beautiful': Torquay and the Introduction of Municipal Refuse Destructors," *Urban History* 34 (2007): 259, on waste stream problems.

20. Morrison, *Excrement*, 130–133.
21. Jonathan Bate comments that "the word is thus removed from the earth and linked to the advance of society. It is, if you will, removed from the country to the city—cultivation comes to mean civility, a word which has its root in the Latin *civilis*, meaning 'of, or pertaining to, the city.'" Jonathan Bate, *Song of the Earth* (Cambridge, MA: Harvard University Press, 2000), 4.
22. Each bodily part is associated with its own excrement: stomach with feces, the liver with urine; sweat, hair, nails, and mucus were likewise viewed as excrements. Michael Schoenfeldt, "Fables of the Belly in Early Modern England," in *The Body in Parts: Fantasies of Corporeality in Early Modern Europe*, ed. David Hillman and Carla Mazzio (New York: Routledge, 1997), 242–243, 245.
23. Norbert Elias, *The Civilizing Process* (Oxford: Blackwell, 1994/2000), 100–102.
24. Caroline Walker Bynum, *Holy Feast and Holy Fast: The Religious Significance of Food to Medieval Women* (Berkeley: University of California Press, 1987), Chapter 2.
25. Elias, *The Civilizing Process*, 90.
26. Lothario Dei Segni (Pope Innocent III), *On the Misery of the Human Condition: De miseria humane conditionis*, ed. Donald R. Howard, trans. Margaret Mary Dietz (Indianapolis: Bobbs-Merrill, 1969), 46.
27. Ibid., 71.
28. Oppenheimer, *Till Eulenspiegel*, 104–106.
29. Ibid., 142–144.
30. Jonathan Dollimore, *Sexual Dissidence: Augustine to Wilde, Freud to Foucault* (Oxford: Clarendon Press, 1991), 111.
31. Oppenheimer, *Till Eulenspiegel*, 164–165.
32. Bate, *Song of the Earth*, 4. Bate writes, "Originally 'culture' was the work done by a labourer in the fields, whereas for [Matthew] Arnold and his successors culture is intellectual, even spiritual, work which serves the moral needs of a society and is set in opposition to the very idea of physical labour" (5).
33. Jonathan Swift, *Gulliver's Travels*, ed. *Philip Pinkus* (Toronto: Macmillan, 1968), 243.
34. Ibid., 244.
35. Slavoj Žižek uses this moment in *The Phantom of Liberty* as a springboard to contemplate toilet construction. "The three basic types of toilet form an excremental correlative-counterpoint to the Lévi-Straussian triangle of cooking (the raw, the cooked and the rotten). In a traditional German toilet, the hole into which shit disappears after we flush is right at the front, so that shit is first laid out for us to sniff and inspect for traces of illness. In the typical French toilet, on the contrary, the hole is at the back, i.e. shit is supposed to disappear as quickly as possible. Finally, the American (Anglo-Saxon) toilet presents a synthesis, a mediation between these opposites: the toilet basin is full of water, so that the shit floats in it, visible, but not to be inspected... It is clear that none of these versions can be accounted for in purely utilitarian terms: each involves

a certain ideological perception of how the subject should relate to excrement." Slavoj Žižek, "Knee-Deep," *London Review of Books*, September 2, 2004: 12.

36. Andrew Martin, "What's Eating Me," *The Guardian Weekend*, January 10, 2004: 36. Alison Johnson, cook and owner of Scarista House on the Isle of Harris, discusses her dislike of serving or consuming conventionally raised animals: "'Ça sent la merde'; each animal's flesh was faintly permeated with the odour of its own dung. Why? Had they been rolling in it? Penned up in the fumes of it?" Alison Johnson, *Scarista Style* (London: Futura, 1987), 14.
37. Seth Lerer "Grendel's Glove, "*ELH* 61 (1994): 721–751, writes how Caedmon "traded on these legends of ritual ingestion and shared poetic performance, contrasting them with the monastic traditions of *ruminatio* and the ideals of a Christian ingestive imagery," no. 19, accessed October 21, 2014, http://muse.jhu.edu/journals/elh/v061/61.4lerer.html. Here the poetic act would be defecation. See also Seth Lerer, "'On fagne flor': the postcolonial *Beowulf*," in *Postcolonial Approaches to the European Middle Ages: Translating Cultures*, ed. Ananya Jahanara Kabir and Deanne Williams (Cambridge: Cambridge University Press, 2005), 77–102.
38. David Inglis, *A Sociological History of Excretory Experience: Defecatory Manners and Toiletry Technologies* (Lewiston, ME: Edwin Mellen Press, 2001), 135.
39. Elias, *The Civilizing Process*, 108.
40. Ibid., 111–112; see also 116.
41. F. Scott Fitzgerald, *The Great Gatsby* (New York: Scribner, 1925/2004), 179.
42. Ibid., 41.
43. Elias, *The Civilizing Process*, 68.
44. Fitzgerald, *The Great Gatsby*, 61–63.
45. Ibid., 154.
46. Ibid., 12.
47. Ibid., 48.
48. Ibid., 98.
49. Ibid., 173.
50. Ibid., 149.
51. Ibid., 171.
52. Ibid., 133.
53. Ibid., 130–1.
54. Dorothy Allison, *Trash* (New York: Plume, 1988/2002), 23.
55. Ibid., 29.
56. Swift, *Gulliver's Travels*, 214.
57. Ibid., 215.
58. Ibid., 217.
59. Ibid., 219.
60. Ibid., 220.
61. Ibid., 253.
62. Ibid., 220.
63. Ibid., 225.
64. Ibid., 227.
65. Ibid., 221.
66. Ibid., 228.

4 Memory and Narrative: Ruins, Nostalgia, and Ghosts

1. John Scanlan, *On Garbage* (London: Reaktion, 2005), 141.
2. William Rathje and Cullen Murphy, *Rubbish! The Archaeology of Garbage* (New York: HarperCollins, 1992), 10–11.
3. Charles Britt Bousman, personal communication, Texas State University, March 6, 2012. Many thanks to Britt for all his stimulating suggestions. See Will Viney, *Waste: A Philosophy of Things* (London: Bloomsbury, 2014), 57–58, where he argues what has distinguished archeologists from garbage collectors have been the rules by which items are interpreted.
4. Patricia Yaeger, "Editor's Column: The Death of Nature and the Apotheosis of Trash; or, Rubbish Ecology," *PMLA* 123 (2008): 335.
5. Robert L. Kelly and David Hurst Thomas, *Archeology*, 6th ed. (Belmont, CA: Wadsworth, 2012,) 108–115.
6. See Michael B. Schiffer, "Toward the Identification of Formation Processes," *American Antiquity* 48.4 (1983): 675–706; and Robert Ascher, "Time's Arrow and the Archaeology of a Contemporary Community," in *Settlement Archaeology*, ed. K. C. Chang (Palo Alto: National Press, 1968), 43–52.
7. Michael B. Schiffer, "Is There a 'Pompeii Premise' in Archaeology?," *Journal of Anthropological Research* 41.1 (1985): 18.
8. Ibid., 24.
9. Ibid., 18.
10. Schiffer, "Toward the Identification," 690.
11. Ibid., 696–697.
12. Kelly and Thomas, *Archeology*, 383, citing Emil Haury.
13. Craig Williamson, *A Feast of Creatures: Anglo-Saxon Riddle Songs* (Philadelphia: University of Pennsylvania Press, 2011), 6.
14. Zygmunt Bauman, *Wasted Lives: Modernity and Its Outcasts* (Cambridge: Polity Press, 2004), 18. See also Walter Moser, "Garbage and Recycling: From Literary Theme to Mode of Production," *Other Voices* 3.1 (2007), accessed April 18, 2012, http://www.othervoices.org/3.1/wmoser/index.php, who argues "recycling . . . is associated with forgetting."
15. Irving Wohlfarth, "Et Cetera? The Historian as Chiffonnier," *New German Critique* 39 (1986): 144–145.
16. Allen J. Frantzen, *Desire for Origins: New Language, Old English, and Teaching the Tradition* (New Brunswick, NJ: Rutgers University Press, 1990), 47. Frantzen warns against eliminating the past, reminding us to be wary of the "desire for origins" to justify our "pure" beginnings.
17. Ardis Butterfield, "Chaucer and the Detritus of the City," in *Chaucer and the City*, ed. Ardis Butterfield (Cambridge: D. S. Brewer, 2006), 6. Butterfield cites Walter Benjamin, *The Arcades Project: Walter Benjamin*, trans. Howard Eiland and Kevin McLaughlin (Cambridge, MA, and London: Belknap Press, 1999), "Translators' Forward," ix.
18. Walter Benjamin, *Illuminations: Essays and Reflections*, ed. Hannah Arendt (New York: Schocken Books, 1968), 255.
19. Scanlan, *On Garbage*, 129.

20. Theodor W. Adorno, "Trying to Understand *Endgame*," trans. Michael T. Jones, *New German Critique, Critical Theory and Modernity* 26 (Spring–Summer 1982), 144, also 125. Richard Halperin comments on the two most common interpretations of *Endgame*: that it depicts a post-nuclear holocaust world or the devastation after a second Noah's flood. Richard Halpern, "Beckett's Tragic Pantry: *Endgame* and the Deflation of the Act." *PMLA* 129.4 (2014): 744.
21. Samuel Beckett, *Endgame* (New York: Grove Press, 1958), 14.
22. Ibid., 15, also 20.
23. Ibid., 18.
24. Ibid., 21.
25. Jorge Luis Borges, "Funes, His Memory," in *Collected Fictions*, trans. Andrew Hurley (New York: Viking, 1998), 135.
26. Ibid., 137.
27. Ibid., 136.
28. Ibid., 137.
29. Ibid.
30. Scanlan, *On Garbage*, 179.
31. See John D. Niles, "Locating *Beowulf* in Literary History," in *The Postmodern Beowulf: A Critical Casebook*, ed. Eileen A. Joy and Mary K. Ramsay (Morgantown: West Virginia University Press, 2006), 153.
32. Alison A. Chapman, "Ophelia's 'Old Lauds': Madness and Hagiography in Hamlet," *Medieval & Renaissance Drama in England* 20 (2007): 131, n. 8.
33. Margaret Aston, "English Ruins and English History: The Dissolution and the Sense of the Past," *Journal of the Warburg and Courtauld Institutes* 36 (1973): 245.
34. Frantzen, *Desire for Origins*, 47.
35. Arthur F. Marotti, "Shakespeare and Catholicism," in *Theatre and Religion: Lancastrian Shakespeare*, ed. Richard Dutton, Alison Findlay, and Richard Wilson (Manchester: Manchester University Press, 2003), 227. Marotti cites Elizabeth Mazzola, *The Pathology of the English Renaissance: Sacred Remains and Holy Ghosts* (Leiden and Boston: Brill, 1998), 3, 7. See also Jacqueline Vanhoutte, "Denmark's Rotting Reconsidered," *Philological Quarterly* 91.3 (2012): 395. "*Hamlet*'s representations of inwardness, often associated with the play's turn to the future, are the product of its engagement with a disease from the past, or, to be more specific, with this disease [leprosy] *as* past." Vanhoutte writes how "the disease of the past assume[s] the additional burden of representing its troubled legacies" (410).
36. See Chapter 8, "Dominant, Residual, and Emergent," in Part II of Raymond Williams, *Marxism and Literature* (Oxford: Oxford University Press, 1977).
37. The residual and emergent are present during the dominant; the "residual" is not the same as archaic. "Any culture includes available elements of its past, but their place in the contemporary cultural process is profoundly variable. I would call the 'archaic' that which is wholly recognized as an element of the past, to be observed, to be examined, or even on occasion to be consciously 'revived', in a deliberately specializing way. What I mean by 'residual' is very different. The residual, by definition, has been effectively formed in the past,

but it is still active in the cultural process, not only and often not at all as an element of the past, but as an effective element of the present." Ibid., 122.

38. Williams, *Marxism and Literature*, 123–124.
39. Aston, "English Ruins and English History," 247.
40. Eamon Duffy, "Bare Ruined Choirs: Remembering Catholicism in Shakespeare's England," in *Theatre and Religion: Lancastrian Shakespeare*, ed. Richard Dutton, Alison Findlay, and Richard Wilson (Manchester: Manchester University Press, 2003), 45.
41. Jennifer Summit, "Leland's *Itinerary* and the Remains of the Medieval Past," in *Reading the Medieval in Early Modern England*, ed. Gordon McMullan and David Matthews (Cambridge: Cambridge University Press, 2007), 168.
42. "The garbage object is always endowed with pastness and thus becomes a vehicle or a trace of the past." Moser, "Garbage and Recycling."
43. Jaś Elsner, "Iconoclasm and the Preservation of Memory," in *Monuments and Memory, Made and Unmade*, ed. Robert S. Nelson and Margaret Olin (Chicago, IL: University of Chicago Press, 2003), 211. In 608, for example, the Pantheon was cleared of "pagan filth," and subsequently rededicated as a Christian church to the Virgin Mary and the martyrs (217).
44. See Duffy, "Bare Ruined Choirs," 40–41. As Susan Dunn-Hensley points out, "the destruction of [Walsingham] and other sites of Marian adoration formed a part of [Shakespeare's] literary consciousness." "Return of the Sacred Virgin: Memory, Loss, and Restoration in Shakespeare's Later Plays," in *Walsingham in Literature and Culture from the Middle Ages to Modernity*, ed. Gary Waller and Dominic Janes (Farnham, England: Ashgate, 2010), 186.
45. Jeffrey Jerome Cohen, *Of Giants: Sex, Monsters, and the Middle Ages* (Minneapolis, MN: University of Minneapolis Press, 1999), 2–3.
46. Scanlan, *On Garbage*, 163.
47. Ibid., 87. "But [the changes in Eastern bloc regimes] is not an exact parallel, for the statues of Stalin (for example) were of recent vintage, and even those who erected them were familiar with the language of the image-breakers and the vulnerability of public icons of this kind." Margaret Aston, "Iconoclasm in England: Official and Clandestine," in *Iconoclasm versus Art and Drama*, ed. Clifford Davidson and Ann Nichols (Kalamazoo, MI: Medieval Institute, 1989), 55, 65.
48. Scanlan, *On Garbage*, 87, 92, 163.
49. Shelly Rambo, "Haunted (by the) Gospel: Theology, Trauma, and Literary Theory in the Twenty-First Century," *PMLA* 125.4 (2010): 938. See Avery F. Gordon's *Ghostly Matters: Haunting and the Sociological Imaginat*ion (Minneapolis: University of Minnesota Press, 1997).
50. Hermann Pálsson and Paul Edwards (trans.), *Eyrbyggja Saga* (London: Penguin, 1989), 136.
51. Ibid., 138.
52. Ibid., 140.
53. Ibid., 141.
54. Gay Hawkins, *The Ethics of Waste: How We Relate to Rubbish* (Lanham, MD: Rowman & Littlefield, 2006), 66, writes "access to privacy [for defecation] also inaugurates a visible citizenship." Such was the case of

an Indian woman who left her new husband until he built her a toilet in the house. He did as she wished and some states have even begun enforcing the "sanitation for all" drive of the government. http://www.bbc.co.uk/news/world-asia-india-17022847, accessed February 28, 2012. This is a problem in China as well, where "potty parity" does not yet exist between men and women. According to a report from 2010, about 45 percent of Chinese "lacked access to improved sanitation facilities that protect users from contact with excrement." Sharon LaFraniere, "For Chinese Women, a Basic Need and Few Places to Attend to It," *The New York Times*, March, 1, 2012, accessed March 5, 2012, http://www.nytimes.com/2012/03/01/world/asia/chinese-women-demand-more-public-toilets.html. As a campaigner says, "I do think the right to go to the bathroom is a basic right."

55. H. G. Adler, *The Journey: A Novel*, trans. Peter Filkins, Afterword by Jeremy Adler (New York: Random House, 2008), 64.
56. Ibid., 67.
57. Ibid.
58. "Leopold's fictional death and disappearance among the refuse is countered in real life by Adler's mining of the factual detritus through the assistance of Leo Baeck's act of preservation. That Adler then uses the 'refuse' of his experience to create vivid factual and fictional renderings of it is his triumph." Leo Baeck was a real-life "spiritual leader of Germany's Jews" who worked the garbage detail and was at Theresienstadt and survived. Filkins, "Introduction," *The Journey*, xx.
59. Ibid., 68–69.
60. Ibid., 102.
61. Ibid., 98.
62. Ibid., 245.
63. Peter Boxall, "'There's No Lack of Void': Waste and Abundance in Beckett and DeLillo," *SubStance* #116, 37.2 (2008): 56.
64. Ibid., 57.
65. F. Scott Fitzgerald, *The Great Gatsby* (New York: Scribner, 1925/2004), 111.
66. Scanlan, *On Garbage*, 92.
67. Fitzgerald, *The Great Gatsby*, 66.
68. Ibid., 67. "Waste is an ironic testimonial to a desire to forget." Myra J. Hird, "Knowing Waste: Towards an Inhuman Epistemology," *Social Epistemology: A Journal of Knowledge, Culture and Policy* 26.3–4 (2012): 455.
69. Fitzgerald, *The Great Gatsby*, 105.
70. Ibid., 99.
71. Ibid., 135.
72. Ibid., 147–148.
73. Ibid., 149.
74. Ibid., 161.
75. Susan Strasser, *Waste and Want: A Social History of Trash* (New York: Henry Holt, 2000), 18. Quoted from Wohlfarth, "Et Cetera?," 151.
76. Wohlfarth, "Et Cetera?," 144.
77. Ibid., 148. See Susan Strasser, *Waste and Want: A Social History of Trash* (New York: Henry Holt, 2000), 18, citing Wohlfarth, "Et Cetera?"; also see Walter

Benjamin, *Charles Baudelaire: A Lyric Poet in the Era of High Capitalism*, trans. Harry Zohn (London: Verso, 1973), 19–20; Barrie M. Ratcliffe, "Perceptions and Realities of the Urban Magic: The Rag Pickers of Paris in the First Half of the Nineteenth Century," *Canadian Journal of History* 27 (August 1992): 205; Jeremy Tambling, "Letters and Litter," in *Rubbish, Waste and Litter: Culture and Its Refuse/als*, ed. Tadeusz Rachwał (Warsaw: Wydawnictwo SWPS Academica, 2008), 36.

78. Wohlfarth, "Et Cetera?," 149.
79. "The historicist is a culture-vulture. He scavenges off the garbage of other times and places, the ruins of Western Civilization, in search of sadly inadequate surrogates for the 'soul' he is in the very process of losing." Ibid., 153.
80. Fitzgerald, *The Great Gatsby*, 176.
81. Ibid., 177–178.
82. Ibid., 57–58.
83. Ibid., 25.
84. Ibid., 110.
85. Ibid., 73. Nick remembered it having happened, "but if I had thought of it at all I would have thought of it as a thing that merely *happened*, the end of some inevitable chain." After Jay's death, Wolfheim tells Nick, "'I raised him up out of nothing, right out of the gutter... We were so thick like that in everything'– he held up two bulbous fingers–'always together.' I wondered if this partnership had included the World's Series transaction in 1919." Ibid., 171.
86. Ibid., 180.

5 Failed Source Reduction: Conspicuous Consumption and the Inability to Minimize

1. Georges Bataille, *Visions of Excess: Selected Writings, 1927–1939*, ed. Allan Stoekl, trans. Allan Stoekl, with Carl R. Lovitt and Donald M. Leslie, Jr. (Minneapolis, MN: University of Minneapolis Press, 1985), 117.
2. "Until the second half of the twentieth century, the great majority of people even in the most developed countries could not afford to discard clothes or household furnishings until they were worn out." Susan Strasser, *Waste and Want: A Social History of Trash* (New York: Henry Holt, 2000), 4.
3. Gay Hawkins, *The Ethics of Waste: How We Relate to Rubbish* (Lanham, MD: Rowman & Littlefield, 2006), 28, using Bill Brown.
4. Strasser, *Waste and Want*, 13.
5. M. Shamsul Haque, "Environmental Discourse and Sustainable Development: Linkages and Limitation," *Ethics and Environment* 5 (2000): 11 [3–21]. See also Alexa Bingham, "Discourse of the Dammed: A Study of the Impacts of Sustainable Development Discourse on Indigenous Peoples in the Brazilian Amazon in the Context of the Proposed Belo Monte Hydroelectric Dam," *POLIS Journal* 4 (2010): 3.
6. Haque, "Environmental Discourse and Sustainable Development," 14.
7. Michel Serres, *Malfeasance: Appropriation Through Pollution?*, trans. Anne-Marie Feenberg-Dibon (Stanford, CA: Stanford University Press, 2011), 45.

8. All references to Thorstein Veblen, *Conspicuous Consumption* (1902), accessed October 18, 2014, http://www.fordham.edu/halsall/mod/1902veblen00.asp.
9. Strasser, *Waste and Want*, 4. As Zsuzsa Gille points out, within capitalism waste lies *outside* production, while in socialism it is perceived as existing *within* production. Zsuzsa Gille, *From the Cult of Waste to the Trash Heap of History: The Politics of Waste in Socialist and Postsocialist Hungary* (Bloomington: Indiana University Press, 2007), 6.
10. Jonathan Swift, *Gulliver's Travels*, ed. Philip Pinkus (Toronto: Macmillan, 1968), 241–242.
11. Scott F. Fitzgerald, *The Great Gatsby* (New York: Scribner, 1925/2004), 95.
12. Michael Thompson, *Rubbish Theory: The Creation and Destruction of Value* (Oxford: Oxford University Press, 1979), 1.
13. Ibid., 2.
14. Indeed, John Clark's work on the increase of incinerators in late nineteenth-century England shows how "the destructor's appetite facilitated a throwaway society." John F. M. Clark, "'The Incineration of Refuse is Beautiful': Torquay and the Introduction of Municipal Refuse Destructors," *Urban History* 34 (2007): 257.
15. Francesco Orlando, *Obsolete Objects in the Literary Imagination: Ruins, Relics, Rarities, Rubbish, Uninhabited Places, and Hidden Treasures*, trans. Gabriel Pihas and Daniel Seidel, with Alessandra Grego (New Haven, CT: Yale University Press, 2006), 21.
16. Georges Bataille, *The Accursed Share: An Essay on General Economy, Vol. II: The History of Eroticism, Vol. III: Sovereignty*, trans. Robert Hurley (New York: Zone Books, 1993), 67.
17. "All things, all objects, begins from nothing (garbage) and eventually return to nothing (garbage again)." John Scanlan, *On Garbage* (London: Reaktion Books, 2005), 98.
18. Fitzgerald, *The Great Gatsby*, 92.
19. As Timothy Morton points out, "Every aesthetic trace, every footprint of an object, sparkles with absence. Sensual things are elegies to the disappearance of objects." *Realist Magic: Objects, Ontology, Causality*, University of Michigan Library (Ann Arbor, MI: Open Humanities Press, 2013), 18; http://openhumanitiespress.org/realist-magic.html.
20. Fitzgerald, *The Great Gatsby*, 39.
21. Ibid., 76.
22. Ibid., 130.
23. Rex Ferguson, "Garbage and Gatsby," in *Aesthetic Fatigue: Modernity and the Language of Waste*, ed. John Scanlan and John F. M. Clark (Newcastle upon Tyne: Cambridge Scholars, 2013), 212.
24. Fitzgerald, *The Great Gatsby*, 39. Compare Gatsby's extravagance to the diet of the poor widow and her daughters in Chaucer's "The Nun's Priest's Tale," who eat a limited diet. "While the widow's diet may seem boring or repetitive, it succeeds in being more ecologically friendly than a more varied and changing diet like that of the hypocritical religious in *Piers Plowman* (XIII.60–93)." Susan Signe Morrison, *Excrement in the Late Middle Ages: Sacred Filth and Chaucer's Fecopoetics* (New York: Palgrave Macmillan, 2008), 124.

25. Fitzgerald, *The Great Gatsby*, 109.
26. Strasser, *Waste and Want*, 9.
27. Fitzgerald, *The Great Gatsby*, 179. "How thin and frail is the line separating a seat of power from a rubbish heap." Zygmunt Bauman, *Wasted Lives: Modernity and Its Outcasts* (Cambridge: Polity Press, 2004), 106.
28. Scanlan, *On Garbage*, 44–46.
29. Fitzgerald, *The Great Gatsby*, 35.
30. Ibid., 36.
31. Ibid., 172.
32. Ibid., 93.
33. Ibid., 149.
34. Ibid.
35. Ibid., 92.
36. Ibid., 93.
37. Scanlan, *On Garbage*, 98.
38. Julian Stallabrass, "Trash," in *The Object Reader*, ed. Fiona Candlin and Raiford Guins (London: Routledge, 2009), 407.
39. Bataille, *Visions of Excess,* 118–119. "Within a neoliberal state, any social investment that does not have a clear end—a projected moment when input value (money, services, care) can be replaced by output value—is not merely economically suspect but morally suspect, no matter the life-enhancing nature of the investment." Elizabeth A. Povinelli, "The Child in the Broom Closet: States of Killing and Letting Die," *South Atlantic Quarterly* 107.3 (2008): 519 [509–530].
40. Bataille, *Visions of Excess,* 123.
41. Dorothy Allison, *Trash* (New York: Plume, 1988/2002), 76.
42. All Middle English references from this edition: William Langland, *The Vision of Piers Plowman* (London and New York: J. M. Dent and E. P. Dutton, 1978), Web, accessed March 30, 2010, http://quod.lib.umich.edu/cgi/t/text/text-idx?c=cme;cc=cme;rgn=div1;view=toc;idno=PPlLan;node=PPlLan%3A1. All Modern English references from this edition: William Langland, *Piers Plowman*, trans. E. Talbot Donaldson, ed. Elizabeth Robertson and Stephen H. A. Shepherd (New York: W. W. Norton, 2006). See Morrison, *Excrement,* 141ff., for a fuller discussion of wasters in both *Piers* and *Wynnere and Wastoure.*
43. "Christianity replaced pagan expenditure prescribed by custom with voluntary alms, either in the form of distributions from the rich to the poor, or (and above all) in the form of extremely significant contributions to churches and later to monasteries." Bataille, *Visions of Excess,* 123–124.
44. Links between gold and excrement occur in discussions about alchemy. John Friedman in a personal communication points to the image of a man defecating in an alchemical context in Munich Staatsbibliothek CLM 25110, folio. 21v, reproduced by H. Buntz, "Europaische Alchimie vom 13 bis zum 18 Jahrhundert," in Emil Ernst Ploss, et al., *Alchimia*: *Ideologie und Technologie* (Munich: Heinz Moos Verlag, 1970), 197. See Munich, Bayerische Staatsbibliothek CLM 455, folio 135v, "Alchemist wiltu werden reich / So thu dein ding klugleich/Isth dreck vnd scheiss golt / So werden dir alle leut

holt." This is a variant of the proverbial image of a man excreting gold coins as depicted on a corbel in Kaiserworth House, Goslar, Germany ca. 1494. See Christa Grössinger, *Humour and Folly in Secular and Profane Prints of Northern Europe, 1430–1540* (London and Turnhout: Harvey Miller, 2002), fig. 2, p. 3. Like the Pardoner, Till Eulenspielgel at one point dresses up as a monk and wanders through the countryside with a skull he dug up from a graveyard, claiming it to be that of Saint Brendan. Then he proclaims that he is to collect money for a new church from all except those women who are adulteresses. "'And whoever here may be such women, let them stand back. For if they offer me something—those who are guilty of adultery—I won't take it, and they will be revealed in shame unto me! So—know yourselves!' Then he offered the people the head to be kissed, a smith's head, maybe, that he had stolen from a church graveyard." Paul Oppenheimern, trans., *Till Eulenspiegel: His Adventures* (New York: Routledge, 2001), 62. Needless to say, none of the women want to be considered adulterous, so they all push forward, donating plentiful money.

45. Bataille, *The Accursed Share*, 81.
46. Sophie Gee, *Making Waste: Leftovers and the Eighteenth-Century Imagination* (Princeton, NJ: Princeton Univerity Press, 2010), 82–83. "Life is a product of putrefaction, and it depends on both death and the dungheap." Bataille, *The Accursed Share*, 80. A legend from 1527 concerns the alchemist Paracelsus, who passed the rumor that he would expose the greatest secret of medicine to the esteemed and learned academics of Basle. The alchemist arrives in besmottered clothes and holds up a dish before the doctors and professors. When revealed, it contained a steaming human turd. Paracelsus mocks them for refusing to acknowledge that it is from decay that all birth originates. Nicolas Flamel (late fourteenth to early fifteenth centuries) links the Resurrection with corruption. The body must decompose in order to (eventually) rise again in glory. "In medieval alchemic thought, putrefaction was considered the necessary source of regeneration." Agostino Paravicini Bagliani, "The Corpse in the Middle Ages: The Problem of the Division of the Body," in *The Medieval World*, ed. Peter Linehan and Janet L. Nelson (London: Routledge, 2001), 341, 339.
47. Alan Dundes, *Life is Like a Chicken Coop Ladder: A Study of German National Character through Folklore* (Detroit, MI: Wayne State University Press, 1984), 13.
48. Oppenheimer, *Till Eulenspiegel*, 185–186.
49. See Keri Fitzgerald's "Maxims and the Waste of Words in Franklin's 'The Way to Wealth,'" Class paper for Honors 3396A, Texas State University, October 13, 2009.
50. Michael Camille, *Image on the Edge: The Margins of Medieval Art* (London: Reaktion, 1992), 36.
51. Bataille, *The Accursed Share*, 69.
52. Paul Fleischman, *Seedfolks* (New York: HarperTeen, 2004), 83.
53. "A swindler is a perpetrator of fraudulent schemes who takes money from behind a trusting person's back (Schwindler is derived from Schwein). The word links feces and money, collapsing the space between them (swine in mud, filthy rich, rolling in dough). The linkage of money with shit was widespread

at the outset of capitalism when the non-aristocratic rich were resented for displaying their fat wealth and piggish greed." Wayne Anderson, *Freud, Leonardo da Vinci, and the Vulture's Tail: A Refreshing Look at Leonardo's Sexuality* (New York: Other Press, 2001), 163.

54. Malia Wollan, "Vandals Create a Pungent Problem in San Francisco," *The New York Times*, January 21, 2009, accessed October 18, 2014, http://www.nytimes.com/2009/01/22/us/22potties.html.
55. Miles Wilson, "Everything," *Line of Fall* (Iowa City: University of Iowa Press, 1989), 21.
56. Ibid., 20. Bennett urges us not to mock the compulsive hoarder who reminds us of our own acquisitiveness, but to see him as one "unusually susceptible to the enchantment-powers of things." Jane Bennett, "Powers of the Hoard: Further Notes on Material Agency," in *Animal, Vegetable, Mineral: Ethics and Objects*, ed. Jeffrey Jerome Cohen (Washington, DC: Oliphaunt Books, 2012), 247. Geoff Sobell's play, *The Object Lesson*, performed at the Brooklyn Academy of Music in November 2014, is ripe for a vibrant materialist reading.
57. Wilson, "Everything," 20.
58. Ibid., 31–32.
59. Ibid., 19.
60. Ibid., 24.
61. Irving Wohlfarth, "Et Cetera? The Historian as Chiffonnier," *New German Critique* 39 (1986): 152.
62. Italo Calvino, "La Poubelle Agréée," in *The Road to San Giovanni*, trans. Tim Parks (New York: Pantheon, 1993), 103.
63. Frederick M. Biggs, "The Politics of Succession in *Beowulf* and Anglo-Saxon Succession," *Speculum* 80.3 (2005): 709–741.

6 Urban Myths: The Civilized and Pristine City-Body

1. Simin Davoudi, "Planning for Waste Management: Changing Discourses and Institutional Relationships," *Progress in Planning* 53 (2000): 170.
2. Ibid., 179.
3. Ibid., 182.
4. Ibid., 179.
5. Lewis Mumford, *The City in History: Its Origins, Its Transformations, and Its Prospects* (New York: Harcourt, Brace & World, 1961), 14.
6. "In human-made ecosystems, the inhabitants of the surrounding villages are the primary producers while city dwellers, despite their cultural sophistication, are mere consumers. Moreover, this parasitic relationship can be reproduced at a larger scale." Manuel De Landa, *A Thousand Years of Nonlinear History* (New York: Swerve Editions, 2000), 106.
7. "All agricultural communities subscribe to the Chinese proverb 'waste is treasure,' in which the least valued is the most valued: in this equation, shit is death which gives life." Martin Pops, "The Metamorphosis of Shit," *Salmagundi* 56 (1982): 50.
8. Donna Landry, "Mud, Blood, and Muck: Country Filth," *Genre* 27 (1994): 328.

9. P. M. Stell and Louise Hampson, *Probate Inventories of the York Diocese 1350–1500* (YML Occasional Series, Dean and Chapter of York, 1998), 227, 241.
10. Jan Karon, *These High, Green Hills* (New York: Penguin, 1996), 127.
11. Brett Clanton, "Alternative Energy Source?," *San Antonio Express News*, October 2, 2009, accessed November 17, 2014, http://www.mysanantonio.com/business/local/article/Alternative-energy-source-858249.php.
12. "Cooper Union Builds Brave New World: The Poetics of Sustainability," *The Cooper Union for the Advancement of Science and Art*, June 1, 2007, quoting Stan Allen, accessed March 19, 2012, http://cooper.edu/about/printed-publications/summer-2007/brave-new-world.
13. Statements made by Coventry council in the fifteenth century, for example, make clear that "the streets and river were overtly understood as a holistic environmental system: what happened upstream in the streets had a direct impact on the downstream river. The government therefore acted to carefully manage both parts of the system by controlling individual behavior." Dolly Jørgensen, "Cooperative Sanitation: Managing Streets and Gutters in Late Medieval England and Scandinavia," *Technology and Culture* 49 (2008): 558. Also see Valerie Allen and Ruth Evans (eds.), *Roadworks: Medieval Roads, Medieval Britain* (Manchester: Manchester University Press, 2015).
14. Ernest L. Sabine, "City Cleaning in Medieval London," *Speculum* 12 (1937): 38. Chaucer may have had some personal experience with sanitation issues. In July 12, 1389 it is recorded: "Writ of aid for Geoffrey Chaucer, appointed clerk of the works at the palace of Westminster, the Tower of London, the castle of Berkhampstede and the manors of Kenyngton, Eltham, Claryndon, Shene, Byflete, Childernelangeley and Feckenham, the lodge of Hathebergh in the New Forest and the lodges within the parks of Claryndon, Eltham, Childernelangeley and Feckenham, and the mews for falcons near Charryncrouch, as well as of the gardens, stanks, mills and enclosures thereof, for which he is to receive 2 *s.* a day." *Calendar of Patent Rolls, Richard II A.D. 1388–1392*, Vol. 4 (London: Mackie, 1902). Now "stanks" comes from French, meaning standing pools of water. In English the aural link between stanks and stinks is too tempting to avoid, as a quotation in the OED confirms. Surely sanitation issues were not unheard of for Chaucer, particularly given this duty. Standing water would be more likely to accumulate smells.
15. Ernest L. Sabine, "Latrines and Cesspools of Mediaeval London," *Speculum* 9 (1934): 321. Ieva Reklaityte's contention that Western Christian cities assumed the worst about people's behavior seems accurate, while, she shows, the medieval Muslim view assumed urbanites would act for the common good. Ieva Reklaityte, "The Hygienic Situation in Medieval Muslim Towns: A Case of al-Andalus," The Medieval Institute, Western Michigan University (May 2008).
16. Jørgensen, "Cooperative Sanitation," 560.
17. Ibid., 564.
18. Gay Hawkins, *The Ethics of Waste: How We Relate to Rubbish* (Lanham, MD: Rowman & Littlefield, 2006), 103.
19. Daniel Lang, "'Give Us the Dumpsters—Or—Give Us Life': *Res Derilictae* and the Trash of Free Trade," *Other Voices* 3.1 (2007: http://www.othervoices.org/3.1

/dlang/index.php (accessed April 18, 2012). These activists do not reject recycling, but "the notion of waste, garbage, or trash on which the concept and practice of 'recycling' is based."

20. Francis X. Clines, "Reimagining What Washes Up," *The New York Times*, March 25, 2010, accessed October 18, 2014, http://www.nytimes.com/2010/03/26/opinion/26fri4.html. "Full social humanity radically excludes the disorder of the senses; it negates its natural principle; it rejects this given and allows only the clean space of a house, of polished floors, furniture, window panes, a space inhabited by venerable persons, at once naive and inviolable, tender and inaccessible." Georges Bataille, *The Accursed Share: An Essay on General Economy, Vol. II: The History of Eroticism, Vol. III: Sovereignty*, trans. Robert Hurley (New York: Zone Books, 1993), 56.
21. John Scanlan, *On Garbage* (London: Reaktion, 2005), 162. The nighttime city lurks in shadow with its invisible garbage collectors and the "spectre of garbage" (157).
22. Ibid., 155.
23. Elizabeth Grosz argues that "the city provides the order and organization that automatically links otherwise unrelated bodies: it is the condition and milieu in which corporeality is socially, sexually, and discursively produced." *Space, Time, and Perversion: Essays on the Politics of Bodies* (New York and London: Routledge, 1995), 104.
24. Eustache Deschamps, "Ballade Showing the Causes and Reasons for the Plague," trans. Anne M. Dropick, *Harper's Magazine*, March 2007: 16.
25. Oxford, Bodleian Library, MS Bodley Rolls 5, membrane 3.
26. Peter Suchsland and Erika Weber, *Deutsche Volksbücher in Drei Bänden. Zweiter Band* (Berlin/Weimar: Aufbau Verlag, 1968), 78–79; Paul Oppenheimer (trans.), *Till Eulenspiegel: His Adventures* (New York: Routledge, 2001), 94–95. In an English translation from the early sixteenth century, Eulenspiegel or Howleglas is said to have "solde turdes for fat" and buys the excrement from "turd farmers." Friedrich W. D. Brie, *Eulenspiegel in England* (Berlin: Mayer & Müller, 1903), 127. See also *Till Eulenspiegel A Merye Jest* [London (1528)], Rpt. Amsterdam: Da Capo Press, 1971; the original printed version has "gong fermers" and "turdes." Eulenspiegel is first mentioned in the OED under "Owlglass"; spellings include Howleglas and Holeglas.
27. Ibid., 116; 144.
28. Ibid., 146; 180.
29. Ibid., 107; 132.
30. Michael Thompson, *Rubbish Theory: The Creation and Destruction of Value* (Oxford: Oxford University Press, 1979), 5, cites Jonathan Swift's "A digression concerning criticks" in *A Tale of a Tub*, reprinted 1949 (London: Nonesuch Press, 1696), 308.
31. "The moment we stopped removing ourselves from that waste, it had to be removed from us. Thus the origins of civilization . . . A civilization that cannot escape its own fecal matter is a civilization in trouble—unless, of course, the uneasy relationship between man and his effluents can evolve." Frederick Kaufman, "Wasteland: A Journey through the American Cloaca," *Harper's Magazine*, February 2008: 46.

32. Susan Strasser, *Waste and Want: A Social History of Trash* (New York: Henry Holt, 2000), 14.
33. In the late medieval city, a rapid increase in urbanization meant that excrement could no longer be harmonized with the environment, though Edward D. Melillo argues for the nineteenth century as the time when urban waste, formerly so vital to rural crops, became waste. Edward D. Melillo, "Nutrient rifts," accessed October 18, 2014, http://discardstudies.com/discard-studies-compendium/.
34. Dominique Laporte, *History of Shit*, trans. N. Benabid and R. El-Khoury (Cambridge, MA: MIT Press, 2000), 15, 17–18.
35. Ibid., 2. Every house must now install a cesspool. This edict suggests how the "disciplinary effects of [the State's] inquisitional gaze" insisted on the removal of "excrement from sight" (63). There was increasingly a general tendency to put noisy and smelly activities far from the political and religious areas. Jean-Pierre Leguay, *La Pollution au Moyen Âge*, 4th ed. (Paris: Editions Jen-Paul Gisserot, 2005), 52.
36. Hawkins, *The Ethics of Waste*, 50.
37. Gay Hawkins, "Down the Drain: Shit and the Politics of Disturbance," in *Culture and Waste: The Creation and Destruction of Value*, ed. Gay Hawkins and Stephen Muecke (Lanham, MD: Rowman & Littlefield, 2003), 40. "In cultures that pride themselves on being technologically 'advanced' catching a glimpse of the brute physicality of waste signals a kind of failure." Hawkins, *The Ethics of Waste*, 1.
38. "All those other spaces for things we don't want to face—prisons, madhouses, hospitals, dumps, *drains*—remind us of the place of secrecy in public knowledge, or the force of the hidden and its role in political authority and social order." Hawkins, "Down the Drain," 41.
39. Hawkins emphasizes the "important place of waste in the formation of the modern subject." *The Ethics of Waste*, 47.
40. Hawkins, "Down the Drain," 45–46. Also see Cyrille Harpet, *De Déchet: Philosophie des Immondices Corps, Ville, Industrie* (Paris: L'Harmattan, 1998), 238.
41. Scanlan, *On Garbage*, 8, 13, 15, 120.
42. Ibid., 9.
43. Italo Calvino, "La Poubelle Agréée," in *The Road to San Giovanni*, trans. Tim Parks (New York: Pantheon, 1993) 101.
44. Ibid., 105.
45. Ibid., 107.
46. Ibid., 97–98.
47. Ibid., 101.
48. Ibid., 114.
49. Ibid., 96.
50. Jane Bennett, "The Force of Things: Steps toward an Ecology of Matter," *Political Theory* 32.3 (June 2004): 350.
51. Ibid., 354.
52. Ibid., 365. Concurring, Matthew Zantingh argues that a "new relationship to things" may help "bodies" hurt by, for example, the toxicity in e-waste.

"When Things Act Up: Thing Theory, Actor-Network Theory, and Toxic Discourse in Rita Wong's Poetry," *Interdisciplinary Studies in Literature and Environment* 20.3 (2013): 638.

53. Elisabeth Malkin, "For Some in Mexico, Trash Is a Treasure Worth Defending," *The New York Times*, February 16, 2012, accessed October 18, 2014, http://www.nytimes.com/2012/02/17/world/americas/for-some-in-mexico-trash-is-a-treasure-worth-defending.html?pagewanted=all.
54. Barrie M. Ratcliffe, "Perceptions and Realities of the Urban Magic: The Rag Pickers of Paris in the First Half of the Nineteenth Century," *Canadian Journal of History* 27 (1992): 215, also 200; 213.
55. Ibid., 201.
56. Ibid., 226.
57. Ibid., 233. Indeed Hubert Zapf suggests the valley of the ashes has become "omnipresent" in Don DeLillo's *Underworld*. "Creative Matter and Creative Mind: Cultural Ecology and Literary Creativity," in *Material Ecocriticism*, ed. Serenella Iovino and Serpil Oppermann (Bloomington: Indiana University Press, 2014), 64.
58. "Slum tourism, or 'poorism,' is catching on." Critics of such tours argue that it "is to make Westerners feel better about their station in life." Others argue that it forces us to acknowledge our—all of our—humanity. Eric Weiner, "Slum Visits: Tourism or Voyeurism?," *The New York Times*, March 9, 2008, accessed October 18, 2014, http://www.nytimes.com/2008/03/09/travel/09heads.html.
59. Zygmunt Bauman, *Wasted Lives: Modernity and Its Outcasts* (Cambridge: Polity Press, 2004), 66.
60. Scanlan, *On Garbage*, 171.
61. Nicholas D. Kristof, "Where Sweatshops Are a Dream," *The New York Times*, January 14, 2009, accessed April 8, 2010, http://www.nytimes.com/2009/01/15/opinion/15kristof.html.
62. H. G. Adler, *The Journey: A Novel*, trans. Peter Filkins, Afterword by Jeremy Adler (New York: Random House, 2008), 231.
63. Ibid., 237.
64. Ibid., 238.
65. Peter Filkins, "Introduction," in *The Journey: A Novel*, trans. Peter Filkins, Afterword by Jeremy Adler (New York: Random House, 2008), xi. Heinrich Böll argues, "It only takes someone like Adler to describe something as seemingly harmless as rubbish collecting in order to reveal the uncanny." Heinrich Böll, *Frankfurt Lecture on Poetics*, quoted by Jeremy Adler in Afterword, Adler, *The Journey*, 291.
66. Adler, 85.
67. Ibid., 186. Serenella Iovino argues that cities "function as if they were huge natural organisms." "Naples 2008, or, the Waste Land: Trash, Citizenship, and an Ethic of Narration," *Neohelicon* 36 (2009): 335. See also to Serenella Iovino, "Bodies of Naples: Stories, Matter, and the Landscapes of Porosity," in *Material Ecocriticism*, ed. Serenella Iovino and Serpil Oppermann (Bloomington: Indiana University Press, 2014), 97–113.
68. Adler, *The Journey*, 66.
69. Ibid., 67.

70. The *zabeleen* in Cairo, who pick up and recycle garbage, are far from anonymous. Peter Hessler, "Tales of the Trash," *The New Yorker*, October 13, 2014, accessed October 15, 2014, http://www.newyorker.com/magazine/2014/10/13/tales-trash.

7 Interiorized Waste: Sin and Metaphysical Meaninglessness

1. Julia Kristeva, *Powers of Horror: An Essay on Abjection*, trans. Leon S. Roudiez (New York: Columbia University Press, 1987), 113.
2. Ibid., 114–115.
3. Lothario Dei Segni (Pope Innocent III), *On the Misery of the Human Condition: De miseria humane conditionis*, ed. Donald R. Howard, trans. Margaret Mary Dietz (Indianapolis: Bobbs-Merrill, 1969), 3.
4. Ibid., 6.
5. Ibid., 8.
6. Ibid., 67–68.
7. Ibid., 27.
8. Ibid., 12.
9. Ibid., 24.
10. Ibid., 61–62.
11. Ibid., 46.
12. Martha Bayless, *Sin and Filth in Medieval Culture: The Devil in the Latrine* (New York: Routledge, 2012), 27.
13. Dei Segni, *On the Misery of the Human Condition*, 85.
14. Guillaume de Deguileville, *The Pilgrimage of Human Life*, trans. Eugene Clasby (New York: Garland, 1992), 44. Compare with Tertullian's passage about the offenses of makeup in Alcuin Blamires, ed., *Woman Defamed and Woman Defended: An Anthology of Medieval Texts* (Oxford: Clarendon Press, 1992), 53.
15. All Chaucer references to Larry D. Benson, *The Riverside Chaucer*, 3rd ed. (Boston: Houghton Mifflin, 1987). See also "The Parson's Tale" in *The Canterbury Tales*, X.839–840 and X.848, 850, in Larry D. Benson, *The Riverside Chaucer*.
16. McAvoy reads the bear as a symbol for priestly hypocrisy and corruption, which transforms "blooms" of virtuous living into sin and corruption. Liz Herbert McAvoy, *Authority and the Female Body in the Writings of Julian of Norwich and Margery Kempe* (Cambridge: D. S. Brewer, 2004), 139, n. 37.
17. "Sin stank; and therefore hell was imagined as an enormous privy." William Ian Miller, *The Anatomy of Disgust* (Cambridge, MA: Harvard University Press, 1997), 151.
18. Durling describes this as "the belly of Hell, [where the] Malebolge . . . represent perhaps the most gigantic case of constipation on record." Robert M. Durling, "Deceit and Digestion in the Belly of Hell," in *Allegory and Representation: Selected Papers from the English Institute, 1979–80*, ed. Stephen J. Greenblatt (Baltimore, MD: Johns Hopkins University Press, 1981), 65.

19. Although waste could be used in the rural economy, getting it there was an issue. "Even in normal times, however, the removal of privy filth from mediaeval London was bound to give rise to frequent nuisances ... Too much was left to the initiative of the individual and too little responsibility was taken in the way of close and constant supervision by the city, to prevent lazy and careless people from frequently dumping privy, and other, filth into out-of-the-way suburban lanes, the city water-courses, or along the banks of the Thames." Ernest L. Sabine, "Latrines and Cesspools of Mediaeval London," *Speculum* 9 (1934): 321.
20. Dolly Jørgensen, "'All Good Rule of the Citee': Sanitation and Civic Government in England, 1400–1600," *Journal of Urban History* 36 (2010): 303.
21. David N. DeVries, "And Away Go Troubles Down the Drain: Late Medieval London and the Poetics of Urban Renewal," *Exemplaria* 8 (1996): 417. See also Dolly Jørgensen, "The Metamorphosis of Ajax," *Early English Studies* 3 (2010): 17. "A well-flushed city is finally a matter of politics as well as hygiene ... Purging itself of rot and ordure, the city maintains itself as a healthy body politic." Paul Strohm, "Sovereignty and Sewage," in *Lydgate Matters: Poetry and Material Culture in the Fifteenth Century*, ed. Lisa H. Cooper and Andrea Denny-Brown (New York: Palgrave Macmillan, 2007), 62.
22. Dei Segni, *On the Misery of the Human Condition*, 50.
23. See Helmut Puff, "The Sodomite's Clothes: Gift-Giving and Sexual Excess in Early Modern Germany and Switzerland," in *The Material Culture of Sex, Procreation, and Marriage in Premodern Europe*, ed. Anne L. McClanan and Karen Rosoff Encarnación (New York: Palgrave, 2002), 252.
24. As Allen Frantzen has pointed out, "The problem with sodomy is not only sex, therefore, but also animality, anality, excrement, the stuff of the Dead Sea—in short, the worst stuff of medieval life." In *Cleanness*, there is an affiliation between "the Sodomites' sexual act with anal filth." Michel Foucault, "The Disclosure of Sodomy in *Cleanness*," *PMLA* 111 (1996): 462.
25. "The anus is seen as the footing on which our dignity depends. It must be secured or everything else built upon it crumbles. For this reason, however, the anus is also a temptation." Tiffany Beechy, "Devil Take the Hindmost: Chaucer, John Gay, and the Pecuniary Anus," *The Chaucer Review* 41.1 (2006): 73. She cites Jeffrey Masten, "Is the Fundament a Grave?" in *The Body in Parts: Fantasies of Corporeality in Early Modern Europe*, ed. David Hillman and Carla Mazzio (New York: Routledge, 1997), 129–145. See also Miller, *The Anatomy of Disgust*, 100–101. "The asshole is the meeting-place of the human and the demonic (in the gross materiality of shit)." Martin Pops, "The Metamorphosis of Shit," *Salmagundi* 56 (1982): 39.
26. Allen J. Frantzen, "The Disclosure of Sodomy in *Cleanness*," *PMLA* 111 (1996): 460–461.
27. Jerome J. Citrome, "Medicine as Metaphor in the Middle English *Cleanness*," *The Chaucer Review* 35 (2001): 266.
28. Middle English, from J. J. Anderson, *Sir Gawain and the Green Knight, Pearl, Cleanness, Patience* (London: Everyman Paperback Classics, 1996). All

translations from *Cleanness*, trans. and ed. Kevin Gustafson (Buffalo, NY: Broadview, 2010).

29. "It is through abolishment of dietary taboos, partaking of food with pagans, verbal and gestural contact with lepers, as well as through its power over impure spirits that the message of Christ is characterized." Kristeva, *Powers of Horror*, 113.
30. While the Old Testament is associated with surgical excision, the New Testament is associated with the higher mode of medical treatment, with Christ himself as the ideal leche. See Citrome, "Medicine as Metaphor," 261.
31. *Mankind* quotes from Douglas Bruster and Eric Rasmussen (eds.), *Everyman and Mankind*, Arden Early Modern Drama (London: Methuen Drama, 2009), 171.
32. "Bodiliness provides access to the sacred." Caroline Walker Bynum, *Fragmentation and Redemption Essays on Gender and the Human Body in Medieval Religion* (New York: Zone Books, 1991), 186.
33. Guillaume de Deguileville, *The Pilgrimage of Human Life*, trans. by Eugene Clasby (New York: Garland, 1992), 32. Lydgate's version has her say: "I am callyd Dame Penaunce. / The cheff wardeyn (who lyst se,) / Off thylkë ylë most secre; / The wych (who espyë kan,) / Ys yhyd with-Inne a man. / I make yt clene (I yow ensure,) / Off allë fylthe & al ordure." John Lydgate, *The Pilgrimage of the Life of Man*, ed. F. J. Furnivall and Katharine B. Locock, *EETS* 77, 83, 92 (1899, 1901, 1904), reprinted as one volume (Woodbridge, Suffolk: Boydell & Brewer, 1996), lines 4054–4060.
34. See Jeremy J. Citrome, *The Surgeon in Medieval English Literature* (New York: Palgrave Macmillan, 2006), 106, 109, concerning excremental language in the confessional exchange.
35. The wine, rather than a sign of consecretion, is one of "desecration." Gerard Kilroy, "Requiem for a Prince: Rites of Memory in *Hamlet*," in *Theatre and Religion: Lancastrian Shakespeare*, ed. Richard Dutton, Alison Findlay, and Richard Wilson (Manchester: Manchester University Press, 2003), 151–152.
36. See Jacqueline Vanhoutte, "Denmark's Rotting Reconsidered," *Philological Quarterly* 91:3 (2012): 409.
37. Elizabeth V. Spelman, "Combing Through Trash: Philosophy Goes Rummaging," *The Massachusetts Review* 52.2 (2011): 323.
38. See Vanhoutte, "Denmark's Rotting," who brilliantly traces the leprous undersores of *Hamlet*.
39. As Hawkins has argued, "Making ourselves clean is ethical work; we are transforming the body in relation to a wider moral ethos." Indeed, "most of us are far cleaner than we need to be." Gay Hawkins, *The Ethics of Waste: How We Relate to Rubbish* (Lanham, MD: Rowman & Littlefield, 2006), 58.
40. John Scanlan, *On Garbage* (London: Reaktion, 2005), 25.
41. Lawrence Babb, *The Elizabethan Malady: A Study of Melancholia in English Literature from 1580 to 1642* (East Lansing: Michigan State College Press, 1951, rpt. 1965), 16. Melancholy, related to *acedia* or sloth, translates into boredom in modern society. Harvie Ferguson, "Exteriority: Boredom, Disgust and the Margins of Humanity," in *Aesthetic Fatigue: Modernity and the*

Language of Waste, ed. John Scanlan and John F. M. Clark (Newcastle upon Tyne: Cambridge Scholars, 2013), 225.

42. Fecal dissidence "unsettles the very opposition between the dominant and the subordinate." Jonathan Dollimore, *Sexual Dissidence: Augustine to Wilde, Freud to Foucault* (Oxford: Clarendon Press, 1991), 21. Here I play with Dollimore's phrase of "sexual dissidence."
43. Peter Boxall, "'There's No Lack of Void': Waste and Abundance in Beckett and DeLillo," *SubStance* 37.2 (2008): 58. Hamm accuses Clov of polluting "the air." Samuel Beckett, *Endgame* (New York: Grove Press, 1958), 3.
44. "What Beckett offers in the way of philosophy he himself also reduces to culture-trash." Theodor W. Adorno, "Trying to Understand *Endgame*," trans. Michael T. Jones, *New German Critique, Critical Theory and Modernity* 26 (Spring–Summer 1982): 119. As Boxall points out in his reading juxtaposing Beckett and DeLillo, a major structure of contemporary life sets up an "opposition between Waste and abundance." Boxall, "There's No Lack of Void," 56. He asks, provocatively, "But what if waste and abundance are in fact no oppositions at all?" (57). Beckett has a long history of fascination with waste, though Édouard Magessa O'Reilly has shown how Beckett toned down the scatological elements originally in works such as *Molloy*, part II, and the radio play *All That Fall* (1956). Édouard Magessa O'Reilly, "*Molloy*, Part II, Where the Shit Hits the Fan," *Genetic Joyce Studies* 6 (2006). http://www.geneticjoycestudies.org/GJS6/GJS6OReilly.htm, accessed April 6, 2010. See also J. D. Esty, "Excremental Postcolonialism," *Contemporary Literature* 40 (1999): 22–59.
45. Beckett, *Endgame*, 4, 8, 9, 13.
46. Boxall, "There's No Lack of Void," 57.
47. Beckett, *Endgame*, 5.
48. Ibid., 44.
49. Ibid., 46. As Mephistopheles observes, "'Why, this is hell, nor am I out of it . . . Hell hath no limits, nor is circumscrib'd / In one self place, but where we [the damned] are is hell, / And where hell is must we ever be.' For those who reject heaven, hell is everywhere, and thus is limitless. For them, even the thought of heaven is hell." Cited in Wendell Berry, "Faustian Economics: Hell Hath No Limits," *Harper's Magazine* May 2008: 38–39, accessed November 16, 2009 at http://www.harpers.org/archive/2008/05/0082022.
50. Beckett, *Endgame*, 26.
51. Ibid., 36.
52. Ibid., 31.
53. Will Viney, *Waste: A Philosophy of Things* (London: Bloomsbury, 2014), 20. As Viney points out, the *tōhû wābōhû* of *Genesis* falls outside of human time amd use (21).
54. "God's creation from Chaos is a creation from waste." Sophie Gee, *Making Waste: Leftovers and the Eighteenth-Century Imagination* (Princeton, NJ: Princeton Univerity Press, 2010), 86.
55. Beckett, *Endgame*, 57.
56. Ibid., 32.
57. Ibid., 77.
58. Ibid., 62.

59. Ibid., 68.
60. Ibid., 69.
61. Ibid., 6.
62. Ibid., 82. As Levinas asserts, nausea "is the affirmation itself of being . . . It is the impotence of pure being, in all its nakedness." Emmanuel Levinas, *On Escape: De l'évasion*, trans. Bettina Bergo (Stanford, CA: Stanford University Press, 2003), 68.
63. Beckett, *Endgame*, 55.
64. Ibid., 83.
65. Ibid., 84.

8 The Toxic Metaphor of Wasted Humans: Those Filthy Cleaners Who Scrub Us Spotless

1. Italo Calvino, "La Poubelle Agréée," in *The Road to San Giovanni*, trans. Tim Parks (New York: Pantheon, 1993), 109.
2. 24 March–31 August 2011. From a plaque at "The Community: New Delhi and Kolkata, 2011." "For man, when he is most particular to be clean, will despise those by whose labours cleanliness is maintained." Reginald Reynolds, *Cleanliness and Godliness* (Garden City, NY: Doubleday, 1946), 31.
3. George Lakoff and Mark Johnson, *Metaphors We Live By*, 2nd ed. (Chicago, IL: University of Chicago Press, 2003), 16.
4. Ibid., 29.
5. Sandra Blakeslee, "Mind Games: Sometimes a White Coat Isn't Just a White Coat," *The New York Times*, April 3, 2012, accessed April 4, 2012, http://www.nytimes.com/2012/04/03/science/clothes-and-self-perception.html.
6. Lakoff and Johnson, *Metaphors We Live By*, 243.
7. Martha C. Nussbaum, *From Disgust to Humanity: Sexual Orientation and Constitutional Law* (Oxford/New York: Oxford University Press, 2010), xiii.
8. Ibid., 24.
9. Martha C. Nussbaum, *Upheavals of Thought: The Intelligence of Emotions* (Cambridge: Cambridge University Press, 2001), 204.
10. Ibid., 205.
11. Nussbaum, *From Disgust to Humanity*, xv. As Dundes points out, it is a common trope to defame minority groups with the insult of filth. Alan Dundes, *Life Is Like a Chicken Coop Ladder: A Study of German National Character through Folklore* (Detroit: Wayne State University Press, 1984), 119.
12. Nussbaum, *Upheavals of Thought*, 347.
13. Mary Douglas, "Jokes," *Implicit Meanings: Essays in Anthropology* (London: Routledge & Kegan Paul, 1975), 102.
14. The show "is a working-class fantasy that bursts its own illusions." Alessandra Stanley, "Undercover Boss: He's Good at Pushing Paper, but Can He Pick Up Trash?" *The New York Times*, February 6, 2010, accessed February 8, 2010, http://tv.nytimes.com/2010/02/06/arts/television/06under.html?_r=1.
15. Calvino, "La Poubelle Agréée," 108. Not limited to Western society, in Yemen the "'Al Akhdam'—the servants"—prefer to be called "Al Muhamascheen"—the

marginalized ones. Only allowed to beg or sweep streets, they live in "fetid slums on the edge of town" and are the hereditary lowest caste. "Degrading myths pursue them: they eat their own dead, and their women are all prostitutes. Worst of all, they are reviled as outsiders in their own country, descendants of an Ethiopian army that is said to have crossed the Red Sea to oppress Yemen before the arrival of Islam." Robert F. Worth, "Despite Caste-Less Society in Yemen, Generations Languish at Bottom of Ladder," *The New York Times,* February 27, 2008, accessed October 18, 2014, http://www.nytimes.com/2008/02/27/world/africa/27iht-27yemen.10456434.html?pagewanted=all.

16. Amelia Gentleman, "Picking Up Trash by Hand, and Yearning for Dignity," *The New York Times*, September 27, 2007, accessed November 23, 2014, http://www.nytimes.com/2007/09/27/world/asia/27ragpickers.html?pagewanted=print.
17. J. F. M. Clark, "'The Incineration of Refuse is Beautiful': Torquay and the Introduction of Municipal Refuse Destructors," *Urban History* 34 (2007): 258, citing Henry Mayhew, *London Labour and London Poor*, ed. Victor Neuburg (London: Penguin, 1985), 230.
18. "Because of their living conditions, their appearance, their clothes, their smell." Because they were "peripatetic" and without a definite location, they were always suspect. One document by "the Prefect of Police" in 1828 writes they are "*sans domicile fixe*." Barrie M. Ratcliffe, "Perceptions and Realities of the Urban Magic: The Rag Pickers of Paris in the First Half of the Nineteenth Century," *Canadian Journal of History* 27 (1992): 204.
19. Irving Wohlfarth, "Et Cetera? The Historian as Chiffonnier," *New German Critique* 39 (1986): 157.
20. Ibid., 151.
21. Ratcliffe, "Perceptions and Realities," 210.
22. Ibid., 203. In order to understand the ragpicker, they were made exotic to tame them, while others wrote about them as "noble savages" (207–208). Mohammed Rafi Arefin criticizes Peter Hessler's piece on Cairo's *zabeleen* for its orientalism. "The Dirty Details: A Response to 'Tales of the Trash,' accessed November 10, 2014, http://discardstudies.com/2014/11/03/the-dirty-details-a-response-to-tales-of-the-trash/.
23. Calvino, "La Poubelle Agréée," 110.
24. John Scanlan, *On Garbage* (London: Reaktion, 2005), 169.
25. H. G. Adler, *The Journey: A Novel*, trans. Peter Filkins, Afterword by Jeremy Adler (New York: Random House, 2008), 66.
26. F. Scott Fitzgerald, *The Great Gatsby* (New York: Scribner, 1925/2004), 99.
27. Calvino, "La Poubelle Agréée," 99.
28. "Will there be enough dustmen, collectors of the garbage which 'our way of life' daily spawns, or—as Richard Rorty asks—a sufficient number of 'people who get their hands dirty cleaning our toilets' while being paid ten times less than we 'who sit behind desks and punch keyboards?'" Zygmunt Bauman, *Wasted Lives: Modernity and Its Outcasts* (Cambridge: Polity Press, 2004), 45.
29. Myra J. Hird, "Waste Flows," accessed November 10, 2014, http://discardstudies.com/discard-studies-compendium/.
30. Ibid.

31. Alexa Bingham, "Discourse of the Dammed: A Study of the Impacts of Sustainable Development Discourse on Indigenous Peoples in the Brazilian Amazon in the Context of the Proposed Belo Monte Hydroelectric Dam," *POLIS Journal* 4 (2010): 28.
32. Ibid., 9, 14.
33. Nigel Clark and Myra J. Hird, "Deep Shit," *O-Zone: A Journal of Object-Oriented Studies* 1 (2014): 45.
34. Zsuzsa Gille, *From the Cult of Waste to the Trash Heap of History: The Politics of Waste in Socialist and Postsocialist Hungary* (Bloomington: Indiana University Press, 2007), 22.
35. John Frow, "A Pebble, a Camera, a Man Who Turns into a Telegraph Pole," *Critical Inquiry* 28.1 (2001): 283.
36. Gille, *From the Cult of Waste*, 23.
37. Bauman, *Wasted Lives*, 44.
38. Nussbaum, *Upheavals of Thought*, 347.
39. Bauman, *Wasted Lives*, 5.
40. Ibid., 12.
41. Ibid., 66.
42. Nussbaum, *From Disgust to Humanity*, 11.
43. Ibid., 20.
44. Ibid., 16.
45. Timothy Morton, "The Liminal Space between Things: Epiphany and the Physical," in *Material Ecocriticism*, ed. Serenella Iovino and Serpil Oppermann (Bloomington: Indiana University Press, 2014), 276.
46. Bauman, *Wasted Lives*, 40–41, citing Stefan Czarnowski.
47. Ibid., 45.
48. Naomi Quinn, "The Cultural Basis of Metaphor," in *Beyond Metaphor: The Theory of Tropes in Anthropology*, ed. James W. Fernandez (Stanford, CA: Stanford University Press, 1991), 58.
49. Ibid., 65, 76–77.
50. See Edward Rothstein, "In a Collection of Memorabilia, Politics at Its Most Boisterous," *The New York Times*, June 28, 2008, accessed October 18, 2014, http://www.nytimes.com/2008/06/28/arts/design/28muse.html?pagewanted=print; and Dan Bilefsky, "With Sharp Satire, Enfant Terrible Challenges Czech Identity," *The New York Times*, September 4, 2009, accessed October 18, 2014, http://www.nytimes.com/2009/09/05/world/europe/05cerny.html.
51. "This gesture and the words that accompany it are based on a literal debasement in terms of the topography of the body, that is, a reference to the bodily lower stratum, the zone of the genital organs. This signifies destruction, a grave for the one who is debased." Mikhael Bakhtin, *Rabelais and His World*, trans. Héléne Iswolsky (Bloomington: Indiana University Press, 1968/1984), 148.
52. "Hanc ad munditiem adde mundiorem, / quod culus tibi purior salillo est, / nec toto decies cacas in anno; / atque id durius est faba et lapillis, / quod tu si manibus terras fricesque, / non umquam digitum inquinare posses." Many thanks to Leo Landrey for this translation. He suggests variant options for the first line "elegance/more elegant": "cleanness/cleaner," "manners/better

mannered," or "taste/more tasteful." All these alternatives suggest a division between what is privileged and disdained in the culture.

53. Ruth Mazo Karras, *From Boys to Men: Formations of Masculinity in Late Medieval Europe* (Philadelphia: University of Pennsylvania Press, 2003), 100–108.
54. Jonathan Swift, *Gulliver's Travels*, ed. Philip Pinkus (Toronto: Macmillan, 1968), 252.
55. Although not that uncommon among animals, coprophagy is generally looked down upon by humans. Ralph A. Lewin, *Merde: Excursions in Scientific, Cultural, and Sociohistorical Coprology* (New York: Random House, 1999), 95.
56. "Group hatred and the oppression of groups is very often based on a failure to individualize... portraying the group as altogether subhuman." Martha C. Nussbaum, *Poetic Justice: The Literary Imagination and Public Life* (Boston: Beacon Press, 1995), 92.
57. Sigmund Freud, *The Standard Edition of the Complete Psychological Works of Sigmund Freud, Totem and Taboo and Other Works*, vol. 13 (London: Hogarth Press, 1957), 19.
58. Chaucer clearly knew Jephthah, citing his tragic vow to God to kill the first being to emerge from his house should he win in battle. That person is none other than his only daughter, whom he then must execute. Chaucer cites this in *The Physician's Tale* (VI.240) in *The Riverside Chaucer*, ed. Larry D. Benson, 3rd ed. (Boston: Houghton Mifflin, 1987).
59. Jonathan Hsy, *Trading Tongues: Merchant, Multilingualism, and Medieval Literature* (Columbus, OH: Ohio State University Press, 2013), 18–19.
60. As John Clark points out, "Recent studies of environmental justice in the United States have found that incinerators are disproportionately overrepresented in areas populated by socially marginalized groups." Clark, "The Incineration," 262.
61. "And just as in the natural body the physiological danger points were at the joints, where member met member, or at the openings where the body could be invaded by harmful influences from without, so in the social body tensions arose at the jointures which linked group to group, class to class; and in the social body too there were the openings through which might pour invasion from without." Mervyn James, "Ritual, Drama and the Social Body in the Late Medieval English Town," *Past & Present* 98 (1983): 7.
62. "In the positioning of medieval hospitals, stigma was used to protect and define the boundaries of towns, and the boundaries between social and economic groups... The stigmatized body, in particular that of the leper, took on a transitional identity appropriate to this liminality... This liminal area was used to control and observe stigmatised groups, such as the poor, aged and infirm, but was in addition used to *display* bodies classified through stigma." Roberta Gilchrist, "Medieval Bodies in the Material World: Gender, Stigma and the Body," in *Framing Medieval Bodies*, ed. Sarah Kay and Miri Rubin (Manchester: Manchester University Press, 1994): 47–49.
63. Susan Signe Morrison, *Excrement in the Late Middle Ages: Sacred Filth and Chaucer's Fecopoetics* (New York: Palgrave Macmillan, 2008), 69–72.

64. Fitzgerald, *The Great Gatsby*, 124. Jennifer Gabrys, "Cité Multimédia: Noise and Contamination in the Information City," Paper delivered at the Visual Knowledges Conference, University of Edinburgh, September 17–20, 2003. "Peripheral zones always exist in complement to centers."
65. Rex Ferguson, "Garbage and Gatsby," in *Aesthetic Fatigue: Modernity and the Language of Waste*, ed. John Scanlan and John F. M. Clark (Newcastle upon Tyne: Cambridge Scholars, 2013), 211.
66. Paul Fleischman, *Seedfolks* (New York: HarperTeen, 2004), 25.
67. Ibid., 26. Trash, of course, never disappears. It may transform, but never vanishes in a magic puff of smoke.
68. Fleischman, *Seedfolks*, 28.
69. Katherine Boo, *Behind the Beautiful Forevers: Life, Death, and Hope in a Mumbai Undercity* (New York: Random House, 2012), 237. See also Rose George, *The Big Necessity: Adventures in the World of Human Waste* (New York: Metropolitan Books, 2008).
70. Richard Jago, "The Scavengers: A Town Eclogue," in *A Book of English Pastoral Verse*, ed. John Barrell and John Bull (New York: Oxford University Press, 1975), 314–316.
71. Fleischman, *Seedfolks*, 33.
72. Ibid., 34.
73. Ibid., 35.
74. Calvino, "La Poubelle Agréée," 110.
75. Ibid., 111.
76. Ibid., 108–110.
77. Ibid., 110.
78. Daniel Lang has argued the legal restrictions on dumpster-diving intend to force "practitioners" to become "consumers." Lang, "'Give Us the Dumpsters—Or—Give Us Life': *Res Derilictae* and the Trash of Free Trade," *Other Voices* 3.1 (2007): http://www.othervoices.org/3.1/dlang/index.php, accessed April 18, 2012.
79. Christine Temko, "Regulation and Refuse Matter in Don DeLillo's *Underworld* and Eugene Marten's *Waste*," *Interdisciplinary Studies in Literature and Environment* 20.3 (2013): 495. She cites Todd McGowen, "The Obsolescence of Mystery and the Accumulation of Waste in Don DeLillo's *Underworld*," *Critique: Studies in Contemporary Fiction* 46.2 (Winter 2005): 136. Waste "has no proper place within the prevailing socioeconomic structure." Temko, "Regulation," 498.
80. Adler, *The Journey*, 96–97.
81. Kristeva, *Powers of Horror*, 8.
82. Cotton Vitellius A XV ff. 98v–106v.
83. Michael Swanton, *Anglo-Saxon Prose* (London: Everyman, 1993), 231–232.
84. See Nussbaum, *Upheavals of Thought*, 203.
85. "Immigrants, and particularly the fresh arrivals among them, exude the faint odour of the waste disposal tip which in its many disguises haunts the nights of the prospective casualties of rising vulnerability. For their detractors and haters, immigrants embody—visibly, tangibly, in the flesh—the inarticulate yet hurtful and painful presentiment of their own disposability." Bauman,

Wasted Lives, 56. This parallels Lennard Davis's point about disabled bodies reminding "able bodied" folks of their inevitable decay and fragmentation.

86. Ibid., 38, quoting Charles Darwin. See also William A. Cohen, "Deep Skin," in *Thinking the Limits of the Body*, ed. Jeffrey J. Cohen and Gail Weiss (Albany: SUNY Press, 2003), 63–82 on the nexus of metropolitan filth and the colonial subject in Victorian England.
87. Bingham, "Discourse of the Dammed," 9, 14. She cites M. Shamsul Haque, "Environmental Discourse and Sustainable Development: Linkages and Limitation," *Ethics and Environment* 5 (2000): 14. For her fascinating history of SD discourse, see pages 8–17.
88. "The energy sustainability and national development concepts that dominate the socioeconomic developmentalist story-line are here shown to contradict socioenvironmental sustainability and development at the local level... This commoditization of nature is condemned by proponents of the sustainability story-line." Bingham, "Discourse of the Dammed," 27.
89. Bauman, *Wasted Lives*, 1–4. See Gay Hawkins, *The Ethics of Waste: How We Relate to Rubbish* (Lanham, MD: Rowman & Littlefield, 2006), 10: "In the quest to purify, which Bruno Latour describes as a typically modern strategy, waste has no generative capacities, only destructive ones." See Bruno Latour, *We Have Never Been Modern* (Cambridge, MA: Harvard University Press, 1993), 10–12.
90. Bauman, *Wasted Lives*, 39.
91. Warwick Anderson, "Excremental Colonialism: Public Health and the Poetics of Pollution," *Critical Inquiry* 21 (1995): 649.
92. Ibid., 644.
93. Ibid., 648. Here he cites Norbert Elias, *The History of Manners*, vol. 1 of *The Civilizing Process*, trans. Edmund Jephcott (New York: Urizen Books, 1978), xiv. Davis points out that "what [Bakhtin's] term [grotesque] has failed to liberate is the notion of actual bodies." Lennard J. Davis, "Enforcing Normalcy: Disability, Deafness, and the Body," in *Visualizing the Disabled Body: The Classical Nude and the Fragmented Torso,* reprinted in *The Norton Anthology of Theory and Criticism*, ed. Vincent B. Leitch (New York: Norton, 2001), 2418.
94. Anderson, "Excremental Colonialism," 667. As Toni Morrison writes in *Tar Baby*, "Although they called it architecture it was in fact elaborately built toilets, decorated toilets, toilets surrounded with and by business and enterprise in order to have something to do in between defecations since waste was the order of the day and the ordering principle of the universe... That was the sole lesson of their world: how to make waste, how to make machines that made more waste, how to make wasteful products, how to talk waste, how to study waste, how to design waste, how to cure people who were sickened by waste so they could be well enough to endure it, how to mobilize waste, legalize waste and how to despise the culture that lived in cloth houses and shit on the ground far away from where they ate." Toni Morrison, *Tar Baby* (New York: Knopf, 1981), cited by Anderson, "Excremental Colonialism," 640.
95. Though he is writing in the postindependence African context, J. D. Esty has pointed out how "the toilet... is a powerful symbol of techological and developmental superiority—one that has the corollary effect of intensifying,

via a newly potent scientific language, the negative valence of shit." J. D. Esty, "Excremental postcolonialism," *Contemporary Literature* 40 (1999): 29. For a criticism of Anderson and Esty, see Dana Phillips, "Excremental Ecocriticism and the Global Sanitation Crisis," in *Material Ecocriticism*, ed. Serenella Iovino and Serpil Oppermann (Bloomington: Indiana University Press, 2014), 176–178.

96. Michel Serres, *Malfeasance: Appropriation Through Pollution?*, trans. Anne-Marie Feenberg-Dibon (Stanford, CA: Stanford University Press, 2011), 46. See Bauman, *Wasted Lives*, 43.
97. Marcus Heeresma, "Dumping Ground," in *The Dedalus Book of Dutch Fantasy*, ed. and trans. Richard Huijing (Sawtry, Cambridgeshire: Dedalus, 1993), 147.
98. Ibid., 142.
99. Ibid., 144.
100. Ibid., 145.
101. Ibid., 146.
102. Ibid., 147.
103. Ibid., 150–151.
104. Ibid., 154.
105. Swift, *Gulliver's Travels*, 260.
106. Ibid., 261.
107. Ibid., 238.
108. Ibid., 247.
109. Ibid., 284.
110. Ibid., 231.
111. Ibid., 235.
112. Ibid., 278.
113. Ibid., 280.
114. Ibid., 271.
115. Ibid., 279.
116. For more on this tale and its excremental anti-Semitism, see Morrison, *Excrement*, 82–88.
117. Quoted in John Victor Tolan, *Petrus Alfonsi and His Medieval Readers* (Gainesville, FL: University Press of Florida, 1993), 117.
118. Martha Bayless, "The Story of the Fallen Jews and the Iconography of Jewish Unbelief," *Viator* 34 (2003): 147.
119. Kathleen Biddick, *The Shock of Medievalism* (Durham, NC: Duke University Press, 1998), 119. As Jews became more visible in the imagination and ideologically, and as they became literally "marked" sartorially through yellow badges, they were literally less visible, because of being expelled, as victims of pogroms, and through ghettoization (114–115).
120. Dei Segni, *On the Misery of the Human Condition*, 30–1.
121. Sharon Achinstein, "John Foxe and the Jews," *Renaissance Quarterly* 54.1 (2001): 86, n. 2.
122. Paul Oppenheimer (trans.), *Till Eulenspiegel: His Adventures* (New York: Routledge, 2001), xxxvii.
123. Ibid., 69.

124. Ibid., 70–71.
125. See Morrison, *Excrement*, Chapters 6 and 7.
126. Fitzgerald, *The Great Gatsby*, 34.
127. Ibid., 170.
128. Michael Thompson, *Rubbish Theory: The Creation and Destruction of Value* (Oxford: Oxford University Press, 1979), 115. See Roger Griffin, *Modernism and Fascism: The Sense of a Beginning under Mussolini and Fascism* (London: Palgrave, 2007), 333.
129. Rachel Nolan, "Refuse Heap Is Archive for Night of Hatred," *The New York Times*, October 28, 2008, accessed October 18, 2014, http://www.nytimes.com /2008/10/28/world/europe/28germany.html?pagewanted=print.
130. Fleischman, *Seedfolks*, 77.
131. As Peter Filkin points out, *der Abfall*—"'rubbish heap' of history"—also means "the Fall" from grace, thus adding tones of metaphysical import to the story. Peter Filkins (trans.), "Introduction," *The Journey: A Novel* (New York: Random House, 2008), xiv.
132. Adler, *The Journey*, 62.
133. Ibid., 63.
134. Ibid., 65.
135. Ibid., 80.
136. Ibid., 262.
137. "The lack of washing facilities at concentration camps prevented prisoners from washing themselves properly, thus turning them literally into what they had been figured as rhetorically: filthy." Morrison, *Excrement*, 147–148. The use of "showers" with which to gas the Jews in death camp was the ideal trope to suggest how Germany could be made clean by their elimination. "We seem to have here a made metaphorical *reductio ad absurdum* in which, through a fiendish triumph of technology, 'dirty' Jewish flesh was melted down in order to transform it into 'clean' soap." Dundes, *Life Is Like a Chicken Coop Ladder*, 127.
138. Adler, *The Journey*, 84.
139. Ibid., 100.
140. Ibid., 87.
141. Ibid., 100.
142. "According to the measure of socially useful labor, which they can no longer perform, old people are superfluous and must be discarded." Adorno, "Trying to Understand *Endgame*," 142.
143. "*Endgame* is the true gerontology [where old people are literally thrown into the trashcan] . . . The national socialists irreparably overturned the taboo of old age. Beckett's trashcans are the emblem of a culture restored after Auschwitz." Adorno, "Trying to Understand *Endgame*," 143.
144. Beckett, *Endgame*, 9.
145. Ibid., 17.
146. Ibid., 16.
147. Adler, *The Journey*, 77.
148. Ibid., 193.
149. Ibid., 194.
150. Ibid., 195.

9 The Secret Life of Objects: The Audacity of Thingness and the Poignancy of Materiality

1. Cole Porter, "You've Got That Thing," in *Fifty Million Frenchmen* (1929).
2. Thanks to Jim Kilfoyle for this enmeshed insight linking the catchphrase Oliver Hardy said to Stan Laurel in many of their films ("Well, here's another nice mess you've gotten me into"), with Timothy Morton's concept of the "mesh," which I discuss later in this chapter.
3. Bill Brown, "Thing Theory," *Critical Inquiry* 28.1 (2001): 7.
4. Jane Bennett, "The Elements," *postmedieval: a journal of medieval cultural studies* 4.1 (2013): 109.
5. Serenella Iovino and Serpil Oppermann, "Introduction," in *Material Ecocriticism*, ed. Serenella Iovino and Serpil Oppermann (Bloomington: Indiana University Press, 2014), 2.
6. Iovino and Oppermann, *Material Ecocriticism*, 75.
7. A variety of thinkers work in these related fields: Bruno Latour, Jane Bennett, Graham Harman, Timothy Morton, Levi Bryant, Iain Hamilton, Jeffrey Jerome Cohen, and Manuel De Landa, to name a few. See Andrew Cole, "The Call of Things: A Critique of Object-Oriented Ontologies," *minnesota review* 80 (2013): 111.
8. Maurizia Boscagli, *Stuff Theory: Everyday Objects, Radical Materialism* (New York: Bloomsbury, 2014).
9. Heather I. Sullivan, "The Ecology of Colors: Goethe's Materialist Optics and Ecological Posthumanism," in *Material Ecocriticism*, ed. Serenella Iovino and Serpil Oppermann (Bloomington: Indiana University Press, 2014), 82.
10. Eileen A. Joy, "You Are Here: A Manifesto," in *Animal, Vegetable, Mineral: Ethics and Objects*, ed. Jeffrey Jerome Cohen (Washington, DC: Oliphaunt Books, 2012), 169.
11. A term associated with De Landa. See Alan Montroso, "Human," in *Inhuman Nature*, ed. Jeffrey Jerome Cohen (Washington, DC: Oliphaunt Books, 2014), 39; and Nigel Clark and Myra J. Hird, "Deep Shit," *O-Zone: A Journal of Object-Oriented Studies* 1 (2014): 51.
12. Clark and Hird, "Deep Shit," 46–7.
13. Iovino and Oppermann, *Material Ecocriticism*, 85.
14. Ian Bogost, *Alien Phenomenology: Or What It's Like to Be a Thing* (Minneapolis, MN: University of Minnesota Press, 2012), 34.
15. Jane Bennett, "Systems and Things: A Response to Graham Harman and Timothy Morton," *New Literary Theory* 43 (2012): 230. See Cole, "The Call of Things," 108, 111.
16. Iovino and Oppermann, "Introduction," 4, quoting Bruno Latour, *Pandora's Hope: Essays on the Reality of Science Studies* (Cambridge, MA: Harvard University Press, 1999), 174.
17. Montroso, "Human," 40. Cole contends that, while these approaches work "hard *not* to project the human into the heart of things, in their attempt to respect the indifference of objects in themselves, they do so anyway by dint of the ancient Logos principle by which things call out to us and speak their being." Cole, "The Call of Things," 106–107.

18. Jeffrey J. Cohen, "Introduction: All Things," in *Animal, Vegetable, Mineral: Ethics and Objects*, ed. Jeffrey Jerome Cohen (Washington, DC: Oliphaunt Books, 2012), 7.
19. Bennett describes this as "the strange ability of ordinary man-made items to exceed their status as objects and to manifest traces of independence or aliveness, constituting the outside of our own experience." Jane Bennett, *Vibrant Matter: A Political Ecology of Things* (Durham: Duke University Press, 2010), xvi.
20. Jane Bennett, "The Force of Things: Steps toward an Ecology of Matter," *Political Theory* 32.3 (June 2004): 350.
21. Bennett, *Vibrant Matter*, 5.
22. Bennett, "The Force of Things," 358.
23. William Rathje and Cullen Murphy, *Rubbish*! *The Archaeology of Garbage* (New York: HarperCollins, 1992), 12.
24. Paul Reyes, "Bleak Houses: Digging through the Ruins of the Mortgage Crisis," *Harper's Magazine,* October 2008: 31.
25. Ibid., 38.
26. Stacy Alaimo, "Oceanic Origins, Plastic Activism, and New Materialism," in *Material Ecocriticism*, ed. Serenella Iovino and Serpil Oppermann (Bloomington: Indiana University Press, 2014), 187–188. In this article Alaimo nuances her concept by pointing out that "although particular strands of thing theory, object-oriented ontology, speculative realisms, new vitalisms, and material feminisms may or may not be particularly posthumanist or environmentally oriented, material ecocriticism, by definition, focuses on material agencies as part of a wider environmentalist ethos that values ecosystems, biodiversity, and nonhuman life" (193).
27. Bennett, *Vibrant Matter*, 2, citing W. J. T. Mitchell, *What Do Pictures Want? The Lives and Loves of Images* (Chicago, IL: University of Chicago Press, 2005), 156–157.
28. Iovino and Oppermann, *Material Ecocriticism*, 77.
29. If the concept of "inert matter helps animate our current practice of aggressively wasteful and planet-endangering consumption, then a materiality experienced as a lively force with agentic capacity could animate a more ecologically sustainable public." Bennett, *Vibrant Matter*, 51.
30. Montroso, "Human," 41, n. 4, quoting Graham Harman.
31. Timothy Morton, *The Ecological Thought* (Cambridge, MA: Harvard University Press, 2010), 130. "We are bequeathing a particular futurity through a projected responsibility for the toxicity, contamination, and resource depletion our epoch created." Jesse Goldstein, "Wastelands," http://discardstudies.com/discard-studies-compendium/.
32. Morton, *The Ecological Thought*, 10–11. In a similar vein, see Slavoj Žižek in *The Examined Life* (Dir. Astra Taylor, 2008). See https://www.youtube.com/watch?v=iGCfiv1xtoU.
33. Christopher Schmidt, *The Poetics of Waste: Queer Excess in Stein, Ashbery, Schuyler, and Goldsmith* (New York: Palgrave Macmillan, 2014), 61. See Timothy Morton, *Ecology without Nature: Rethinking Environmental Aesthetics* (Cambridge, MA and London: Harvard University Press, 2007).

34. Michel Serres, *Malfeasance: Appropriation Through Pollution?*, trans. Anne-Marie Feenberg-Dibon (Stanford, CA: Stanford University Press, 2011), 53.
35. Myra J. Hird, "Waste Flows," accessed November 10, 2014. http://discard studies.com/discard-studies-compendium/.
36. Ibid.
37. Ruth Evans, "Lacan's *belles-lettres*: On Difficulty and Beauty," in *On Style: An Atelier*, ed. Eileen A. Joy and Anna Kłosowska (Brooklyn, NY: punctum books, 2013), 20.
38. Morton, *The Ecological Thought*, 132.
39. Ibid., 17. Robertson similarly aspires "to help brown, blacken and mottle environmental discourse—adding poop to the party, so to speak." Eric Robertson, "Volcanoes, Guts and Cosmic Collisions: The Queer Sublime in *Frankenstein* and *Melancholia*," *Green Letters: Studies in Ecocriticism* 18.1 (2014): 65.
40. Morton, *Ecology without Nature*, 17. "Human society used to define itself by excluding dirt and pollution. We cannot now endorse this exclusion, nor can we believe in the world it produces. This is literally about realizing where your waste goes. Excluding pollution is part of performing Nature as pristine, wild, immediate, and pure." Timothy Morton, "Queer Ecology," *PMLA* 125.2 (2010): 274.
41. A. R. Ammons, *Garbage* (New York: W. W. Norton, 2002), 74.
42. Ibid., 75.
43. Ibid., 90.
44. Ibid., 115.
45. Sullivan, "The Ecology of Colors," 90.
46. Walt Whitman, "When I Heard the Learn'd Astronomer," in *Literature: A Pocket Anthology*, ed. R. S. Gwynn, 5th ed. (Boston, MA: Longman, 2012), 566.
47. Serpil Oppermann, "From Ecological Postmodernism to Material Ecocriticism: Creative Materiality and Narrative Agency," in *Material Ecocriticism*, ed. Serenella Iovino and Serpil Oppermann (Bloomington: Indiana University Press, 2014), 24, quoting Charles Birch, "The Postmodern Challenge to Biology," in *The Reenchantment of Science: Postmodern Proposals*, ed. David Ray Griffin (Albany: Statue University of New York Press, 1988), 70–71. Also Iovino and Oppermann, *Material Ecocriticism*, 80.
48. Bennett, *Vibrant Matter*, 6. She cites Robert Sullivan, *The Meadowlands: Wilderness Adventures on the Edge of a City* (New York: Doubleday, 1998) 96–97.
49. Oppermann, "From Ecological Postmodernism," 30, 34. He builds here from the idea that by "proposing that we can read the world as matter endowed with stories, material ecocriticism speaks of a new mode of description designated as 'storied matter,' or 'material expressions' constituting an agency with signs and meaning" (21).
50. Wendy Wheeler, "Natural Play, Natural Metaphor, and Natural Stories: Biosemiotic Realism," in *Material Ecocriticism*, ed. Serenella Iovino and Serpil Oppermann (Bloomington: Indiana University Press, 2014), 78.
51. Morton, *The Ecological Thought*, 15.
52. Ibid., 7.

53. Bennett, "The Force of Things," 367.
54. Kellie Robertson, "Exemplary Rocks," in *Animal, Vegetable, Mineral: Ethics and Objects*, ed. Jeffrey Jerome Cohen (Washington, DC: Oliphaunt Books, 2012), 100.
55. Morton, *The Ecological Thought*, 50.
56. Wayne C. Booth, "Metaphor as Rhetoric: The Problem of Evaluation," in *On Metaphor*, ed. Sheldon Sacks (Chicago, IL: University of Chicago Press, 1979), 54–55.
57. Morton, *The Ecological Thought*, 87.
58. Ibid., 94.
59. Bennett, *Vibrant Matter*, 4.
60. Viney's work on ruins is instructive here, where "imagining the ruins of the future... provides an opportunity to re-examine the present." Will Viney, *Waste: A Philosophy of Things* (London: Bloomsbury, 2014), 154. This is certainly applicable to the many instances when future ruins are invoked in *Beowulf.*
61. For Beowulf's revision of the fight with Grendel's Mother and her origins, see Dana M. Oswald, "'Wigge under Wætere': Beowulf's Revision of the Fight with Grendel's Mother," *Exemplaria* 21 (2009): 63–82.
62. See Seth Lerer "Grendel's Glove," *English Literary History* 61 (1994): 732, writes about "corporeal poetics," the way in which "each body contains all the parts that may describe the world; in turn, the things of the world may describe in full all the body." Grendel as a "disabled" other must be controlled by the "normal" hero. "The disability becomes a power derived from its otherness, its monstrosity, in the eyes of the 'normal' person. The disability must be decapitated and then contained in a variety of magic wallets." Lennard J. Davis, "Enforcing Normalcy: Disability, Deafness, and the Body," in *Visualizing the Disabled Body: The Classical Nude and the Fragmented Torso*, Reprinted in *The Norton Anthology of Theory and Criticism*, ed. Vincent B. Leitch (New York: Norton, 2001), 2405. Here, the magic wallet is displaced onto the glove Beowulf ascribes to Grendel.
63. Bennett, *Vibrant Matter*, ix. See Jeffrey Jerome Cohen, "Introduction: Ecostitial," in *Inhuman Nature*, ed. Jeffrey Jerome Cohen (Washington, DC: Oliphaunt Books, 2014), vi. Compare to the "noble body" of the pebble held in the hand of the speaker of Zbigniew Herbert's poem "Pebble." John Frow, "A Pebble, a Camera, a Man Who Turns into a Telegraph Pole," *Critical Inquiry* 28.1 (2001): 271. Orlando writes, "Time wears things out *or* lends them dignity: it wears things out *and* lends them dignity." Francesco Orlando, *Obsolete Objects in the Literary Imagination: Ruins, Relics, Rarities, Rubbish, Uninhabited Places, and Hidden Treasures*, trans. Gabriel Pihas and Daniel Seidel with Alessandra Grego (New Haven, CT: Yale University Press, 2006), 11–12.
64. Frow, "A Pebble," 276, 273.
65. Timothy Morton, *Realist Magic: Objects, Ontology, Causality* (University of Michigan Library, Ann Arbor: Open Humanities Press, 2013), 18. http://openhumanitiespress.org/realist-magic.html.
66. Richard Conniff, "Useless Creatures," *The New York Times*, September 14, 2014, accessed October 18, 2014, http://opinionator.blogs.nytimes.com/2014/09/13/useless-creatures/?_php=true&_type=blogs&_r=0.

67. Heather I. Sullivan, "Dirt Theory and Material Ecocriticism," *Interdisciplinary Studies in Literature and Environment* 19.3 (Summer 2012): 528. "[Catriona] Sandilands and [Timothy] Morton prompt us to ask what exactly is wrong with *pollution*, a word that comes freighted with a history of disciplining *sexual* as well as environmental deviance? For Sandilands, 'environmental governmentality [is] a particular technology of abjection, a discourse organizing, both symbolically and somatically, myriad practices of ingestion and excretion, desire and revulsion. It is an irrevocably social process, linking a desire for internal corporeal order with the expulsion of disorderly and terrifying substances, disorderly and terrifying bodies.'" Greg Garrard, "Nature Cures? Or How to Police Analogies of Personal and Ecological Health," *Interdisciplinary Studies in Literature and Environment* 19.3 (Summer 2012): 503, quoting Catriona Sandilands, "Eco Homo: Queering the Ecological Body Politic," *Social Philosophy Today* 19 (2004): 31.
68. Ralph A. Lewin, *Merde: Excursions in Scientific, Cultural, and Sociohistorical Coprology* (New York: Random House, 1999), 141.
69. As Laporte asks, "Who will write the history of Saint Jerome, advisor to the ladies of Rome from 382 to 385, who warned against the practice of smearing one's face with shit to preserve a youthful complexion? How could he know that the Church itself would later sanctify women who—surpassing common semen-swallowers and rivaling Sadean heroines—went so far as to ingest it?" Dominique Laporte, *History of Shit*, trans. N. Benabid and R. El-Khoury (Cambridge, MA: MIT Press, 2000), 102.
70. Michael Thompson, *Rubbish Theory: The Creation and Destruction of Value* (Oxford: Oxford University Press, 1979), 11.
71. Gay Hawkins, *The Ethics of Waste: How We Relate to Rubbish* (Lanham, MD: Rowman & Littlefield, 2006), 78, discussing Thompson, *Rubbish Theory.*
72. Thompson, *Rubbish Theory*, 2.
73. Ibid., 103; see also 114–115.
74. Ibid., 7–9.
75. "But usefulness is precisely the argument other people put forward to justify destroying or displacing wildlife, and they generally bring a larger and more persuasive kind of green to the argument." Conniff, "Useless Creatures."
76. See chapter 6.
77. Marcus Hensel, "The Gift of Good Land?: Settled Lands and Wastelands in Anglo-Saxon Thought," Talk given at BABEL Working Group, November 2010.
78. "The impossibility of getting out of the game and of giving back to things their toy-like uselessness heralds the precise instant at which infancy comes to an end, and defines the very notion of seriousness." Emmanuel Levinas, *On Escape De l'évasion*, trans. Bettina Bergo (Stanford, CA: Stanford University Press, 2003), 52.
79. Hawkins, *The Ethics of Waste*, 21–22. For more on *American Beauty*, see Boscagli, *Stuff Theory*, 244–251.
80. As Hawkins points out, pieces of trash are objects weighted with history and can carry a "sensuous presence." Gay Hawkins, "Waste in Sydney: Unwelcome Returns." *PMLA* 122 (2007): 351. Alaimo takes Hawkins to task. Alaimo, "Oceanic Origins," 194–195.

81. Ariel Sabar, "One Man's Trash Is Another Man's Two-Bit 'Trashball,'" *The New York Times*, April 29, 2007, accessed October 18, 2014, http://www.nytimes.com/2007/04/29/us/29trash.html. See his blog: guyclinch.blogspot.com. See also David Scott Diffrient, "Stories That Objects Might Live to Tell: The 'Hand-Me-Down' Narrative in Film," *Other Voices* 3.1 (2007), accessed April 18, 2012, http://www.othervoices.org/3.1/sdiffrient/index.php.
82. Alaimo, "Oceanic Origins," 200. Alaimo points out that "ostensibly discrete entities such as plastic bottle caps are, in a sense, already part of who we are, as human diets ontologically entangle us with the plastic seas" (198).
83. Paul Fleischman, *Seedfolks* (New York: HarperTeen, 2004), 38.
84. Ibid., 44.
85. Hawkins, *The Ethics of Waste*, 85.
86. Ibid., 86.
87. Ibid., 88.
88. Italo Calvino, "La Poubelle Agréée," in *The Road to San Giovanni*, trans. Tim Parks (New York: Pantheon, 1993), 109.
89. H. G. Adler, *The Journey: A Novel*, trans. Peter Filkins, Afterword by Jeremy Adler (New York: Random House, 2008), 268.
90. Ibid., 283.
91. Hawkins, *The Ethics of Waste*, 98; Susan Strasser, *Waste and Want: A Social History of Trash* (New York: Henry Holt, 2000), 21.
92. Music by Sammy Fain, lyrics by Irving Kahal.
93. Jane Bennett, "Powers of the Hoard: Further Notes on Material Agency," in *Animal, Vegetable, Mineral: Ethics and Objects*, ed. Jeffrey Jerome Cohen (Washington, DC: Oliphaunt Books, 2012), 240, 238, 239.
94. T. Tyler, *Ciferae: A Bestiary in Five Fingers* (Minneapolis, MN: University of Minnesota Press, 2012), 63. As Alain Robbe-Grillet points out, "In practically all our contemporary literature these anthropomorphic analogies are too insistently, too coherently, repeated, not to reveal a whole metaphysical system." Alain Robbe-Grillet, "Nature, Humanism and Tragedy" (1958), in *Snapshots and Towards a New Novel*, trans. Barbara Wright (London: Calder & Boyars, 1965), 78, quoted in Sophie Ratcliffe, *On Sympathy* (Oxford: Oxford University Press, 2008), 32. "The world is neither meaningful nor absurd. It is . . . quite simply. But it is not the world of the humanist text and anthropomorphic metaphor. The world is . . . the strange." Quoted in Raylene Ramsay, *Robbe-Grillet & Modernity: Science, Sexuality, & Subversion* (Gainesville, FL: University Press of Florida, 1992), 245.
95. Juliet Fleming, "Scraping by: Towards a Pre-Historic Criticism," *postmedieval: a journal of medieval cultural studies* 3 (2012): 121–122; citing Barbara Johnson, *The Feminist Difference: Literature, Psychoanalysis, Race, and Gender* (Cambridge, MA: Harvard University Press, 1998); also Bennett, *Vibrant Matter*, 120.
96. Morton, *The Ecological Thought*, 76. "The contradictions within each of these new philosophies—it is and it is not anthropocentrism, anthropocentrism is and is not a bad thing." Cole, "The Call of Things," 107.
97. Scott Slovic, "Editor's Note," *Interdisciplinary Studies in Literature and Environment* 20.3 (2013): 456, quoting Catherine Diamond, "Hiking in

Yangming Mountain: 'Listening,'" *Interdisciplinary Studies in Literature and Environment* 20.3 (2013): 684.

98. Iovino and Oppermann, *Material Ecocriticism*, 82. See Bennett, *Vibrant Matter*, 120; Iovino and Oppermann, "Introduction," 8; Oppermann, "From Ecological Postmodernism," 29; and Hubert Zapf, "Creative Matter and Creative Mind: Cultural Ecology and Literary Creativity," in *Material Ecocriticism*, ed. Serenella Iovino and Serpil Oppermann (Bloomington: Indiana University Press, 2014), 52. Carolynn Van Dyke writes about "the poetics of transmorphism" in this same vein. Carolynn Van Dyke, "Touched by an Owl? An Essay in Vernacular Ethology," *postmedieval: a journal of medieval cultural studies* 7 (2014): 15.
99. Morton, *Realist Magic*, 17. After all, Morton points out, "I can't help anthropomorphizing everything I handle... Just as I fail to avoid anthropomorphizing everything, so all entities whatsoever constantly translate other objects into their own terms... Everything else is doing the same thing." Timothy Morton, "An Object-Oriented Defense of Poetry," *New Literary History* 43 (2012): 207.
100. Emmanuel Levinas, *Totality and Infinity: An Essay on Exteriority*, trans. Alphonso Lingis (Dordrecht: Kluwer, 1991), 139–140. Harman contends Levinas does not go quite far enough in his argument, remaining centered on the human. Yet Levinas sets up object-oriented sympathies, laying "the groundwork for a strange new form of realism." Graham Harman "Levinas," 408. Levinas respects the object, where its matter provides its meaning. Ibid., 411, quoting Levinas, *Totality and Infinity*, 82, 192–193.
101. Levinas, *Totality and Infinity*, 140.
102. Emmanuel Levinas, *Humanism of the Other*, trans. Nidra Poller, Introduction by Richard A. Cohen (Urbana: University of Illinois Press, 2006), 11.
103. Emmanuel Levinas, *Ethics and Infinity: Conversations with Philippe Nemo*, trans. R. A. Cohen (Pittsburgh: Duquesne University Press, 1985), 88; Robert Eaglestone, *Ethical Criticism: Reading after Levinas* (Edinburgh: Edinburgh University Press, 1997), 141.
104. Levinas, *Humanism of the Other*, 44.
105. Ibid., 41.
106. "[Eckhart] prescribes mindfulness toward things that, to our ear, will remain 'quite dumb' unless we enter into a 'face to face' encounter with the object and open ourselves to it" in a Levinasian encounter. Cole, "The Call of Things," 109. Levinasian affinity has, in turn, affinity with Oppermann's observation: "The liveliness of matter has liberating effects of moving the human vision from the language of otherness to that of differential coemergence." Oppermann, "From Ecological Postmodernism," 35. Even the poop emoticon wears a face complete with eyes and smile, thus defying social norms.
107. Calvino, "La Poubelle Agréée," 94–95.
108. Ibid., 111.
109. The quote continues, "as that which is *said*—reduced to fixed identity or synchronized presence—is an ontological closure to the other." Emmanuel Levinas and Richard Kearney, "Dialogue with Emmanuel Levinas," in *Face to Face with Levinas*, ed. Richard A. Cohen (Albany, NY: SUNY Press, 1986), 29.

110. Eaglestone, *Ethical Criticism*, 166.
111. François Rabelais, *Gargantua and Pantagruel*, trans. Sir Thomas Urquhart and Peter Motteux (Chicago, IL: Encyclopaedia Britannica, 1952), 17.
112. He continues, "[that marks] interspaces where identities are formed through negotiation, interaction, and engagement," and "contact zones . . . wherein dialectic relations of self and other . . . [are] always charged with ambivalence . . . between extremes of attraction and repulsion, of mastery and anxiety." Michael Uebel, *Ecstatic Transformation: On the Uses of Alterity in the Middle Ages* (New York: Palgrave Macmillan, 2005), 36, 41.
113. Ibid., 43.
114. As Teresa Mack argues, even if presented from the point of view of the colonizing discourse, "the trace of the constructed Other . . . nevertheless deposits a residue of the ambivalence, ambiguity." Teresa Mack, "Toward an Understanding of the 'Fiction of the Space Between': An(O)ther Kind of Agency," April 2010, Unpublished paper.
115. Leo Lionni, *little blue and little yellow* (New York: McDowell, Obolensky, 1959), Unpaginated.
116. All quotes from Heather O'Neill, "The Secret Life of Our Trash Can," *The New York Times Magazine*, March 14, 2014, accessed November 28, 2014, http://www.nytimes.com/2014/03/16/magazine/the-secret-life-of-our-trash-can.html.
117. Bennett, "Powers of the Hoard," 244.
118. Julian Yates, "Sheep-Tracks—A Multi-Species Impression," in *Animal, Vegetable, Mineral: Ethics and Objects*, ed. Jeffrey Jerome Cohen (Washington, DC: Oliphaunt Books, 2012), 179 [173–209].
119. Lowell Duckert, "Speaking Stones, John Muir, and a Slower (Non) Humanities," in *Animal, Vegetable, Mineral: Ethics and Objects*, ed. Jeffrey Jerome Cohen (Washington, DC: Oliphaunt Books, 2012), 278. See Joy, "You Are Here," 163.
120. Bennett, "Powers of the Hoard," 240, 267.
121. Bennett, "The Force of Things," 360.
122. Bennett, *Vibrant Matter*, 13.
123. http://www.youtube.com/watch?v=VF9-sEbqDvU.
124. See also the incinerator scene in *Toy Story 3* (2010), where the doughty band of toys holds hands, inexorably dragged to their fiery demise, before a last-minute rescue by the aliens.
125. The line "a dollie for Sue" suggests this doll was abandoned by a child named Sue. *Toy Story 2* includes the heartrending "When Somebody Loved Me," sung by Jesse, the abandoned cowgirl doll.
126. Compare to Timothy Morton, "Treating Objects Like Women: Feminist Ontology and the Question of Essence," in *International Perspectives in Feminist Ecocriticism*, ed. Greta Gaard, Simon C. Estok, and Serpil Oppermann (New York: Routledge, 2013), 56–69.
127. Mark Atherton (trans.), *Hildegard von Bingen: Selected Writings* (London: Penguin, 2001), 107.
128. Ibid., 108.
129. Robertson, "Exemplary Rocks," 92.

130. Even minerals and rocks, "these apparently inert strata," contain "traces of bygone biospheres." Bennett, *Vibrant Matter*, 8, citing Lynn Margulis and Dorion Sagan, *What is Life?* (Berkeley: University of California Press, 2000), 50.
131. Atherton, *Hildegard von Bingen*, 109.
132. "An alliance between human beings and primordial stone can loosen the temporal fixedness of one and the spatial immobility of the other." Jeffrey Jerome Cohen, "Stories of Stone," *postmedieval: a journal of medieval cultural studies* 1 (2010): 61.
133. Robertson, "Exemplary Rocks," 100.
134. Atherton, *Hildegard von Bingen*, 111–112.
135. Perhaps, then, it's not a matter of equalizing the playing field between human and nonhuman, not a "flat morality but one of infinite, incommensurable hierarchies." Karl Steel, "With the World, or Bound to Face the Sky: The Postures of the Wolf-Child of Hesse," in *Animal, Vegetable, Mineral: Ethics and Objects*, ed. Jeffrey Jerome Cohen (Washington, DC: Oliphaunt Books, 2012): 33.
136. Thanks to Deanna Rodriguez for this perceptive insight.
137. Robertson, "Exemplary Rocks," 100.
138. Ibid., 105.
139. Ibid., 106.
140. Bogost, *Alien Phenomenology*, 6.
141. Ibid., 9. My emphasis.
142. S. A. J. Bradley (trans. and ed.), *Anglo-Saxon Poetry* (London: J. M. Dent & Sons, 1982), 404.
143. Lowell Duckert, "Glacier," *postmedieval: a journal of medieval cultural studies* 4.1 (2013): 70.
144. See Bennett, *Vibrant Matter*, 11, referencing Michael De Landa, *A Thousand Years of Nonlinear History* (New York: Swerve Editions, 1997), 26–27.
145. Ibid., 58. My emphasis.
146. Duckert, "Speaking Stones," 274. Citing Bruno Latour, *Reassembling the Social: An Introduction to Actor-Network-Theory* (Oxford: Oxford University Press, 2005), 25. For more on slowness see Bennett, "Powers of the Hoard," 254, and Joy, "You Are Here," 172.
147. Bennett, *Vibrant Matter*, 116.
148. All references to the translation by Roy M. Liuzza, "Two Old English Elegies from the Exeter Book: *The Wanderer* and *The Ruin*," accessed October 19, 2014, https://web.utk.edu/~rliuzza/401/Elegies.pdf.
149. "Objects in ruins speak back." Tim Edenson, "Waste Matter: The Debris of Industrial Ruins and the Disordering of the Material World," *Journal of Material Culture* 10.3 (2005): 317. Viney discusses the temporality of ruins, pointing out "the ruin is anchored to a use-time that has passed." *Waste*, 140.
150. Edenson, "Waste Matter," 311.
151. Liuzza, "Two Old English Elegies."
152. A form of excess, grace is "what is gratuitous," yet which "[abounds]." Steven Connor, "The Poorest Things Superfluous: On Redundancy," Talk at Rubbish

Symposium, Birkbeck College, London, July 30, 2011, accessed October 21, 2014, http://stevenconnor.com/redundancy.html.

153. Thanks to Ray Stockstad for catalyzing the focus on garbage as riddle for me.
154. "Knowing waste is rendering the indeterminate determinate." Myra J. Hird, "Knowing Waste: Towards an Inhuman Epistemology," *Social Epistemology: A Journal of Knowledge, Culture and Policy* 26.3–4 (2012): 454.
155. "The riddle endows things with history" and makes history a riddle. Daniel Tiffany, "Lyric Substance: On Riddles, Materialism, and Poetic Obscurity," *Critical Inquiry* 28.1 (2001): 74.
156. Thanks to Shannon Shaw for this quip.
157. Bennett, *Vibrant Matter*, 358.
158. Ibid., 360.
159. Naomi Shihab Nye, "The Traveling Onion," in *Literature: A Pocket Anthology*, ed. R. S. Gwynn, 5th ed. (Boston, MA: Longman, 2012), 778–779.
160. See Bradley, *Anglo-Saxon Poetry*, 370–372.
161. Quotes from Peter Campion (trans.), "Who Is So Smart, So Crafty-Spirited?" in *The Word Exchange: Anglo-Saxon Poems in Translation*, ed. Greg Delanty and Michael Matto (New York: W. W. Norton, 2011), 66–73; see also Bradley, *Anglo-Saxon Poetry*, 370–372.

10 Trash Meditation: The Arts of Transience and Proximity

1. According to legend, Flaubert evidently claimed, "Madame Bovary, c'est moi."
2. "While sustainable development discourse is seen as both enabling and constraining for indigenous peoples, the socioeconomic priorities that characterize the dominant story-line are reproduced in the Brazilian context and legitimize policies that are detrimental to indigenous interests and sustainability objectives." Alexa Bingham, "Discourse of the Dammed: A Study of the Impacts of Sustainable Development Discourse on Indigenous Peoples in the Brazilian Amazon in the Context of the Proposed Belo Monte Hydroelectric Dam," *POLIS Journal* 4 (2010): 1.
3. Bingham, "Discourse of the Dammed," 31.
4. Ibid., 33.
5. M. Shamsul Haque, "Environmental Discourse and Sustainable Development: Linkages and Limitation," *Ethics and Environment* 5 (2000): 4. Patricia Yaeger cites Garrett Hardin's phrase, "tragedy of the commons." "Freedom in a commons brings ruin to all." Garrett Hardin, "The Tragedy of the Commons," *Science* 162, December 13, 1968, 1244.
6. Haque, "Environmental Discourse and Sustainable Development," 12–13. Ferraro and Reid call homo economicus bad for the environment and "responsible for the current unsustainability of the Planet." Emilia Ferraro and Louise Reid, "On Sustainability and Materiality. Homo Faber, a New Approach," *Ecological Economics* 96 (2013): 127. As Morton points out, we cannot accommodate "ecological thinking . . . to postmodern consumer capitalism." Timothy

Morton, *The Ecological Thought* (Cambridge, MA: Harvard University Press, 2010), 16.

7. John Scanlan, *On Garbage* (London: Reaktion, 2005), 182. See Tim Edenson, "Waste Matter: The Debris of Industrial Ruins and the Disordering of the Material World," *Journal of Material Culture* 10.3 (2005); and Slavoj Žižek, *The Ticklish Subject* (London and New York: Verso, 2000), 12. Jane Bennett privileges vital materialism over an environmentalism in which nonhumans are passive, making us more removed from "Nature." Jane Bennett, *Vibrant Matter: A Political Ecology of Things* (Durham: Duke University Press, 2010), 111.
8. Wendell Berry, "The Necessity of Agriculture," *Harper's Magazine*, December 2009, 16.
9. "Something else must happen [besides separation and mastery] to nurture an ethos of positive engagement with waste, to trigger a relation of openness and care, and to encourage the cultivation of new habits." Gay Hawkins, *The Ethics of Waste: How We Relate to Rubbish* (Lanham, MD: Rowman & Littlefield, 2006), 121; see also Jhan Hochman, "Green Cultural Studies: An Introductory Critique of an Emerging Discipline," *Mosaic* 30.1 (1997): 93, and Gay Hawkins, "Down the Drain: Shit and the Politics of Disturbance," in *Culture and Waste: The Creation and Destruction of Value*, ed. Gay Hawkins and Stephen Muecke (Lanham, MD: Rowman & Littlefield, 2003), 50.
10. Wendell Berry, "Faustian Economics: Hell Hath No Limits," *Harper's Magazine*, May 2008, 35, 36, 41, 36, accessed November 16, 2009. http://www.harpers.org/archive/2008/05/0082022.
11. Michael Pollan, "The Vegetable-Industrial Complex," *The New York Times Magazine*, October 15, 2006, accessed October 18, 2014, http://www.nytimes.com/2006/10/15/magazine/15wwln_lede.html. On the contribution of late medieval deforestation to the fourteenth-century ecological crisis, see Michael De Landa, *A Thousand Years of Nonlinear History* (New York: Swerve Editions, 1997), 122.
12. Hawkins, *The Ethics of Waste*, 29.
13. "Seeing waste, instead, as a temporary set of things in between forms of life, rather than between disorder and order, would arguably serve to foster a better environmental politics." Joshua Ozias Reno, "Toward a New Theory of Waste: From 'Matter out of Place' to Signs of Life," *Theory Culture Society* 31.6 (2014): 20.
14. Walter Benjamin, *Illuminations: Essays and Reflections*, ed. Hannah Arendt (New York: Schocken Books, 1968), 256.
15. Ibid., 257–258.
16. Scanlan, *On Garbage*, 14.
17. Joseph Carroll, *Literary Darwinism: Evolution, Human Nature, and Literature* (New York: Routledge, 2004), 88; also 116.
18. Emmanuel Levinas, "Ethics of the Infinite," in Richard Kearney, *States of Mind: Dialogues with Contemporary Thinkers* (New York: New York University Press, 1995), 189.
19. *WALL-E*, directed by Andrew Stanton (2008).
20. See Christopher Schmidt, *The Poetics of Waste: Queer Excess in Stein, Ashbery, Schuyler, and Goldsmith* (New York: Palgrave Macmillan, 2014), preface and

afterword, in which he argues *Wall-E* is not innovative in terms of gender politics. Also Morton, *The Ecological Thought*, 2.

21. Daniel B. Smith, "Is There an Ecological Unconscious?" *The New York Times*, January 31, 2010, accessed October 18, 2014, http://www.nytimes.com/2010/01/31/magazine/31ecopsych-t.html?pagewanted=all.
22. Scanlan, *On Garbage*, 33.
23. Ibid., 180.
24. "Garbage provides a shadow history of modern life." Ibid., 36.
25. Will Viney, *Waste: A Philosophy of Things* (London: Bloomsbury, 2014), 2, 9–10. Michelle Coyne's work on dumpster diving addresses this issue. Food is often labelled with "sell by" or "use by" dates that do not reflect the healthiness of the food but require the "wasting" of it. "Waste reclamation," she argues, "redefines the space and meaning of waste itself." Michelle Coyne, "From Production to Destruction to Recovery: Freeganism's Redefinition of Food Value and Circulation," *Iowa Journal of Cultural Studies* 10/11 (2009), accessed March 9, 2012, http://www.uiowa.edu/~ijcs/waste/coyne.htm.
26. Viney, *Waste*, 21.
27. H. G. Adler, *The Journey: A Novel*, trans. Peter Filkins, Afterword by Jeremy Adler (New York: Random House, 2008), 78.
28. Ibid., 79.
29. Ibid., 280.
30. Ibid., 281.
31. Emmanuel Levinas, *Humanism of the Other*, trans. Nidra Poller, Introduction by Richard A. Cohen (Urbana: University of Illinois Press, 2006), 32.
32. Ibid., 33.
33. Ibid., 37.
34. Adler, *The Journey*, 281.
35. Ibid., 284.
36. Eric Leake, "Humanizing the Inhumane: The Value of Difficult Empathy," in *Rethinking Empathy Through Literature*, ed. Meghan Marie Hammond and Sue J. Kim (New York: Routledge, 2014), 175.
37. Ibid., 176.
38. Ibid.
39. Ibid., 178.
40. Ibid., 184.
41. See especially Adler, *The Journey*, 77–87.
42. Peter Filkins, "Introduction," *The Journey*, xv.
43. Scanlan, *On Garbage*, 163.
44. Elizabeth A. Povinelli, "The Child in the Broom Closet: States of Killing and Letting Die," *South Atlantic Quarterly* 107.3 (2008): 511.
45. Ibid., 516.
46. "Empathy asks us to put ourselves in someone else's shoes...And yet, this very act—the ethical gesture—initiates a separation between you and me." Ibid., 520.
47. Povinelli, "The Child in the Broom Closet," 528.
48. Berry, "Faustian Economics," 39.

49. Quoted by Filkins, "Introduction," *The Journey*, xx, quoting W. G. Sebald, "An Attempt at Restitution," in *Campo Santo*, ed. Sven Meyer and trans. Anthea Bell (London: Hamish Hamilton, 2005); also W. G. Sebald, "An Attempt at Restitution," *The New Yorker* December 20 and 27, 2004, accessed October 19, 2014, http://www.newyorker.com/magazine/2004/12/20/an-attempt-at-restitution.
50. "Recuperation is a loving gesture." Schmidt, *The Poetics of Waste*, 159.
51. A. R. Ammons, *Garbage* (New York: W. W. Norton, 2002), 112.
52. Eli Clare movingly suggests the term "restoration," an ongoing process that would enable a "radical valuing of disabled and chronically ill bodies... The bodies of both disabled and chronically ill people and restored prairies resist the impulse toward and the reality of monocultures." Eli Clare, "Meditations on Natural Worlds, Disabled Bodies, and a Politics of Cure," in *Material Ecocriticism*, ed. Serenella Iovino and Serpil Oppermann (Bloomington: Indiana University Press, 2014), 214–215.
53. Gerald Vizenor, *Landfill Meditation: Crossblood Stories* (Hanover, NH: Wesleyan University Press/University Press of New England, 1991), 104.
54. Ibid., 99. See also Lowell Duckert, "Speaking Stones, John Muir, and a Slower (Non)Humanities," in *Animal, Vegetable, Mineral: Ethics and Objects*, ed. Jeffrey Jerome Cohen (Washington, DC: Oliphaunt Books, 2012), 279.
55. Vizenor, *Landfill Meditation*, 104–105.
56. Ian Bogost, *Alien Phenomenology: Or What It's Like to Be a Thing* (Minneapolis, MN: University of Minnesota Press, 2012), 34.
57. Morton, *The Ecological Thought*, 127.
58. Ibid., 125.
59. Richard Panek, "Out There," *The New York Times Magazine*, March 11, 2007, accessed October 18, 2014, http://www.nytimes.com/2007/03/11/magazine/11dark.t.html?pagewanted=all.
60. After *Hell* was destroyed in a fire, the brothers recreated it as *Fucking Hell* (2008). There is a great deal written about trash art, including good anthologies like *The Sublime*, ed. Simon Morley (London and Cambridge, MA: Whitechapel Gallery/MIT Press, 2010) and *Ruins*, ed. Brian Dillon (London and Cambridge, MA: Whitechapel Gallery/MIT Press, 2011).
61. This dance has now been filmed as a documentary, *Trash Dance*, Choreographer: Allison Orr (2012). Also see http://trashdancemovie.com/.
62. Robert Faires, "Corps de Garbage," *The Austin Chronicle*, September 11, 2009, accessed November 10, 2014, http://www.austinchronicle.com/arts/2009-09-11/837971/. The *Handshake Ritual* of Mierle Ukeles similarly breaks down the distance between us and those who pick up our trash. See Mark B. Feldman, "Inside the Sanitation System: Mierle Ukeles, Urban Ecology, and the Social Circulation of Garbage," *Iowa Journal of Cultural Studies* 10.11 (2009), accessed March 9, 2012, http://www.uiowa.edu/~ijcs/waste/feldman.html. Calvino comments on the "musical rhythm" of "kitchen life," and the "art of tying an overfull bag." Italo Calvino, "La Poubelle Agréée," in *The Road to San Giovanni*, trans. Tim Parks (New York: Pantheon, 1993), 94, 123.

11 Waste Aesthetics: Puns, Litter-ature, and Intertextuality

*Ruth Evans, "Lacan's *belles-lettres*: On Difficulty and Beauty," in *On Style: An Atelier*, ed. Eileen A. Joy and Anna Kłosowska (Brooklyn, NY: punctum books, 2013), 19. Viney discusses types of textual waste as well; Will Viney, *Waste: A Philosophy of Things* (London: Bloomsbury, 2014), 80.

1. John G. Bourke, *The Scatalogic Rites of All Nations* (Washington, DC: W. H. Lowdermilk, 1891): 3–4; David Rhode, "Coprolites from Hidden Cave, Revisted: Evidence for Site Occupation History, Diet and Sex of Occupants," *Journal of Archaeological Science* 30 (2003): 915, n. 3; Penny Van Esterik, "No Free Lunch," *Agriculture and Human Values* 22 (2005): 207–208.
2. Bourke, *The Scatalogic Rites*, 3.
3. W. W. Newcomb, *The Indians of Texas: From Prehistoric to Modern Times* (Austin, TX: University of Texas Press, 1961/2002), 41.
4. Evans, "Lacan's *belles-lettres*," 19.
5. Ibid., 20.
6. Even reviews of my book, *Excrement*, insist on puns: "Morrison's approach is thus hands-on or, more precisely, pants-down. Her book . . . rubs our noses in the midden of medieval poetry." Peter J. Smith, Review, *Times Higher Education*, January 15, 2009.
7. John Kelso, "These Residents Just Aren't Going with the Flow," *Austin American-Statesman*, March 28, 2008.
8. John Scanlan, *On Garbage* (London: Reaktion, 2005), 9, 98. See Irving Wohlfarth, "Et Cetera? The Historian as Chiffonnier," *New German Critique* 39 (1986): 157: "The rag-and-bone man, a cousin of the Shakespearean grave-digger, is, like the destructive character, an allegorist in action—the grave-digger of the bourgeois world."
9. Gordon C. F. Bearn shows how "more meanings of a word are in play than those we have intentionally put into play." Gordon C. F. Bearn, "Notes and Fragments: The Possibility of Puns: A Defense of Derrida," *Philosophy and Literature* 19.2 (1995): 331.
10. References to *The Riverside Shakespeare*, ed. G. Blakemore Evans (Boston: Houghton Mifflin, 1974).
11. All references to the translations in Nancy Mason Bradbury, "Rival Wisdom in the Latin *Dialogue of Solomon and Marcolf*," *Speculum* 83 (2008): 348.
12. Ibid., 349. Marcolf "turns all Solomon's turgid truths into turds." Michael Camille, *Image on the Edge: The Margins of Medieval Art* (London: Reaktion, 1992), 27.
13. Bradbury, "Rival Wisdom," 349.
14. Ibid., 357.
15. Ibid.
16. Ibid., 362.
17. Bearn, "Notes and Fragments," 334.
18. Jonathan Swift, *Gulliver's Travels*, ed. Philip Pinkus (Toronto: Macmillan, 1968), 240.
19. It has been suggested that many words dealing with anality appear in German folklore. For an extensive and fascinating study of German character as

reflected in folklore, see Alan Dundes, *Life Is Like a Chicken Coop Ladder: A Study of German National Character through Folklore* (Detroit, MI: Wayne State University Press, 1984), 9.

20. Paul Oppenheimer (trans.), *Till Eulenspiegel: His Adventures* (New York: Routledge, 2001), 97, 106, 115, 136.
21. Ibid., 22–23.
22. Mary Douglas's work on jokes suggests that we could read Till as a "minor mystic. Though only a mundane and border-line type, [the joker] is one of those people who pass beyond the bounds of reason and society and give glimpses of a truth which escapes through the mesh of structured concepts." Mary Douglas, "Jokes," *Implicit Meanings: Essays in Anthropology* (London: Routledge & Kegan Paul, 1975), 108.
23. See Gary Snyder, *A Place in Space* (Washington, DC: Counterpoint, 1995), 173–175.
24. François Rabelais, *Gargantua and Pantagruel*, trans. Sir Thomas Urquhart and Peter Motteux (Chicago, IL: Encyclopaedia Britannica, 1952), 10.
25. Ibid., 15.
26. Ibid., 16, also 17.
27. Ibid., 17. See Thomas DiPiero, "Shit Happens: Rabelais, Sade, and the Politics of Popular Fiction," *Genre* 27 (1994): 304–305. For DiPiero, shit appears in Rabelais to undermine the distinction between the classes, the biological and the social, and high and low culture.
28. Rictor Norton, *Early Eighteenth-Century Newspaper Reports: A Sourcebook*, "Silvia and Tom Turd-man," last modified December 16, 2003, accessed April 1, 2010, http://rictornorton.co.uk/grubstreet/waterman.htm.
29. See Michael Snediker, "To Peach or Not to Peach: Style and the Interpersonal," in *On Style: An Atelier*, ed. Eileen A. Joy and Anna Kłosowska (Brooklyn, NY: punctum books, 2013), 52–53, about bodice rippers, "trashy reading," and "the trash of stylistic difficulty."
30. Christine Neufeld, "Always Accessorize: In Defense of Scholarly *Cointise*," in *On Style: An Atelier*, ed. Eileen A. Joy and Anna Kłosowska (Brooklyn, NY: punctum books, 2013), 88.
31. "Without selection there would be no story." Zygmunt Bauman, *Wasted Lives: Modernity and Its Outcasts* (Cambridge: Polity Press, 2004), 17.
32. Hermann Pálsson and Paul Edwards (trans.), *Eyrbyggja Saga* (London: Penguin, 1989), 31, 68, 128.
33. Michael Camille, "Glossing the Flesh: Scopophilia and the Margins of the Medieval Book," in *The Margins of the Text*, ed. D. C. Greetham (Ann Arbor: University of Michigan Press, 1997), 261, 263.
34. "*Ulysses* is packed to overflowing with [excreted things]." Tom McCarthy, "'Ulysses' and Its Wake," *London Review of Books* June 19, 2014: 39. "*Ulysses* matters most, because it makes matter of everything." McCarthy, "'Ulysses' and Its Wake," 40. We might even see this style as a form of "camp." As Guy Schaffer explains, "Camp offers a mode of celebrating, reappropriating, and rendering waste visible, without pretending that waste has stopped being waste." Guy Schaffer, "Camp," http://discardstudies.com/discard-studies-compendium/.

35. Lawrence Sterne, *Tristram Shandy*, ed. James Aiken Work (Indianapolis: The Odyssey Press/Bobbs-Merrill, 1940/1977), 3.
36. Ibid., 7.
37. Ibid., 8.
38. Ibid., 10.
39. Ibid., 23.
40. Ibid., 32–34.
41. Ibid., 35–36.
42. Ibid., 36–37.
43. Ibid., 37.
44. Ibid., 285–286.
45. Anna Kłosowska, "Style as Third Element," in *On Style: An Atelier*, ed. Eileen A. Joy and Anna Kłosowska (Brooklyn, NY: punctum books, 2013), 26.
46. Keri Fitzgerald, "The Creation of Meaning through Metafictional Digression in Sterne's *Tristram Shandy*," Paper for Honors 3396A, November 17, 2009, Texas State University.
47. Sterne, *Tristram Shandy*, 36.
48. Italo Calvino, "La Poubelle Agréée," in *The Road to San Giovanni*, trans. Tim Parks (New York: Pantheon, 1993), 125–126.
49. The "meaning we attach to the world is constituted through a continual *recycling of ideas and matter*, in the resurrection and re-ordering of all kinds of garbage that furnishes the actual physical 'being-ness' of existence." Scanlan, *On Garbage*, 112.
50. Marilyn Randall explores the metaphor of recycling and responds to Walter Moser's work on cultural recycling. "Recycling Recycling, or *plus ça change*." *Other Voices* 3.1 (2007): accessed April 18, 2012, http://www.othervoices.org/3.1/mrandall/index.php. Randall writes, "The concept of *intertextuality* (followed closely by *bricolage*, *appropriation* and *recycling*) subsequently fulfilled a theoretical need to explain the fact of aesthetic repetition . . . As opposed to other metaphors, the process invoked by *recycling* is one which explicitly negates its origins, the new creation implying the elimination of the old."
51. Will Viney, "The Future of Ruins," in *Aesthetic Fatigue: Modernity and the Language of Waste*, ed. John Scanlan and John F. M. Clark (Newcastle upon Tyne: Cambridge Scholars, 2013), 144–145. Viney explores Eliot's allusions as a way to make what was redundant "available for reuse." Viney, *Waste*, 94ff. For his discussion of "Ozymandias," see Viney, *Waste*, 161ff.
52. Francesco Orlando, *Obsolete Objects in the Literary Imagination: Ruins, Relics, Rarities, Rubbish, Uninhabited Places, and Hidden Treasures*, trans. Gabriel Pihas and Daniel Seidel with Alessandra Grego (New Haven, CT: Yale University Press, 2006), 205.
53. Viney, *Waste*, 82–83.
54. Tim Armstrong, "Eliot's Waste Paper," *Modernism, Technology, and the Body: A Cultural Study* (Cambridge: Cambridge University Press, 1998), 70. Reprinted in *The Waste Land: A Norton Critical Edition*, ed. Michael North (New York: W. W. Norton, 2001), 275–280.
55. Armstrong, "Eliot's Waste Paper," 70–71.

56. Ibid., 73.
57. Gerald Vizenor, *Landfill Meditation: Crossblood Stories* (Hanover, NH: Wesleyan University Press/University Press of New England, 1991), 155.
58. Ibid., 158.
59. Ibid., 161.
60. H. G. Adler, *The Journey: A Novel,* trans. Peter Filkins, Afterword by Jeremy Adler (New York: Random House, 2008), 171.
61. Ibid., 172.
62. Ibid., 173.
63. Ibid., 178.
64. Peter Filkins, "Introduction," *The Journey: A Novel,* trans. Peter Filkins, Afterword by Jeremy Adler (New York: Random House, 2008), xii, xiii, xiv.
65. Véronique Bragard, "Sparing Words in the Wasted Land: Garbage, Texture, and Écriture Blanche in Auster's *In the Country of Last Things* and McCarthy's *The Road,*" *Interdisciplinary Studies in Literature and Environment* 20.3 (2013): 491. I am applying her assessment of Cormac McCarthy to Beckett. She argues for a Beckettian "poetics of diminishment" as a sign of waste style in Auster and McCarthy (488).
66. Samuel Beckett, *Endgame* (New York: Grove Press, 1958), 2.
67. Ibid., 22–23.
68. Ibid., 48.
69. Ibid., 50.
70. Ibid., 38.
71. Ibid., 53.
72. Ibid., 54.
73. Ibid., 61.
74. Ibid., 77.
75. Ibid., 78.
76. Ibid., 81.
77. Ibid., 44.
78. Ibid., 70.
79. Ibid., 72.
80. Ibid., 79.
81. Ibid., 56.
82. Ibid., 23.
83. Ibid., 58.

12 Gleaning Aesthetics: Poetry as Communal Salvage

1. Quoted by Lisa J. Kiser, "The Garden of St. Francis: Plants, Landscape, and Economy in Thirteenth Century Italy," *Environmental History* 8 (2003): 232, accessed October 19, 2014, http://rbedrosian.com/Gardens/Garden_Italy_Saint_Francis.pdf.
2. André Lacoeque, *The Feminine Unconventional: Four Subversive Figures in Israel's Tradition* (Minneapolis, MN: Fortress Press, 1990), 85.
3. Moabites are wasted humans from the Jewish perspective, hence the extraordinary heritage of David. And a female Moabite lies even lower than male

Moabites, since they "attempted to corrupt the Israelites coming from Egypt on their way to Canaan." Ibid., 85.

4. Lacoeque, *The Feminine Unconventional*, 106.
5. "'All that is mine is yours, and all that is yours remains yours.' Martin Buber's characterization of Hasidic love." Lacoeque, *The Feminine Unconventional*, 96.
6. John Scanlan, *On Garbage* (London: Reaktion, 2005), 25.
7. Gregory M. Lamb, "The Need to Feed Hungry Families Cultivates New Interest in Gleaning," *Christian Science Monitor*, November 2, 2009, accessed November 20, 2014, http://www.csmonitor.com/World/Making-a-difference/2009/1102/p07s01-lign.html.
8. Allan Stoekl, *Bataille's Peak: Energy, Religion, and Postsustainability* (Minneapolis, MN: University of Minnesota Press, 2007), xix, 145–148. See Daniel Lang, "'Give Us the Dumpsters—Or—Give Us Life': *Res Derilictae* and the Trash of Free Trade," *Other Voices* 3.1 (2007): http://www.othervoices.org/3.1/dlang/index.php, accessed April 18, 2012.
9. Gay Hawkins, *The Ethics of Waste: How We Relate to Rubbish* (Lanham, MD: Rowman & Littlefield, 2006), 107.
10. Ibid., 110.
11. Ibid., 114.
12. Ibid., 115.
13. Italo Calvino, "La Poubelle Agréée," in *The Road to San Giovanni*, trans. Tim Parks (New York: Pantheon, 1993), 124.
14. Even the filthiest of places has the potential for resurrection and revival. Royte observes that in Fresh Kills, "the graded road wrapped around an enormous shoulder of grass, and Queen Anne's lace, dandelions, black-eyed Susan, cinquefoil, yellow-blossomed mugwort, blue chicory, and pink multiflora roses covered its 145-food-high plateau." Elizabeth Royte, *Garbage Land: On the Secret Trail of Trash* (New York: Little, Brown, 2005), 88. Royte points out how Fresh Kills has even become sacred with the rubble from the World Trade Center Towers: "hallowed ground . . . the final resting place of heroes" (100–101).
15. Paul Fleischman, *Seedfolks* (New York: HarperTeen, 2004), 65.
16. Ibid., 9.
17. Ibid., 10.
18. Ibid., 3–4.
19. Ibid., 15–16.
20. Ibid., 18.
21. Ibid., 22.
22. Ibid., 42–43.
23. Ibid., 72.
24. Ibid., 30.
25. Ibid., 31.
26. Hawkins, *The Ethics of Waste*, 87. "Mending and restoring objects often require even more creativity than original production." Susan Strasser, *Waste and Want: A Social History of Trash* (New York: Henry Holt, 2000), 10.

27. Strasser, *Waste and Want*, 11, citing Claude Lévi-Strauss, *The Savage Mind* (Chicago, IL: University of Chicago Press, 1966), 17–18, 33.
28. Fleischman, *Seedfolks*, 33.
29. Ibid., 61.
30. Ibid., 65.
31. Ibid., 74.
32. Ibid., 81.
33. Ibid., 87.

13 Compost Aesthetics: The Poet[h]ics of Metaphor

1. John Berger, *Keeping a Rendezvous* (New York: Vintage, 1991), 42.
2. Martha C. Nussbaum, *Poetic Justice: The Literary Imagination and Public Life* (Boston, MA: Beacon Press, 1995), 36.
3. Martha C. Nussbaum, *Upheavals of Thought: The Intelligence of Emotions* (Cambridge: Cambridge University Press, 2001), 661.
4. Ibid., 665.
5. Berger, *Keeping a Rendezvous*, 39.
6. Nussbaum, *Upheavals of Thought*, 221.
7. Martha C. Nussbaum, *From Disgust to Humanity: Sexual Orientation and Constitutional Law* (Oxford and New York: Oxford University Press, 2010), xvii.
8. Ibid., xix.
9. Ibid., 208.
10. See John Keats, "Letter to George and Tom Keats, 21, 27 (?) December 1817," in *The Letters of John Keats*, ed. Robert Gittings (Oxford: Oxford University Press, 1970/1982), 43. "Seeing yourself from another point of view is the beginning of ethics and politics." Timothy Morton, *The Ecological Thought* (Cambridge, MA: Harvard University Press, 2010), 14.
11. Sophocles, *Philoctetes*, trans. Carl Phillips (Oxford: Oxford University Press, 2003), 65, line 789.
12. Ibid., 90, line 1437.
13. "Mindfulness is political because the restoration of our senses is also a restoration of our ties to the material world and to the consequences of our actions." Linda Holler, *Erotic Morality: The Role of Touch in Moral Agency* (New Brunswick, NJ: Rutgers University Press, 2002), 171, 97, 172.
14. Gay Hawkins, *The Ethics of Waste: How We Relate to Rubbish* (Lanham, MD: Rowman & Littlefield, 2006), 48.
15. Eileen A. Joy, "You Are Here: A Manifesto," in *Animal, Vegetable, Mineral: Ethics and Objects*, ed. Jeffrey Jerome Cohen (Washington, DC: Oliphaunt Books, 2012), 161.
16. "But it is also very valuable to extend this literary understanding by seeking out literary experiences in which we do identify sympathetically with individual members of marginalized or oppressed groups within our own society, learning both to see the world, for a time, through their eyes and then reflecting as spectators on the meaning of what we have seen." Nussbaum, *Poetic Justice*, 92.

17. Ottmar Ette, "Literature as Knowledge for Living, Literary Studies as Science for Living," trans. Vera M. Kutzinski, *PMLA* 125 (2010): 987.
18. Nussbaum, *Upheavals of Thought*, 677.
19. All quotes from Walt Whitman, *Song of Myself (1892 version)*, accessed October 25, 2014. http://www.poetryfoundation.org/poem/174745.
20. Nussbaum, *Upheavals of Thought*, 351.
21. Ette, "Literature as Knowledge," 988.
22. Wellcome Collection, May 14, 2011, interview of Rob Smith, Chief Flusher, Thames Water, and facilitator Rosie Cox.
23. The poem continues, "And wheeling to it, storming up the slope, / I think of the angle of repose the manure / pile assumes, how sparrows come to pick / the redelivered grain." Maxine Kumin, "The Excrement Poem," in *After Frost: An Anthology of Poetry from New England*, ed. Henry Lyman (Amherst, MA: University of Massachusetts Press, 1996), 201.
24. "Eating constitutes a series of mutual transformations between human and nonhuman materials . . . In the eating encounter, all bodies are shown to be but temporary congealments of a materiality that is a process of becoming." Jane Bennett, *Vibrant Matter: A Political Ecology of Things* (Durham: Duke University Press, 2010), 40, 49.
25. *The Poetry Paper*, Issue Seven 2010/11, Editor Dean Parkin, The Poetry Trust, 13.
26. Susan Stanford Friedman, "Why Not Compare?," *PMLA* 126.3 (May 2011): 757.
27. Vilashini Cooppan, "Net Work: Area Studies, Comparison, and Connectivity," *PMLA* 128.3 (2013): 615.
28. See Thomas Claviez, "Done and Over With—Finally? Otherness, Metonymy, and the Ethics of Comparison," *PMLA* 128.3 (2013): 613, on metonymy.
29. As Joy and Ramsey have contested, echoing W. G. Sebald (2004). E. A. Joy and M. K. Ramsey. "Introduction: Liquid *Beowulf*," in *The Postmodern Beowulf: A Critical Casebook*, ed. E. A. Joy and M. K. Ramsey (Morgantown, WV: West Virginia Press, 2006): lv. See W. G. Sebald, "An Attempt at Restitution," *The New Yorker*, December 20 and 27, 2004, accessed October 19, 2014, http://www.newyorker.com/magazine/2004/12/20/an-attempt-at-restitution.
30. Erich Auerbach, "Philology and *Weltliteratur*," trans. Marie and Edward Said, *The Centennial Review* 13.1 (1969): 17.
31. Ette, "Literature as Knowledge," 989.
32. Ibid., 990.
33. As Katherine Hayles articulates, her dream "is a version of the posthuman . . . that understands human life [as] embedded in a material world of great complexity, one on which we depend for our continued survival." N. Katherine Hayles, *How We Became Posthuman: Virtual Bodies in Cybernetics, Literature, and Informatics* (Chicago, IL: University of Chicago Press, 1999), 5.
34. A. R. Ammons, *Garbage* (New York: W. W. Norton, 2002), 56.
35. Ibid., 103–104.
36. Ibid., 106.

14 Poetry as Homeopathy: The Poet as Ragpicker

1. Emmanuel Levinas, "Ethics of the Infinite," in *States of Mind: Dialogues with Contemporary Thinkers*, ed. Richard Kearney (New York: New York University Press, 1995), 190, 192.
2. See Karl Steel, "With the World, or Bound to Face the Sky: The Postures of the Wolf-Child of Hesse," in *Animal, Vegetable, Mineral: Ethics and Objects*, ed. Jeffrey Jerome Cohen (Washington, DC: Oliphaunt Books, 2012), 25.
3. Cited by Martha C. Nussbaum, *Upheavals of Thought: The Intelligence of Emotions* (Cambridge: Cambridge University Press, 2001), 659.
4. Myra J. Hird, "Feminist Matters: New Materialist Considerations of Sexual Difference," *Feminist Theory* 5 (2004): 224.
5. Ibid., 228.
6. Michael Swanton (trans. and ed.), *Anglo-Saxon Prose* (London: Everyman/J. M. Dent, 1993), 230.
7. Michael Uebel's visit to "Waste Studies: Cultural Refuse/als in Medieval Literature," Texas State University, February 11, 2010.
8. Levinas, "Ethics of the Infinite," 190, 192.
9. Walter Benjamin, *Illuminations: Essays and Reflections*, ed. Hannah Arendt (New York: Schocken Books, 1968), 256.
10. As O'Brien O'Keefe argues, "By trying to assign Grendel a fixed essence, we reassure ourselves, for by doing so we provide him with limits. Indeed, when Grendel is lurking in the fens and marches, he is not so very frightening, for we understand what he is . . . Grendel is at his most terrifying not in the marches but in the place of men." Katherine O'Brien O'Keefe, "*Beowulf*, Lines 702b–836: Transformations and the Limits of the Human," *Texas Studies in Literature and Language* 23.4 (1981): 492.
11. Sianne Ngai, "Raw Matter: A Poetics of Disgust," in *Telling It Slant*, ed. Mark Wallace and Steven Marks (Tuscaloosa, AL: University of Alabama Press, 2002): 168.
12. Ibid., 184.
13. Ibid., 186.
14. See Jeffrey Jerome Cohen, *Of Giants: Sex, Monsters, and the Middle Ages* (Minneapolis, MN: University of Minneapolis Press, 1999), 4.
15. Linda Holler, *Erotic Morality: The Role of Touch in Moral Agency* (New Brunswick, NJ: Rutgers University Press, 2002), 169.
16. Zygmunt Bauman, *Wasted Lives: Modernity and Its Outcasts* (Cambridge: Polity Press, 2004), 128.
17. David Sandner, "Tracking Grendel: The Uncanny in *Beowulf*," *Extrapolation* 40 (1999): 167. Indeed, "Grendel stands as an uncanny double for Beowulf, for humanity, and for the modern reader" (175).
18. Bauman, *Wasted Lives*, 106. "The memory of Æschere's body having been both ingested and also discarded, almost as trash, along the tracks of the stateless forest, serves as a frightening rebuke to the idea that anyone could ever be safe, *at home*, from the enemy." Eileen A. Joy, "'In his eyes stood a light, not beautiful': Levinas, Hospitality, *Beowulf*," in *Levinas and Medieval Literature:*

The "Difficult Reading" of English and Rabbinic Texts, ed. Ann Astell and Justin Jackson (Pittsburgh, PA: Duquesne University Press, 2009), 80. As Eli Clare writes, "Defects are disposable and *abnormal*, bodies to eradicate." Eli Clare, "Meditations on Natural Worlds, Disabled Bodies, and a Politics of Cure," in *Material Ecocriticism*, ed. Serenella Iovino and Serpil Oppermann (Bloomington: Indiana University Press, 2014), 205.

19. Quoted by T. S. Eliot in *The Waste Land*, http://www.poetryfoundation.org/poem/176735. Accessed October 8, 2014.
20. Myra J. Seaman, "Becoming More (than) Human: Affective Posthumanisms, Past and Future," *Journal of Narrative Theory* 37.2 (2007): 246.
21. Ibid., 269.
22. Ibid., 247; see also Jeffrey Jerome Cohen, *Medieval Identity Machines* (Minneapolis, MN: University of Minnesota Press 2003), xiii.
23. Lennard J. Davis, "Enforcing Normalcy: Disability, Deafness, and the Body," in *Visualizing the Disabled Body: The Classical Nude and the Fragmented Torso*, reprinted in *The Norton Anthology of Theory and Criticism*, ed. Vincent B. Leitch (New York: Norton, 2001), 2414.
24. As Bennett points out, the "association of matter with passivity still haunts us today." Jane Bennett, *Vibrant Matter: A Political Ecology of Things* (Durham: Duke University Press, 2010), 65.
25. Véronique Bragard, "Introduction: Languages of Waste: Matter and Form in Our Garb-age," *Interdisciplinary Studies in Literature and Environment* 20.3 (2013): 462.
26. Bennett, *Vibrant Matter*, 65. "Nature is not the passive, inert, or 'dead' matter that Frankenstein imagines." Anne K. Mellor, "Possessing Nature: The Female in *Frankenstein*," reprinted in Mary Shelley, *Frankenstein A Norton Critical Edition*, 2nd ed., ed. J. Paul Hunter (New York: W. W. Norton, 2012): 363, from *Romanticism and Feminism*, ed. Anne K. Mellor (Bloomington: Indiana University Press, 1988).
27. "The idea of touch always initiates a dialectic of attraction and repulsion, of fear, hatred, or erotic attraction . . . So, the fragmented body is hacked up, exploded, into the fragments that make it up." Davis, "Enforcing Normalcy," 2415.
28. Holler, *Erotic Morality*, 113.
29. See Timothy Morton, *The Ecological Thought* (Cambridge, MA: Harvard University Press, 2010), 112.
30. Michael Wines, "China's Impolitic Artist, Still Waiting to be Silenced," *The New York Times*, November 28, 2009, accessed October 18, 2014, http://www.nytimes.com/2009/11/28/world/asia/28weiwei.html.
31. Mark Twain, *The Adventures of Huckleberry Finn* (New York: Bantam, 1981), 84.
32. Norbert Elias, *The Civilizing Process* (Oxford: Blackwell, 1994/2000), 69.
33. Martin Halliwell and Andy Mousley, *Critical Humanisms: Humanist/Anti-Humanist Dialogues* (Edinburgh: Edinburgh University Press, 2003), 2.
34. Elizabeth Alvilda Petroff, *Medieval Women's Visionary Literature* (New York: Oxford University Press, 1986), 157.
35. Ibid., 158.
36. Mark Atherton (trans.), *Hildegard von Bingen: Selected Writings* (London: Penguin, 2001), 166–167.
37. Ibid., 120.

38. Calvert Watkins, *The American Heritage Dictionary of Indo-European Roots*, 3rd ed. (Boston, MA: Houghton Mifflin Harcourt, 2011), 52.
39. Thanks to Allison Estrada-Carpenter for suggesting this.
40. Atherton, *Hildegard von Bingen*, 118.
41. See Donna J. Haraway, "A Cyborg Manifesto: Science, Technology, and Socialist-Feminism in the Late Twentieth Century," *Simians, Cyborgs, and Women: The Reinvention of Nature* (New York: Routledge, 1991), 155–156.
42. Ibid., 157, mentions "effective" affinities, while "elective" affinities comes from Goethe's novel of that title, *Die Wahlverwandtschaften*. See also Steven Shaviro, *Passion & Excess: Blanchot, Bataille, and Literary Theory* (Tallahassee: Florida State University Press, 1990), 98, where he argues that the communal moment for Bataille is "not one of fusion . . . but one of explosion, when the boundaries of exclusion (and also, therefore, the constraints of self-definition) are swept away."
43. William Ian Miller, *The Anatomy of Disgust* (Cambridge, MA: Harvard University Press, 1997), 284. n. 37.
44. For another valence in this vein, see a discussion on "God himself" as the original pervert, which modernity repudiates and attempts to contain. Jonathan Dollimore, *Sexual Dissidence: Augustine to Wilde, Freud to Foucault* (Oxford: Clarendon Press, 1991), 27.
45. These include homo sustinens, homo reciprocans, homo politicus, or homo ecologicus. Emilia Ferraro and Louise Reid, "On Sustainability and Materiality: Homo Faber, a New Approach," *Ecological Economics* 96 (2013): 127 [125–131].
46. Ibid., 128. Their framework "emphasises the interdependence of materiality and practice by focusing on the processes through which humans are makers rather than profiteers."
47. Citing Bennett, they acknowledge that the focus on materiality helps us to see "humans not as separate and superior to other living organisms." Ferraro and Reid, "On Sustainability and Materiality," 129. They emphasize "living" as opposed to any thing/object, living or inert.
48. As Jane Bennett agrees, eloquently arguing, "Texts are bodies that can light up, by rendering human perception more acute, those bodies whose favored vehicle of affectivity is less wordy: plants, animals, blades of grass, household objects, trash . . . Poetry can help us feel more of the liveliness hidden in such things and reveal more of the threads of connection binding our fate to theirs." Jane Bennett, "Systems and Things: A Response to Graham Harman and Timothy Morton," *New Literary Theory* 43 (2012): 232.
49. Compare what Cleanth Brooks writes about the urn holding the "ashes of the phoenix" in Donne's "Canonization": "The urn to which we are summoned . . . is the poem itself." Cleanth Brooks, *The Well Wrought Urn: Studies in the Structure of Poetry* (New York: Harcourt Brace Jovanovich, 1947), 20–21.
50. See quotes from Ann Murphy, "The Political Significance of Shame," *Borderlands e-journal* 3 (2004), accessed October 21, 2014, http://www.borderlands.net.au/vol3no1_2004/murphy_shame.htm.
51. Emmanuel Levinas, *Autrement qu'être ou au-delà l'essence*, *Phaenomenologica*, vol. 54 (The Hague: Martinus Nijhoff, 1974), trans. Alphonso Lingis,

Otherwise than Being or Beyond Essence (The Hague: Martinus Nijhoff, 1981), 124. Cited in Murphy, "The Political Significance of Shame."

52. Murphy, "The Political Significance of Shame."
53. Twain, *Huckleberry Finn*, 84.
54. A. R. Ammons, *Garbage* (New York: W. W. Norton, 2002), 52.
55. "Dirt can purify by drawing dirt unto itself, the idea that the unclean can have a special cleansing power, and the belief that the voluntary embrace of filth can have a prophylactic power against the polluting effects of filth." Heinrich Von Staden, "Women and Dirt," *Helios* 19 (1992): 16. Boscagli writes about Walter Benjamin's homeopathic use of fetishism "to unmake the fetishism of the commodity." Maurizia Boscagli, *Stuff Theory: Everyday Objects, Radical Materialism* (New York: Bloomsbury, 2014), 47.
56. "Organisms like the millions of bacteria, viruses and especially worms that enter the body along with 'dirt' spur the development of a healthy immune system." Jane E. Brody, "Babies Know: A Little Dirt is Good for You," *The New York Times*, January 26, 2009, accessed October 18, 2014, http://www.nytimes.com/2009/01/27/health/27brod.html. See Mary Ruebush, *Why Dirt is Good: 5 Ways to Make Germs Your Friends* (New York: Kaplan, 2009). See also http://uk.news.yahoo.com/22/20110119/tsc-oukoe-uk-faeces-transplants-011ccfa.html.
57. In the early nineteenth century, criteria for charges of obscenity focused on three aspects: the intent of the accused ("a scandalous and evil disposed person"), the corruption of youth ("contriving, devising and intending, the morals as well of youth as of other good citizens of said commonwealth to debauch and corrupt, and to raise and create in their minds inordinate and lustful desire"), and disturbance of the peace. These quotes come from the Holmes Case, Massachusetts, 1821, concerning *Memoirs of a Woman of Pleasure*. Felice Flanery Lewis, *Literature, Obscenity, & the Law* (Carbondale, IL: Southern Illinois University Press, 1976), 5.
58. Lewis, *Literature, Obscenity, & the Law*, 7.
59. Edward de Grazia, *Girls Lean Back Everywhere: The Law of Obscenity and the Assault on Genius* (New York: Random House, 1992), 4.
60. Donna Dennis, *Licentious Gotham: Erotic Publishing and Its Prosecution in Nineteenth-Century New York* (Cambridge, MA: Harvard University Press, 2009), 265–266.
61. Grazia, *Girls Lean Back Everywhere*, 4.
62. Lewis, *Literature, Obscenity, & the Law*, 41.
63. James C. N. Paul and Murray L. Schwartz, *Federal Censorship: Obscenity in the Mail* (New York: Free Press of Glencoe, 1961), 47–48.
64. Grazia, *Girls Lean Back Everywhere*, 33.
65. Ibid., 37.
66. After a meeting at which advocates on both sides showed up and spoke in force, the city council delayed action and a new library board was chosen, chaired by an outspoken opponent of book-branding. Gerald Ashford, "Book Branding Issue Dead in Texas So Far," *The Daily Republic*, 4, Mitchell, South Dakota, Friday July 24, 1953, accessed October 19, 2014, http://newspaperarchive.com/mitchell-daily-republic/1953-07-24/page-4.

67. Nicholas J. Karolides, Margaret Bald, and Dawn B. Sova, *100 Banned Books: Censorship Histories of World Literature* (New York: Checkmark Books, 1999), 360.
68. Ibid., 361.
69. In *Virgil v. School Board of Columbia County*, 677 F. Supp. 1547, 1551–51 (M.D. Fla. 1988); see Dana B. Sova, *Banned Books: Literature Suppressed on Social Grounds*, revised edition (New York: Facts on File, 2006), 77.
70. This argument was based on a decision made in 1982 concerning Richard Wright's *Black Boy*; the court decided in that case that, in arbitrarily removing books, the school board in Island Trees Union Free School District of New York violated the First Amendment rights of students. Karolides, Bald, and Sova, *100 Banned Books*, 361.
71. From www.udel.edu/educ/whitson/897s05/files/Virgil.pdf, accessed October 22, 2014.
72. Karolides, Bald, and Sova, *100 Banned Books*, 361.
73. From the decision of the United States Court of Appeals, Eleventh Circuit, *Virgil v. School Board of Columbia County*, Florida, January 12, 1989, accessed October 22, 2014, http://openjurist.org/862/f2d/1517/virgil-v-school-board-of-columbia-county-florida.
74. Ibid.
75. Ibid.
76. Ibid.
77. Ibid.
78. www.udel.edu/educ/whitson/897s05/files/Virgil.pdf.
79. *Virgil v. School Board of Columbia County*, January 12, 1989.
80. www.udel.edu/educ/whitson/897s05/files/Virgil.pdf.
81. *Virgil v. School Board of Columbia County*, January 12, 1989.
82. Removal of books is legitimate when it is "'reasonably related to legitimate pedagogical concerns' of denying students exposure to "potentially sensitive topics." *Virgil v. School Board of Columbia County*, January 12, 1989.
83. Thanks to Ruth Evans for this insight as a respondant to my talk, "Dirty Chaucer? Poetry as Homeopathy," at the "Dirty Chaucer" MLA session, January 3, 2013, in Boston.
84. Watkins, *The American Heritage Dictionary*, 79.
85. Thomas Aquinas, *Summa Theologica*, III Qu.82 a.5, accessed May 16, 2013, http://www.newadvent.org/summa/4082.htm.
86. "Pourtant de nombreux textes apocryphes, condamnés au Ve siècle par le Pape Gélase, interdiction levée au XVIe siècle, comprennent un ensemble d'épisodes dans lesquels les *excreta* du Christ paraissent avoir le caractère de panacée et être dotés de propriétés miraculeuses. La simple eau dans laquelle Jésus-Christ enfant aurait été toiletté, aurait dispensé ses miracles et le Premier Évangile de l'Enfance de Jésus-Christ aurait été reçu pour canonique par les gnostiques du second siècle." Cyrille Harpet, *De Déchet: Philosophie des Immondices Corps, Ville, Industrie* (Paris: L'Harmattan, 1998), 203. "The inference is that the excreta of Christ were believed, as in many other instances, to have the character of a panacea, as well as generally miraculous properties." John G. Bourke, *The Scatalogic Rites of All Nations* (Washington, DC: W. H. Lowdermilk, 1891), 56.

87. "Ecological thinking . . . is therefore akin to metaphorical thinking, and because metaphor is used in the most intense, complex, and self-reflexive ways in poetic language, the discourse of ecology and the discourse of poetry and literature are intrinsically related to each other through the shared relevance of metaphor." Hubert Zapf, "Creative Matter and Creative Mind: Cultural Ecology and Literary Creativity," in *Material Ecocriticism*, ed. Serenella Iovino and Serpil Oppermann (Bloomington: Indiana University Press, 2014), 54. Bursting and linking, these metaphorical devices work in complementary ways. Metaphor and metonymy "are the real means by which both natural and cultural semiosis drives natural and cultural evolution and development." Wendy Wheeler, "Natural Play, Natural Metaphor, and Natural Stories: Biosemiotic Realism," in *Material Ecocriticism*, ed. Serenella Iovino and Serpil Oppermann (Bloomington: Indiana University Press, 2014), 70.
88. Zapf, "Creative Matter," 55; see also Wheeler, "Natural Play," 79, quoting Paul Ricoeur, *The Rule of Metaphor: The Creation of Meaning in Language*, trans. Robert Czerny, Kathleen McLaughlin, and John Costello, S. J. (London: Routledge, 2003), 235.
89. Graham Harman, *Guerrilla Metaphysics: Phenomenology and the Carpentry of Things* (Chicago, IL: Open Court, 2005), 112. Harman quotes Jacques Derrida, "White Mythology," in *Margins of Philosophy*, trans. A. Bass (Chicago, IL: University of Chicago Press, 1982), 247, who in turn is analyzing Aristotle's *Poetics*. Levinas, Harman tells us, "has no wish to return to an old-fashioned theory of natural lumps of substance whose relations would only be accidental . . . Objects always lie partly beyond the sensible." Graham Harman, "Levinas and the Triple Critique of Heidegger," *Philosophy Today* 53 (2009): 409.
90. José Ortega y Gasset's "An Essay in Esthetics by Way of a Preface" can be found in Ortega's *Phenomenology and Art*, trans. P. Silver (New York: Norton, 1975), 127–160.
91. Harman, *Guerrilla Metaphysics*, 103.
92. Ibid., 104, quoting Ortega, *Phenomenology and Art*, 134.
93. Ibid., 105.
94. Ibid., 106.
95. Ibid., 106–107.
96. Ibid., 109.
97. Ibid., 110.
98. Watkins, *The American Heritage Dictionary*, see *sta* 86–87 and *kad* 37.
99. See Thomas G. Long, "Chronicle of a Death We Can't Accept," *The New York Times*, November 1, 2009, accessed October 18, 2014, http://www.nytimes.com/2009/11/01/opinion/01long.html.
100. Hiroko Tabuchi, "Japan Finds Story of Hope in Undertaker Who Offered Calm Amid Disaster," *The New York Times*, March 11, 2012, accessed March 19, 2012, http://www.nytimes.com/2012/03/11/world/asia/a-year-later-undertakers-story-offers-japan-hope.html.
101. Emmanuel Levinas, *Humanism of the Other*, trans. Nidra Poller, Introduction by Richard A. Cohen (Urbana, IL: University of Illinois Press, 2006), 16.
102. Paul Fleischman, *Seedfolks* (New York: HarperTeen, 2004), 47, 50.

103. Bauman asserts that our morality needs to be the long-term ethics of "self-limitation." Bauman, *Wasted Lives*, 101.
104. Zygmunt Bauman, *Postmodern Ethics* (Oxford: Blackwell, 1993), 220.
105. A. R. Ammons, "Still," accessed October 26, 2014, http://www.poets.org/poetsorg/poem/still.
106. For example, R. Allen Shoaf, *Chaucer's Body: The Anxiety of Circulation in the Canterbury Tales* (Gainesville, FL: University Press of Florida, 2001), 59–61; Lee Patterson, "Perpetual Motion: Alchemy and the Technology of the Self," *Studies in the Age of Chaucer* 15 (1993): 55, 56; and C. David Benson, "Literary Contests and London Records in the *Canterbury Tales*," in *Chaucer and the City*, ed. Ardis Butterfield (Cambridge: D. S. Brewer, 2006), 143.
107. Bauman, *Wasted Lives*, 22.
108. See Wohlfarth, on the analogy between poet and rag-man for whom refuse is a concern. The *Lumpensammler* feeds his refuse back into the jaws of Industry, while "his literary counterpart seeks . . . to save *his* 'treasure' *from* the capitalist order of things in order to construct objects that will help upset its digestive system." Irving Wohlfarth, "Et Cetera? The Historian as Chiffonnier," *New German Critique* 39 (1986): 152; also 148, 151, n. 9; Barrie M. Ratcliffe, "Perceptions and Realities of the Urban Magic: The Rag Pickers of Paris in the First Half of the Nineteenth Century," *Canadian Journal of History* 27 (1992): 233.
109. "Yes, these people harassed by domestic worries, / Ground down by their work, distorted by age, / Worn-out, and bending / beneath a load of debris, / The commingled vomit of enormous Paris." From Charles Baudelaire, "Le Vin de chiffonniers," trans. William Aggeler, *The Flowers of Evil* (Fresno, CA: Academy Library Guild, 1954), accessed May 16, 2013, http://fleursdumal.org/poem/193.
110. Eliot, *The Waste Land*.
111. Czeslaw Milosz, "Ruins and Poetry," in *The Witness of Poetry* (Cambridge: Harvard University Press, 1984), 97.
112. Gay Hawkins, *The Ethics of Waste: How We Relate to Rubbish* (Lanham, MD: Rowman & Littlefield, 2006), 88.
113. Ammons, *Garbage*, dedication page.
114. Nigel Clark and Myra J. Hird, "Deep Shit," *O-Zone: A Journal of Object-Oriented Studies* 1 (2014): 45.
115. As Clark and Hird have argued, "What some of us have lately taken to calling the Anthropocene, is a human signature, a superficial flourish, on what remains, indelibly, a bacterially-orchestrated biosphere." Clark and Hird, 51.
116. Ammons, *Garbage*, 18.
117. Ibid., 20–21.
118. Timothy Morton, "An Object-Oriented Defense of Poetry," *New Literary History* 43 (2012): 215.
119. Wohlfarth, "Et Cetera?," 152.
120. Ibid., 50–51.
121. "One must write and / rewrite till one writes it right." Ibid., 22. See also Italo Calvino, "La Poubelle Agréée," in *The Road to San Giovanni*, trans. Tim Parks (New York: Pantheon, 1993), 109, 124–125.

122. Ammons, *Garbage*, 108–109.
123. Ibid., 42.
124. Ibid., 28.
125. Ibid., 32.
126. Calvino, "La Poubelle Agréée," 105.
127. Ibid., 106.
128. Ibid., 121.
129. Ammons, *Garbage*, 33–34.
130. Ibid., and 63.
131. Writing as part of a movement to "degrow" the global economy, Erik Assadourian distinguishes between "degrowth" and "decline" and urges the decoupling of growth from material consumption. Erik Assadourian, "The Path to Degrowth in Overdeveloped Countries," in *2012 State of the World: Moving toward Sustainable Prosperity*, ed. Matt Richmond (Washington, DC: Island Press, 2012): 23, accessed May 7, 2012, http://blogs.worldwatch.org/sustainableprosperity/sow2012/. Serres takes this even further by arguing that pollution stems from ownership and "*appropriation takes place through dirt.*" Michel Serres, *Malfeasance: Appropriation Through Pollution?*, trans. Anne-Marie Feenberg-Dibon (Stanford, CA: Stanford University Press, 2011), 3. "Pollution is the sign of the world's appropriation by the *species*." Serres, *Malfeasance*, 53. The only solution to the overwhelming degradation of the plant is that we become tenants, not owners, of the world. According to Serres, the "dispossession of the world" is our only hope. Serres, *Malfeasance*, 72–73.
132. Ottmar Ette, "Literature as Knowledge for Living, Literary Studies as Science for Living," trans. Vera M. Kutzinski, *PMLA* 125 (2010): 986.
133. In the documentary, *The Examined Life*, Slavoj Žižek wanders around garbage as he philosophizes. "The difficult thing is to find poetry, spirituality in this dimension. To recreate, if not beauty, then aesthetic dimension in things like this, in trash itself. That's the true love of the world . . . Love is not idealization . . . The true ecologist loves all this." He then gestures at trash. Slavoj Žižek, in *The Examined Life*, (Dir.) Astra Taylor (2008). See https://www.youtube.com/watch?v=iGCfiv1xtoU. Many thanks to Eric Leake for drawing my attention to this segment.
134. See Walt Whitman, *Leaves of Grass: The First (1855) Edition*, ed. Malcolm Cowley (Harmondsworth: Penguin, 1984), 5.
135. Calvert Watkins, *How to Kill a Dragon: Aspects of Indo-European Poetics* (Oxford: Oxford University Press, 1995), 521.
136. Ibid., 534.
137. Serpil Oppermann, "From Ecological Postmodernism to Material Ecocriticism: Creative Materiality and Narrative Agency," in *Material Ecocriticism*, ed. Serenella Iovino and Serpil Oppermann (Bloomington: Indiana University Press, 2014), 27.

Bibliography

Primary Literary Works

Adler, H. G. *The Journey: A Novel.* Translated by Peter Filkins. Afterword by Jeremy Adler. New York: Random House, 2008.

Allison, Dorothy. *Trash.* New York: Plume, 1988/2002.

Ammons, A. R. *Garbage.* New York: W. W. Norton, 2002.

———. "Still." Accessed October 26, 2014. http://www.poets.org/poetsorg/poem/still.

Anderson, J. J. *Sir Gawain and the Green Knight, Pearl, Cleanness, Patience.* London: Everyman Paperback Classics, 1996.

Aquinas, Thomas. *Summa Theologica.* III Qu.82 a.5. Web. Accessed May 16, 2013. http://www.newadvent.org/summa/4082.htm.

Atherton, Mark, trans. *Hildegard von Bingen: Selected Writings.* London: Penguin, 2001.

Babylonian Talmud: Tractate Berakoth. Accessed October 23, 2014. http://www.come-and-hear.com/berakoth/berakoth_23.html.

Barrell, John, and John Bull, eds. *A Book of English Pastoral Verse.* New York: Oxford University Press, 1975.

Barthelme, Donald. "Nothing: A Preliminary Account." *Sixty Stories,* 245–248. New York: G. P. Putnam's Sons, 1981.

Baudelaire, Charles. "Le Vin de chiffonniers." Translated by William Aggeler. *The Flowers of Evil.* Fresno, CA: Academy Library Guild, 1954. Accessed May 16, 2013. http://fleursdumal.org/poem/193.

Beckett, Samuel. *Endgame.* New York: Grove Press, 1958.

Bede. *The Ecclesiastical History of the English People.* Translated by Leo Sherley-Price. Harmondsworth: Penguin, 1990.

Berger, John. *Keeping a Rendezvous.* New York: Vintage, 1991.

Bird, Caroline. "Joy Is Like a Hungry Pig." *The Poetry Paper,* Issue Seven 2010/11, Editor Dean Parkin, The Poetry Trust, 13.

Blamires, Alcuin, ed. *Woman Defamed and Woman Defended: An Anthology of Medieval Texts.* Oxford: Clarendon Press, 1992.

Böll, Heinrich. "In Defense of 'Rubble Literature.'" In *Stories, Political Writings, and Autobiographical Works,* edited by Martin Black, translated by Leila Vennewitz, 269–273. New York: Continuum, 2006.

———. "In Defense of Washtubs." In *Stories, Political Writings, and Autobiographical Works*, edited by Martin Black, translated by Leila Vennewitz, 274–275. New York: Continuum, 2006.

Borges, Jorge Luis. "The Analytical Language of John Wilkins" (1941). Accessed December 3, 2009. http://www.alamut.com/subj/artiface/language/johnWilkins.html.

———. *Collected Fictions*. Translated by Andrew Hurley. New York: Viking, 1998.

Bradley, S. A. J., trans. and ed. *Anglo-Saxon Poetry*. London: J. M. Dent & Sons, 1982.

Brie, Friedrich W. D. *Eulenspiegel in England*. Berlin: Mayer & Müller, 1903.

Bruster, Douglas, and Eric Rasmussen, eds. *Everyman and Mankind*. Arden Early Modern Drama. London: Methuen Drama, 2009.

Calendar of Patent Rolls, Richard II A. D. 1388–1392, Vol. IV. London: Mackie, 1902.

Campion, Peter. trans. "Who Is So Smart, So Crafty-Spirited?" In *The Word Exchange: Anglo-Saxon Poems in Translation*, edited by Greg Delanty and Michael Matto, 66–73. New York: W. W. Norton, 2011.

Calvino, Italo. "La Poubelle Agréée." In *The Road to San Giovanni*. Translated by Tim Parks, 91–126. New York: Pantheon, 1993.

Chaucer, Geoffrey. *The Riverside Chaucer*, edited by Larry D. Benson. 3rd ed. Boston: Houghton Mifflin, 1987.

Dahl, Roald. *The BFG*. New York: Scholastic, 1982.

Deguileville, Guillaume de. *The Pilgrimage of Human Life*. Translated by Eugene Clasby. New York: Garland, 1992.

Delanty, Greg, and Michael Matto, eds. *The Word Exchange: Anglo-Saxon Poems in Translation*. New York: W. W. Norton, 2011.

DeLillo, Don. *Underworld*. London: Picador, 1997.

Deschamps, Eustache. "Ballade Showing the Causes and Reasons for the Plague." Translated by Anne M. Dropick. *Harper's Magazine* March 2007: 16.

Deuteronomy. Accessed February 9, 2010. http://web.archive.org/web/20080512012616/http://etext.lib.virginia.edu/toc/modeng/public/KjvDeut.html.

Diamond, Catherine. "Hiking in Yangming Mountain: 'Listening.'" *Interdisciplinary Studies in Literature and Environment* 20.3 (2013): 683–684.

DuVal, John, trans. *Cuckolds, Clerics, and Countrymen: Medieval French Fabliaux*. Fayetteville: University of Arkansas Press, 1982.

Eliot, T. S. *The Waste Land*. Accessed October 8, 2014. http://www.poetryfoundation.org/poem/176735.

Fitzgerald, F. Scott. *The Great Gatsby*. New York: Scribner, 1925/2004.

Fleischman, Paul. *Seedfolks*. New York: HarperTeen, 2004.

Franklin, Ben. *The Way to Wealth*. 1758.

Fulk, R. D., Bobert E. Bjork, and John D. Niles, eds. *Klaeber's Beowulf and the Fight at Finnsburg*. 4th ed. Toronto: University of Toronto Press, 2008.

Green, Monica Helen, trans. *The Trotula: An English Translation of the Medieval Compendium of Women's Medicine*. Philadelphia: University of Pennsylvania Press, 2002.

Gustafson, Kevin, trans. and ed. *Cleanness*. Buffalo, NY: Broadview, 2010.

Gwynn, R. S., ed. *Literature: A Pocket Anthology*. 5th ed. Boston: Longman, 2012.

Harington, John. *A New Discourse of a Stale Subject, Called the Metamorphosis of Ajax*, edited by Elizabeth Story Donno. New York: Columbia University Press, 1962.

Heeresma, Marcus. "Dumping Ground." In *The Dedalus Book of Dutch Fantasy*, edited and translated by Richard Huijing, 142–154. Sawtry, Cambridgeshire: Dedalus, 1993.

Hildegard of Bingen. *Scivias*. Translated by Mother Columba Hart and Jane Bishop. New York: Paulist Press, 1990.

Isidore of Seville. *Isidori Hispalensis Episcopi, Etymologiarum sive Originum*, edited by W. M. Lindsay. Oxford: Clarendon Press, 1962.

Jago, Richard. "The Scavengers: A Town Eclogue." In *A Book of English Pastoral Verse*, edited by John Barrell and John Bull, 314–316. New York: Oxford University Press, 1975.

Jelinek, Elfriede. "Services." *Gate Biennale*. Translated by Nick Grindell. London: Methuen, 1996.

Joyce, James. *Ulysses*. New York: Vintage, 1986.

Karon, Jan. *These High, Green Hills*. New York: Penguin, 1996.

Keats, John. *The Letters of John Keats*, edited by Robert Gittings. Oxford: Oxford University Press, 1970/1982.

Kumin, Maxine. "The Excrement Poem." In *After Frost: An Anthology of Poetry from New England*, edited by Henry Lyman, 201. Amherst, MA: University of Massachusetts Press, 1996.

Langland, William. *Piers Plowman*, edited by Elizabeth Robertson and Stephen H. A. Shepherd, translated by E. Talbot Donaldson. New York: W. W. Norton, 2006.

———. *The Vision of Piers Plowman*. London and New York: J. M. Dent and E. P. Dutton, 1978. Accessed March 30, 2010. http://quod.lib.umich.edu/cgi/t/text/text-idx?c=cme;cc=cme;rgn=div1;view=toc;idno=PPlLan;node=PPlLan%3A1.

Lazarus, Emma. "The New Colossus." Accessed October 8, 2014. http://www.poetryfoundation.org/poem/175887.

Leviticus quotes from http://etext.virginia.edu/toc/modeng/public/KjvLevi.html. Accessed February 9, 2010.

Lionni, Leo. *little blue and little yellow*. New York: McDowell, Obolensky, 1959.

Liuzza, Roy M. "Two Old English Elegies from the Exeter Book: *The Wanderer* and *The Ruin*." Accessed October 19, 2014. https://web.utk.edu/~rliuzza/401/Elegies.pdf.

Lothario Dei Segni (Pope Innocent III). *On the Misery of the Human Condition: De miseria humane conditionis*, edited by Donald R. Howard, translated by Margaret Mary Dietz. Indianapolis: Bobbs-Merrill, 1969.

Lydgate, John. *The Pilgrimage of the Life of Man*, edited by F. J. Furnivall and Katharine B. Locock, *EETS* 77, 83, 92 (1899, 1901, 1904). Reprinted as one volume. Woodbridge, Suffolk: Boydell & Brewer, 1996.

McNeill, John T., and Helena M. Gamer, eds. and trans. *Medieval Handbooks of Penance: A Translation of the Principle Libri Poenitentiales*. New York: Columbia University Press, 1938/1990.

Milosz, Czeslaw. "Ruins and Poetry." *The Witness of Poetry*. Cambridge: Harvard University Press, 1984: 79–97.

Mitchell, Bruce, and Fred C. Robinson, eds. *Beowulf: An Edition*. Oxford: Blackwell, 1998.

Norton, Rictor. "Silvia and Tom Turd-man." *Early Eighteenth-Century Newspaper Reports: A Sourcebook*. Last modified December 16, 2003. Accessed April 1, 2010. http://rictornorton.co.uk/grubstreet/waterman.htm.

Oates, Joyce Carol. "Landfill." *The New Yorker*, October 9, 2006. Accessed November 23, 2014. http://www.newyorker.com/magazine/2006/10/09/landfill.

O'Neill, Heather. "The Secret Life of Our Trash Can." *New York Times Magazine*, March 14, 2014. Accessed November 28, 2014. http://www.nytimes.com/2014/03/16/magazine/the-secret-life-of-our-trash-can.html.

Oppenheimer, Paul, trans. *Till Eulenspiegel: His Adventures*. New York: Routledge, 2001.

Pálsson, Hermann, and Paul Edwards, trans. *Eyrbyggja Saga*. London: Penguin, 1989.

Plath, Sylvia. "Lady Lazarus." Accessed October 8, 2014. http://www.poetryfoundation.org/poem/178961.

Rabelais, François. *Gargantua and Pantagruel*. Translated by Sir Thomas Urquhart and Peter Motteux. Chicago: Encyclopaedia Britannica, 1952.

Shakespeare, William. *The Arden Edition of the Works of William Shakespeare: Hamlet*, edited by Ann Thompson and Neil Taylor. London: Thomson, 2006.

———. *The Arden Edition of the Works of William Shakespeare: Hamlet*, edited by Harold Jenkins. London: Methuen, 1982.

———. *The Riverside Shakespeare*, edited by G. Blakemore Evans. Boston: Houghton Mifflin, 1974.

Sophocles. *Philoctetes*. Translated by Carl Phillips. Oxford: Oxford University Press, 2003.

Sterne, Lawrence. *Tristram Shandy*, edited by James Aiken Work. Indianapolis: Odyssey Press/Bobbs-Merrill, 1940/1977.

Suchsland, Peter, and Erika Weber. *Deutsche Volksbücher in Drei Bänden. Zweiter Band*. Berlin/Weimar: Aufbau Verlag, 1968.

Swanton, Michael, trans. and ed. *Anglo-Saxon Prose*. London: Everyman/J. M. Dent, 1993.

Swift, Jonathan. *Gulliver's Travels*, edited by Philip Pinkus. Toronto: Macmillan, 1968.

Thoreau, Henry. *The Writings of Henry David Thoreau: Walden*, edited by J. Lyndon Shanley. Princeton, NJ: Princeton Univeristy Press, 1973.

Till Eulenspiegel A Merye Jest. London (1528). Rpt. Amsterdam: Da Capo Press, 1971.

Twain, Mark. *The Adventures of Huckleberry Finn*. New York: Bantam, 1981.

Twelfth Ecumenical Council: Lateran IV 1215. Medieval Sourcebook. Accessed February 15, 2010. http://www.fordham.edu/halsall/basis/lateran4.html.

Vizenor, Gerald. *Landfill Meditation: Crossblood Stories*. Hanover, NH: Wesleyan University Press/University Press of New England, 1991.

Whitman, Walt. *Leaves of Grass: The First (1855) Edition*, edited by Malcolm Cowley. Harmondsworth: Penguin, 1984.

———. *Song of Myself (1892 version)*. Accessed October 25, 2014. http://www.poetryfoundation.org/poem/174745.

Wilbur, Richard. "Junk." Accessed October 8, 2014. http://www.poetryfoundation.org/poem/171790.

Williamson, Craig. *A Feast of Creatures: Anglo-Saxon Riddle Songs*. Philadelphia: University of Pennsylvania Press, 2011.

Wilson, Miles. "Everything." *Line of Fall*, 17–32. Iowa City: University of Iowa Press, 1989.

Films and Television

Ball of Fire. Directed by Howard Hawks (1941).

The Best Years of Our Lives. Directed by William Wyler (1946).

The Examined Life. Directed by Astra Taylor (2008).

Judgment at Nuremberg. Directed by Stanley Kramer (1961).

Les glaneurs et la glaneuse (*The Gleaners and I*). Directed by Agnès Varda (2000).

The Phantom of Liberty. Directed by Luis Buñuel (1974).

WALL-E. Directed by Andrew Stanton (2008).

Secondary Works Cited

Achinstein, Sharon. "John Foxe and the Jews." *Renaissance Quarterly* 54.1 (2001): 86–120.

Ackroyd, Peter. *London: A Biography*. New York: Doubleday, 2000.

Adichie, Chimamanda Ngozi. "The Danger of a Single Story." *TED Talks*. October 2009. Accessed November 27, 2014. http://www.ted.com/speakers/chimamanda_ngozi_adichie.

Adorno, Theodor W. "Trying to Understand *Endgame*." Translated by Michael T. Jones. *New German Critique. Critical Theory and Modernity* 26 (Spring–Summer 1982): 119–150.

Agamben, Giorgio. *Homo Sacer: Sovereign Power and Bare Life*. Translated by Daniel Heller-Roazen. Stanford, CA: Stanford University Press, 1998.

Agus, Irving A. *Rabbi Meir of Rothenburg: His Life and His Works as Sources for the Religious, Legal, and Social History of the Jews of Germany in the Thirteenth Century*. New York: Ktav, 1970.

Alaimo, Stacy. "Oceanic Origins, Plastic Activism, and New Materialism." In *Material Ecocriticism*, edited by Serenella Iovino and Serpil Oppermann, 186–203. Bloomington: Indiana University Press, 2014.

Allen, Charlotte. "A Dark Age for Medievalists: At Their Annual Congress in Kalamazoo, It's No Longer Your Grandfather's Middle Ages." *The Weekly Standard* 13.36 (June 2, 2008). Accessed October 17, 2014. http://www.weeklystandard.com/Content/Public/Articles/000/000/015/146etleh.asp.

Allen, Valerie. "Commentary on Susan Signe Morrison's Waste Studies—A New Paradigm for Literary Analysis: Something Is Rotten in the Denmark of *Beowulf* and *Hamlet*." The 2009 Compass Interdisciplinary Virtual Conference. Accessed

October 18, 2014. http://compassconference.files.wordpress.com/2009/10/civc-commentary-valerie-allen-john-jay-college-cuny-on-waste-studies-a-new-paradigm-for-literary-analysis-susan-signe-morrison.pdf.

———. *On Farting: Language and Laughter in the Middle Ages*. New York: Palgrave Macmillan, 2007.

Allen, Valerie, and Ruth Evans, eds. *Roadworks: Medieval Roads, Medieval Britain*. Manchester: Manchester University Press, 2015.

Anderson, Warwick. "Excremental Colonialism: Public Health and the Poetics of Pollution." *Critical Inquiry* 21 (1995): 640–669.

Anderson, Wayne. *Freud, Leonardo da Vinci, and the Vulture's Tail: A Refreshing Look at Leonardo's Sexuality*. New York: Other Press, 2001.

Appadurai, Arjun. "Deep Democracy." *Public Culture* 14 (2002): 21–47.

Arefin, Mohammed Rafi. "The Dirty Details: A Response to 'Tales of the Trash.'" Accessed November 10, 2014. http://discardstudies.com/2014/11/03/the-dirty-details-a-response-to-tales-of-the-trash/.

Armstrong, Tim. "Eliot's Waste Paper." In *Modernism, Technology, and the Body: A Cultural Study*, 69–74. Cambridge: Cambridge University Press, 1998. Reprinted in *The Waste Land: A Norton Critical Edition*, edited by Michael North, 275–280. New York: W. W. Norton, 2001.

Ascher, Robert. "Time's Arrow and the Archaeology of a Contemporary Community." In *Settlement Archaeology*, edited by K. C. Chang, 275–280. Palo Alto: National Press, 1968.

Ashenburg, Katherine. *The Dirt on Clean: An Unsanitized History*. New York: North Point Press, 2007.

Ashford, Gerald. "Book Branding Issue Dead in Texas So Far." *The Daily Republic*, p. 4. Mitchell, South Dakota, Friday, July 24, 1953. Accessed October 19, 2014. http://newspaperarchive.com/mitchell-daily-republic/1953–07–24/page-4.

Assadourian, Erik. "The Path to Degrowth in Overdeveloped Countries." In *2012 State of the World: Moving Toward Sustainable Prosperity*, edited by Matt Richmond, 22–37. Washington, DC: Island Press, 2012. Accessed May 7, 2012. http://blogs.worldwatch.org/sustainableprosperity/sow2012/.

Astell, Ann, and Justin Jackson. "Before the Face of the Book: A Levinasian Preface." In *Levinas and Medieval Literature: The "Difficult Reading" of English and Rabbinic Texts*, edited by Ann W. Astell and J. A. Jackson, 1–13. Pittburgh: Duquesne University Press, 2009.

Aston, Margaret. "English Ruins and English History: The Dissolution and the Sense of the Past." *Journal of the Warburg and Courtauld Institutes* 36 (1973): 231–255.

———. "Iconoclasm in England: Official and Clandestine." In *Iconoclasm versus Art and Drama*, edited by Clifford Davidson and Ann Nichols, 47–91. Kalamazoo, MI: Medieval Institute, 1989.

Attridge, Derek. "What Does it Make You Feel? Responding Affectively to Literature." Literary and Critical Theory Seminar, Institute of English Studies, School of Advanced Study, London. February 14, 2011.

Auerbach, Erich. "Philology and *Weltliteratur*." Translated by Marie and Edward Said. *The Centennial Review* 13.1 (1969): 1–17.

Babb, Lawrence. *The Elizabethan Malady: A Study of Melancholia in English Literature from 1580 to 1642*. East Lansing, Michigan State College Press, 1951, rpt. 1965.

Baegert, J. *Observations in Lower California*. Translated by M. Brandenburg and C. L. Baumann, 909–922. Berkeley, CA: University of California Press, 1952.

Bagliani, Agostino Paravicini. "The Corpse in the Middle Ages: The Problem of the Division of the Body." In *The Medieval World*, edited by. Peter Linehan and Janet L. Nelson, 327–341. London: Routledge, 2001.

Bakhtin, Mikhail. *Rabelais and His World*. Translated by Héléne Iswolsky. Bloomington: Indiana University Press, 1968/1984.

Banerjee, Neela. "Imprisoned Rabbi Sues Over Space for Prayer." *The New York Times*, February 16, 2008. Accessed October 18, 2014. http://www.nytimes.com/2008/02/16/us/16prison.html?pagewanted=print.

Barley, Nigel. "The Letter to Brother Edward." *Neuphilologische Mitteilungen* 79 (1978): 23–24.

Barney, Richard A. "Filthy Thoughts, or, Cultural Criticism and the Ordure of Things." *Genre* 27 (1994): 275–293.

Bashford, Alison. *Imperial Hygiene: A Critical History of Colonialism, Nationalism and Public Health*. New York: Palgrave Macmillan, 2004.

Bataille, Georges. *The Accursed Share: An Essay on General Economy, Vol. II: The History of Eroticism, Vol. III: Sovereignty*. Translated by Robert Hurley. New York: Zone Books, 1993.

———. *Visions of Excess: Selected Writings, 1927–1939*, edited by Allan Stoekl, translated by Allan Stoekl, with Carl R. Lovitt and Donald M. Leslie, Jr. Minneapolis, MN: University of Minneapolis Press, 1985.

Bate, Jonathan. *Song of the Earth*. Cambridge, MA: Harvard University Press, 2000.

Battaglia, Frank. "The Germanic Earth Goddess in *Beowulf*?" *The Mankind Quarterly* 31 (1997): 415–446.

Baudrillard, Jean. *Simulacra and Simulation*. Translated by Sheila Faria Glaser. Ann Arbor: University of Michigan Press, 1994.

Baugh, Albert C., and Thomas Cable. *A History of the English Language*. 6th ed. Upper Saddle River, NJ: Prentice Hall, 2013.

Bauman, Zygmunt. *Postmodern Ethics*. Oxford: Blackwell, 1993.

———. *Wasted Lives: Modernity and Its Outcasts*. Cambridge: Polity Press, 2004.

Bayless, Martha. *Sin and Filth in Medieval Culture: The Devil in the Latrine*. New York: Routledge, 2012.

———. "The Story of the Fallen Jews and the Iconography of Jewish Unbelief." *Viator* 34 (2003): 142–156.

Bearn, Gordon C. F. "Notes and Fragments: The Possibility of Puns: A Defense of Derrida." *Philosophy and Literature* 19.2 (1995): 330–335.

Beechy, Tiffany. "Devil Take the Hindmost: Chaucer, John Gay, and the Pecuniary Anus." *The Chaucer Review* 41.1 (2006): 71–85.

Bell, Michael Mayerfeld. "Deep Fecology: Mikhail Bakhtin and the Call of Nature. *Capitalism, Nature, Socialism* 5.4 (1994) 65–84.

Benjamin, Walter. *The Arcades Project: Walter Benjamin*. Translated by Howard Eiland and Kevin McLaughlin. Cambridge, MA, and London: Belknap Press, 1999.

———. *Charles Baudelaire: A Lyric Poet in the Era of High Capitalism*. Translated by Harry Zohn. London: Verso, 1973: 19–20.

———. *Illuminations: Essays and Reflections*, edited by Hannah Arendt. New York: Schocken Books, 1968.

Bennett, Jane. "The Elements." *postmedieval: a journal of medieval cultural studies* 4.1 (2013): 105–111.

———. "The Force of Things: Steps toward an Ecology of Matter." *Political Theory* 32.3 (June 2004): 347–372.

———. "Powers of the Hoard: Further Notes on Material Agency." In *Animal, Vegetable, Mineral: Ethics and Objects*, edited by Jeffrey Jerome Cohen, 237–269. Washington, DC: Oliphaunt Books, 2012.

———. "Systems and Things: A Response to Graham Harman and Timothy Morton." *New Literary Theory* 43 (2012): 225–233.

———. *Vibrant Matter: A Political Ecology of Things*. Durham: Duke University Press, 2010.

Benson, C. David. "Literary Contests and London Records in the *Canterbury Tales*." In *Chaucer and the City*, edited by Ardis Butterfield, 129–144. Cambridge: D. S. Brewer, 2006.

Berry, Wendell. "Faustian Economics: Hell Hath No Limits." *Harper's Magazine*, May 2008: 35–42. Accessed November 16, 2009. http://www.harpers.org/archive/2008/05/0082022.

———. "The Necessity of Agriculture." *Harper's Magazine*, December 2009: 15–16.

Biddick, Kathleen. *The Shock of Medievalism*. Durham, NC: Duke University Press, 1998.

Biggs, Frederick M. "The Politics of Succession in *Beowulf* and Anglo-Saxon Succession." *Speculum* 80.3 (2005): 709–741.

Bilefsky, Dan. "With Sharp Satire, Enfant Terrible Challenges Czech Identity." *The New York Times*, September 4, 2009. Accessed October 18, 2014. http://www.nytimes.com/2009/09/05/world/europe/05cerny.html.

Bingham, Alexa. "Discourse of the Dammed: A Study of the Impacts of Sustainable Development Discourse on Indigenous Peoples in the Brazilian Amazon in the Context of the Proposed Belo Monte Hydroelectric Dam." *POLIS Journal* 4 (2010): 1–47.

Biow, Douglas. *The Culture of Cleanliness in Renaissance Italy*. Ithaca, NY and London: Cornell University Press, 2006.

Birch, Charles. "The Postmodern Challenge to Biology." In *The Reenchantment of Science: Postmodern Proposals*, edited by David Ray Griffin, 69–78. Albany, NY: State University of New York Press, 1988.

Blakeslee, Sandra. "Mind Games: Sometimes a White Coat Isn't Just a White Coat." *The New York Times*, April 3, 2012. Accessed April 4, 2012. http://www.nytimes.com/2012/04/03/science/clothes-and-self-perception.html.

Bogost, Ian. *Alien Phenomenology: Or What It's Like to Be a Thing*. Minneapolis, MN: University of Minnesota Press, 2012.

Boo, Katherine. *Behind the Beautiful Forevers: Life, Death, and Hope in a Mumbai Undercity*. New York: Random House, 2012.

Booth, Wayne C. "Metaphor as Rhetoric: The Problem of Evaluation." In *On Metaphor*, edited by Sheldon Sacks, 47–70. Chicago, IL: University of Chicago Press, 1979.

Boscagli, Maurizia. *Stuff Theory: Everyday Objects, Radical Materialism*. New York: Bloomsbury, 2014.

Bourke, John G. *The Scatalogic Rites of All Nations*. Washington, DC: W. H. Lowdermilk, 1891.

Boxall, Peter. "'There's No Lack of Void': Waste and Abundance in Beckett and DeLillo." *SubStance* 37.2 (2008): 56–70.

Bradbury, Nancy Mason. "Rival Wisdom in the Latin *Dialogue of Solomon and Marcolf*." *Speculum* 83 (2008): 331–365.

Bragard, Véronique. "Introduction: Languages of Waste: Matter and Form in Our Garb-age." *Interdisciplinary Studies in Literature and Environment* 20.3 (2013): 459–463.

———. "Sparing Words in the Wasted Land: Garbage, Texture, and Écriture Blanche in Auster's *In the Country of Last Things* and McCarthy's *The Road*." *Interdisciplinary Studies in Literature and Environment* 20.3 (2013): 479–493.

Brody, Jane E. "Babies Know: A Little Dirt Is Good for You." *The New York Times*, January 26, 2009. Accessed October 18, 2014. http://www.nytimes.com/2009/01/27/health/27brod.html.

Brooks, Cleanth. *The Well Wrought Urn: Studies in the Structure of Poetry*. New York: Harcourt Brace Jovanovich, 1947.

Brown, Bill. "All Thumbs." *Critical Inquiry* 30 (2004): 452–457.

———. "Thing Theory." *Critical Inquiry* 28.1 (2001): 1–22.

Brown, Norman O. *Life Against Death: The Psychoanalytical Meaning of History*. Middletown, CT: Wesleyan University Press, 1959.

Bruccoli, Matthew Joseph, ed. *F. Scott Fitzgerald's The Great Gatsby: A Literary Reference*. New York: Carroll & Graf, 2000.

Bryant, Levi R. "Substantial Powers, Active Affects: The Intentionality of Objects." *Deleuze Studies* 6.4 (2012): 529–543.

Butterfield, Ardis, ed. *Chaucer and the City*. Cambridge: D. S. Brewer, 2006.

———. "Chaucer and the Detritus of the City." In *Chaucer and the City*, edited by Ardis Butterfield, 3–22. Cambridge: D. S. Brewer, 2006.

Bynum, Caroline Walker. *Fragmentation and Redemption: Essays on Gender and the Human Body in Medieval Religion*. New York: Zone Books, 1991.

———. *Holy Feast and Holy Fast: The Religious Significance of Food to Medieval Women*. Berkeley: University of California Press, 1987.

———. The *Resurrection of the Body in Western Christianity, 200–1336*. New York: Columbia University Press, 1995.

———. "Why All the Fuss about the Body? A Medievalist's Perspective." *Critical Inquiry* 22 (1995): 1–33.

Camille, Michael. "Glossing the Flesh: Scopophilia and the Margins of the Medieval Book." In *The Margins of the Text*, edited by D. C. Greetham, 245–267. Ann Arbor: University of Michigan Press, 1997.

———. *Image on the Edge: The Margins of Medieval Art*. London: Reaktion, 1992.

Campkin, Ben, and Rosie Cox, eds. *Dirt: New Geographies of Cleanliness and Contamination*. London: I. B. Tauris, 2007.

Candlin, Fiona, and Raiford Guins. *The Object Reader*. London: Routledge, 2009.

Carroll, Joseph. *Literary Darwinism: Evolution, Human Nature, and Literature*. New York: Routledge, 2004.

Caviness, Madeline H. *Visualizing Women in the Middle Ages: Sight, Spectacle, and Scopic Economy.* Philadelphia: University of Pennsylvania Press, 2001.

Chapman, Alison. "Ophelia's 'Old Lauds': Madness and Hagiography in *Hamlet*." *Medieval and Renaissance Drama in England* 20 (2007): 111–135.

Citrome, Jerome J. "Medicine as Metaphor in the Middle English *Cleanness*." *The Chaucer Review* 35 (2001) 260–280.

———. *The Surgeon in Medieval English Literature.* New York: Palgrave Macmillan, 2006.

Clanton, Brett. "Alternative Energy Source?" *San Antonio Express News*. October 2, 2009. Accessed November 17, 2014. http://www.mysanantonio.com/business/local/article/Alternative-energy-source-858249.php.

Clare, Eli. "Meditations on Natural Worlds, Disabled Bodies, and a Politics of Cure." In *Material Ecocriticism*, edited by Serenella Iovino and Serpil Oppermann, 204–218. Bloomington: Indiana University Press, 2014.

Clark, J. F. M. "'The Incineration of Refuse is Beautiful': Torquay and the Introduction of Municipal Refuse Destructors." *Urban History* 34 (2007): 255–277.

Clark, Nigel, and Myra J. Hird. "Deep Shit." *O-Zone: A Journal of Object-Oriented Studies* 1 (2014): 44–52.

Clark, Samantha. "Strange Strangers and Uncanny Hammers: Morton's *The Ecological Thought* and the Phenomenological Tradition." *Green Letters: Studies in Ecocriticism* 17.2 (2013): 98–108.

Claviez, Thomas. "Done and Over With—Finally? Otherness, Metonymy, and the Ethics of Comparison." *PMLA* 128.3 (2013): 608–614.

Clines, Francis X. "Reimagining What Washes Up." *The New York Times*, March 25, 2010. Accessed October 18, 2014. http://www.nytimes.com/2010/03/26/opinion/26fri4.html.

Cohen, Jeffrey Jerome. *Animal, Vegetable, Mineral: Ethics and Objects.* Washington, DC: Oliphaunt Books, 2012.

———. ed. *Inhuman Nature.* Washington, DC: Oliphaunt Books, 2014.

———. "Introduction: Midcolonial." In *The Postcolonial Middle Ages*, edited by Jeffrey Jerome Cohen, 1–17. New York: St. Martin's Press, 2001.

———. *Medieval Identity Machines.* Minneapolis, MN: University of Minnesota Press, 2003.

———. *Of Giants: Sex, Monsters, and the Middle Ages.* Minneapolis, MN: University of Minneapolis Press, 1999.

———. ed. *Prismatic Ecology: Ecotheory Beyond Green.* Minneapolis, MN: University of Minnesota Press, 2013.

———. "Stories of Stone." *postmedieval: a journal of medieval cultural studies* 1 (2010): 56–63.

Cohen, Richard A., ed. *Face to Face with Levinas.* Albany, NY: State University of New York Press, 1986.

Cohen, Ted. *Thinking of Others: On the Talent for Metaphor.* Princeton, NJ: Princeton University Press, 2012.

Cohen, William A. "Deep Skin." In *Thinking the Limits of the Body*, edited by Jeffrey J. Cohen and Gail Weiss, 63–82. Albany, NY: State University of New York Press, 2003.

Cohen, William A., and Ryan Johnson. *Filth: Dirt, Disgust and Modern Life*. Minneapolis, MN: Minnesota University Press, 2005.

Cole, Andrew. "The Call of Things: A Critique of Object-Oriented Ontologies." *minnesota review* 80 (2013): 106–118.

Conniff, Richard. "Useless Creatures." *The New York Times*, September 14, 2014. Accessed October 18, 2014. http://opinionator.blogs.nytimes.com/2014/09/13/useless-creatures/?_php=true&_type=blogs&_r=0.

Connor, Steven. "The Poorest Things Superfluous: On Redundancy." Talk at Rubbish Symposium, Birkbeck College, London, July 30, 2011. Accessed October 21, 2014. http://stevenconnor.com/redundancy.html.

———. "Exhaust: On Aerial Rejectamenta." In *Aesthetic Fatigue: Modernity and the Language of Waste*, edited by John Scanlan and John F. M. Clark, 71–80. Newcastle upon Tyne: Cambridge Scholars, 2013.

Cooppan, Vilashini. "Net Work: Area Studies, Comparison, and Connectivity." *PMLA* 128.3 (2013): 615–621.

Coyne, Michelle. "From Production to Destruction to Recovery: Freeganism's Redefinition of Food Value and Circulation." *Iowa Journal of Cultural Studies* 10/11 (2009). Accessed March 9, 2012. http://www.uiowa.edu/~ijcs/waste/coyne.htm.

Cramer, Sarah. "Let Your Narrator Be Your Guide: Intrusive Narrators as the Voice of Morality in Children's Fiction." Ethics and Children's Literature Symposium, The Prindle Institute of Ethics, DePauw University. September 15, 2012.

Crang, Mike. "The Death of Great Ships: Photography, Politics, and Waste in the Global Imaginary." *Environment and Planning A* 42.5 (2010): 1084–1102. Accessed October 19, 2014. http://dro.dur.ac.uk/6901/1/6901.pdf.

Crawford, Leslie. "The Bowel Movement. What is Your Poo Telling You about Your Health? It's the Burning Question That has Everybody's Head in the Toilet These Days." *Salon.com*. March 12, 2008. http://www.salon.com/life/feature/2008/03/12/poo. Accessed April 1, 2010.

Culler, Jonathan. *Framing the Sign: Criticism and Its Institutions*. Norman and London: University of Oklahoma Press, 1988.

Dane, Gabrielle. "Reading Ophelia's Madness." *Exemplaria* 10 (1998): 405–423.

Darst, David. "Witchcraft in Spain: The Testimony of Martín de Castañega's Treatise on Superstition and Witchcraft (1529)." *Proceedings of the American Philosophical Society* 123 (1979): 298–322.

Davis, Lennard J. "Enforcing Normalcy: Disability, Deafness, and the Body." In *Visualizing the Disabled Body: The Classical Nude and the Fragmented Torso*. Reprinted in *The Norton Anthology of Theory and Criticism*, edited by Vincent B. Leitch, 2398–2421. New York: Norton, 2001.

Davoudi, Simin. "Planning for Waste Management: Changing Discourses and Institutional Relationships." *Progress in Planning* 53 (2000): 165–216.

de Grazia, Edward. *Girls Lean Back Everywhere: The Law of Obscenity and the Assault on Genius*. New York: Random House, 1992.

De Landa, Michael. *A Thousand Years of Nonlinear History*. New York: Swerve Editions, 1997.

Deloughrey, Elizabeth. "Heavy Waters: Waste and Atlantic Modernity." *PMLA* 125.3 (2010): 703–712.

Dennis, Donna. *Licentious Gotham: Erotic Publishing and Its Prosecution in Nineteenth-Century New York*. Cambridge, MA: Harvard University Press, 2009.

Derrida, Jacques. *Margins of Philosophy*. Translated by A. Bass. Chicago, IL: University of Chicago Press, 1982.

DeVries, David N. "And Away Go Troubles Down the Drain: Late Medieval London and the Poetics of Urban Renewal." *Exemplaria* 8 (1996): 415 (401–418).

Diffrient, David Scott. "Stories That Objects Might Live to Tell: The 'Hand-Me-Down' Narrative in Film." *Other Voices* 3.1 (2007): Accessed April 18, 2012. http://www.othervoices.org/3.1/sdiffrient/index.php.

Dillon, Brian, ed. *Ruins*. London and Cambridge, MA: Whitechapel Gallery/MIT Press, 2011.

DiPiero, Thomas. "Shit Happens: Rabelais, Sade, and the Politics of Popular Fiction." *Genre* 27 (1994): 295–314.

Discard Studies. "About Discard Studies." Accessed November 22, 2014. http://discardstudies.com/about/.

———. Bibliography of Resources, including books, articles, and films. Accessed November 22, 2014. http://discardstudies.com/resources/.

Discard Studies Compendium. "A List of Critical Key Terms in Discard Studies." Accessed November 22, 2014. http://discardstudies.com/discard-studies-compendium/.

Dollimore, Jonathan. "Art in Time of War: Towards a Contemporary Aesthetic." In *The New Aestheticism*, edited by John J. Joughin and Simon Malpas, 36–50. Manchester: Manchester University Press, 2003.

———. *Sexual Dissidence: Augustine to Wilde, Freud to Foucault*. Oxford: Clarendon Press, 1991.

Douglas, Mary. "Jokes." *Implicit Meanings: Essays in Anthropology*. London: Routledge & Kegan Paul, 1975: 90–114.

———. *Purity and Danger: An Analysis of the Concepts of Pollution and Taboo*. London: Routledge, 1966/2002.

Duckert, Lowell. "Glacier." *postmedieval: a journal of medieval cultural studies* 4.1 (2013): 68–79.

———. "Speaking Stones, John Muir, and a Slower (Non)Humanities." In *Animal, Vegetable, Mineral: Ethics and Objects*, edited by Jeffrey Jerome Cohen, 273–279. Washington, DC: Oliphaunt Books, 2012.

Duffy, Eamon. "Bare Ruined Choirs: Remembering Catholicism in Shakespeare's England." In *Theatre and Religion: Lancastrian Shakespeare*, edited by Richard Dutton, Alison Findlay, and Richard Wilson, 40–57. Manchester: Manchester University Press, 2003.

Dundes, Alan. *Life Is Like a Chicken Coop Ladder: A Study of German National Character through Folklore*. Detroit, MI: Wayne State University Press, 1984.

Dunn-Hensley, Susan. "Return of the Sacred Virgin: Memory, Loss, and Restoration in Shakespeare's Later Plays." In *Walsingham in Literature and Culture from the Middle Ages to Modernity*, edited by Gary Waller and Dominic Janes, 185–197. Farnham, England: Ashgate, 2010.

Durling, Robert M. "Deceit and Digestion in the Belly of Hell." In *Allegory and Representation: Selected Papers from the English Institute, 1979–80*, edited by

Stephen J. Greenblatt, 61–93. Baltimore, MD: Johns Hopkins University Press, 1981.

Dutton, Richard, Alison Findlay, and Richard Wilson, eds. *Theatre and Religion: Lancastrian Shakespeare*. Manchester: Manchester University Press, 2003.

Duvall, J. Michael. "One Man's Junk: Material and Social Waste in Frank Norris's *McTeague*." *Studies in American Naturalism* 4 (2009): 132–151.

Eaglestone, Robert. *Ethical Criticism: Reading after Levinas*. Edinburgh: Edinburgh University Press, 1997.

Earl, James. "The Forbidden *Beowulf*: Haunted by Incest." *PMLA* 125.2 (2010): 289–305.

Edenson, Tim. "Waste Matter: The Debris of Industrial Ruins and the Disordering of the Material World." *Journal of Material Culture* 10.3 (2005): 311–332.

Edmundson, Mark. "Against Readings." *The Chronicle of Higher Education*, April 24, 2009. B7–B10.

Elias, Norbert. *The Civilizing Process*. Oxford: Blackwell, 1994/2000.

———. *The History of Manners*, Vol. 1 of *The Civilizing Process*, translated by Edmund Jephcott. New York: Urizen Books, 1978.

Elsner, Jaś. "Iconoclasm and the Preservation of Memory." In *Monuments and Memory, Made and Unmade*, edited by Robert S. Nelson and Margaret Olin, 209–231. Chicago, IL: University of Chicago Press, 2003.

Engler, Mira. *Designing America's Waste Landscapes*. Baltimore, MD: Johns Hopkins University Press, 2004.

Esty, J. D. "Excremental Postcolonialism." *Contemporary Literature* 40 (1999): 22–59.

Ette, Ottmar. "Literature as Knowledge for Living, Literary Studies as Science for Living." Translated by Vera M. Kutzinski. *PMLA* 125 (2010): 983–993.

Evans, Ruth. "Lacan's *belles-lettres*: On Difficulty and Beauty." In *On Style: An Atelier*, edited by Eileen A. Joy and Anna Kłosowska, 15–25. Brooklyn, NY: punctum books, 2013.

Faires, Robert. "Corps de Garbage." *The Austin Chronicle*, September 11, 2009. Accessed November 10, 2014. http://www.austinchronicle.com/arts/2009-09-11/837971/.

Fauconnier, Gilles, and Mark Turner. *The Way We Think: Conceptual Blending and the Mind's Hidden Complexities*. New York: Basic, 2002.

Feldman, Mark B. "Inside the Sanitation System: Mierle Ukeles, Urban Ecology, and the Social Circulation of Garbage." *Iowa Journal of Cultural Studies* 10/11 (2009). Accessed March 9, 2012. http://www.uiowa.edu/~ijcs/waste/feldman.html.

Ferguson, Harvie. "Exteriority: Boredom, Disgust and the Margins of Humanity." In *Aesthetic Fatigue: Modernity and the Language of Waste*, edited by John Scanlan and John F. M. Clark, 220–250. Newcastle upon Tyne: Cambridge Scholars, 2013.

Ferguson, Rex. "Garbage and Gatsby." In *Aesthetic Fatigue: Modernity and the Language of Waste*, edited by John Scanlan and John F. M. Clark, 206–219. Newcastle upon Tyne: Cambridge Scholars, 2013.

Ferraro, Emilia, and Louise Reid. "On Sustainability and Materiality. Homo Faber, a New Approach." *Ecological Economics* 96 (2013): 125–131.

Fienberg, Nona. "Thematics of Value in *The Book of Margery Kempe.*" *Modern Philology* 87 (1989): 132–141.

Fisher, Will. "The Renaissance Beard: Masculinity in Early Modern Europe." *Renaissance Quarterly* 54 (2001): 155–187.

Fitzgerald, Keri. "The Creation of Meaning through Metafictional Digression in Sterne's *Tristram Shandy.*" Paper for Honors 3396A, November 17, 2009. Texas State University, San Marcos.

———. "Maxims and the Waste of Words in Franklin's 'The Way to Wealth.'" Class paper for Honors 3396A, October 13, 2009. Texas State University, San Marcos.

Fleming, Juliet. "Scraping by: Towards a Pre-Historic Criticism." *postmedieval: a journal of medieval cultural studies* 3 (2012): 119–133.

Foucault, Michel. *The Order of Things: An Archaeology of the Human Sciences.* New York: Vintage Books, 1973.

Frantzen, Allen J. *Desire For Origins: New Language, Old English, and Teaching the Tradition.* New Brunswick, NJ: Rutgers University Press, 1990.

———. "The Disclosure of Sodomy in *Cleanness.*" *PMLA* 111 (1996): 451–464.

Freedman, Paul. *Images of the Medieval Peasant.* Stanford, CA: Stanford University Press, 1999.

Freud, Sigmund. *The Standard Edition of the Complete Psychological Works of Sigmund Freud. Totem and Taboo and Other Works.* Vol. 13. London: Hogarth Press, 1957.

Friedman, Susan Stanford. "Why Not Compare?" *PMLA* 126.3 (May 2011): 753–762.

Frow, John. "A Pebble, a Camera, a Man Who Turns into a Telegraph Pole." *Critical Inquiry* 28.1 (2001): 270–285.

Gaard, Greta, Simon C. Estok, and Serpil Oppermann, eds. *International Perspectives in Feminist Ecocriticism.* New York: Routledge, 2013.

Gabrys, Jennifer. "Cité Multimédia: Noise and Contamination in the Information City." Paper delivered at the Visual Knowledges Conference, The University of Edinburgh, September 17–20, 2003.

Gagnier, Regenia. "Why Interdisciplinarity?" The 2009 Compass Interdisciplinary Virtual Conference. Accessed June 4, 2010. http://compassconference.wordpress.com/2009/10/19/gagnier/.

Garrard, Greg. "Nature Cures? Or How to Police Analogies of Personal and Ecological Health." *Interdisciplinary Studies in Literature and Environment* 19.3 (Summer 2012): 494–514.

Gee, Sophie. *Making Waste: Leftovers and the Eighteenth-Century Imagination.* Princeton, NJ: Princeton Univerity Press, 2010.

Gentleman, Amelia. "Picking Up Trash by Hand, and Yearning for Dignity." *The New York Times*, September 27, 2007. Accessed November 23, 2014. http://www.nytimes.com/2007/09/27/world/asia/27ragpickers.html?pagewanted=print.

George, Rose. *The Big Necessity: Adventures in the World of Human Waste.* New York: Metropolitan Books, 2008.

Gilchrist, Roberta. "Medieval Bodies in the Material World: Gender, Stigma and the Body." In *Framing Medieval Bodies*, edited by Sarah Kay and Miri Rubin, 43–61. Manchester: Manchester University Press, 1994.

Gille, Zsuzsa. *From the Cult of Waste to the Trash Heap of History: The Politics of Waste in Socialist and Postsocialist Hungary*. Bloomington: Indiana University Press, 2007.

Gordon, Avery F. *Ghostly Matters: Haunting and the Sociological Imagination*. Minneapolis, MN: University of Minnesota Press, 1997.

Gravdal, Kathryn. *Vilain and Courtois: Transgressive Poetry in French Literature of the Twelfth and Thirteenth Centuries*. Lincoln: University of Nebraska Press, 1989.

Griffin, David Ray. *The Reenchantment of Science: Postmodern Proposals*. Albany, NY: State University of New York Press, 1988.

Griffin, Roger. *Modernism and Fascism: The Sense of a Beginning under Mussolini and Fascism*. London: Palgrave, 2007.

———. "The Rainbow Bridge." The 2009 Compass Interdisciplinary Virtual Conference. Accessed June 4, 2010. http://compassconference.wordpress.com/2009/10/19/griffin/.

Grössinger, Christa. *Humour and Folly in Secular and Profane Prints of Northern Europe, 1430–1540*. London and Turnhout: Harvey Miller, 2002.

Grosz, Elizabeth. *Space, Time, and Perversion: Essays on the Politics of Bodies*. New York and London: Routledge, 1995.

Halliwell, Martin, and Andy Mousley. *Critical Humanisms: Humanist/Anti-Humanist Dialogues*. Edinburgh: Edinburgh University Press, 2003.

Halpern, Richard. "Beckett's Tragic Pantry: *Endgame* and the Deflation of the Act." *PMLA* 129.4 (2014): 742–750.

Haque, M. Shamsul. "Environmental Discourse and Sustainable Development: Linkages and Limitation." *Ethics and Environment* 5 (2000): 3–21.

Haraway, Donna J. "A Cyborg Manifesto: Science, Technology, and Socialist-Feminism in the Late Twentieth Century." *Simians, Cyborgs, and Women: The Reinvention of Nature*, 149–182. New York: Routledge, 1991.

Hardin, Garrett. "The Tragedy of the Commons." *Science* 162, December 13, 1968: 1243–1248.

Harman, Graham. "Gold." In *Prismatic Ecology: Ecotheory Beyond Green*, edited by Jeffrey Jerome Cohen, 106–123. Minneapolis, MN: University of Minnesota Press, 2013.

———. *Guerrilla Metaphysics: Phenomenology and the Carpentry of Things*. Chicago, IL: Open Court, 2005.

———. "Levinas and the Triple Critique of Heidegger." *Philosophy Today* 53 (2009): 407–413.

———. *Tool-Being: Heidegger and the Metaphysics of Objects*. Peru, IL: Open Court, 2002.

———. "The Well-Wrought Broken Hammer: Object-Oriented Literary Criticism." *New Literary History* 43.2 (2012): 183–203.

Harpet, Cyrille. *De Déchet: Philosophie des Immondices Corps, Ville, Industrie*. Paris: L'Harmattan, 1998.

Hawkes, Terence. *Shakespeare in the Present*. London: Routledge, 2002.

Hawkins, Gay. "Down the Drain: Shit and the Politics of Disturbance." In *Culture and Waste: The Creation and Destruction of Value*, edited by Gay Hawkins and Stephen Muecke, 39–52. Lanham, MD: Rowman & Littlefield, 2003.

———. *The Ethics of Waste: How We Relate to Rubbish*. Lanham, MD: Rowman & Littlefield, 2006.

———. "Waste in Sydney: Unwelcome Returns." *PMLA* 122 (2007): 348–351.

Hawkins, Gay, and Stephen Muecke, eds. *Culture and Waste: The Creation and Destruction of Value*. Lanham, MD: Rowman & Littlefield, 2003.

Hayles, N. Katherine. *How We Became Posthuman: Virtual Bodies in Cybernetics, Literature, and Informatics*. Chicago, IL: University of Chicago Press, 1999.

Hensel, Marcus. "The Gift of Good Land?: Settled Lands and Wastelands in Anglo-Saxon Thought." Talk: BABEL Working Group. November 2010.

Hessler, Peter. "Tales of the Trash." *The New Yorker*, October 13, 2014. Accessed October 15, 2014. http://www.newyorker.com/magazine/2014/10/13/tales-trash.

Hird, Myra J. "Feminist Matters: New Materialist Considerations of Sexual Difference." *Feminist Theory* 5 (2004): 223–232.

———. "Knowing Waste: Towards an Inhuman Epistemology." *Social Epistemology: A Journal of Knowledge, Culture and Policy* 26.3–4 (2012): 453–469.

———. "Waste Flows." Accessed November 10, 2014. http://discardstudies.com/discard-studies-compendium/.

Hochman, Jhan. "Green Cultural Studies: An Introductory Critique of an Emerging Discipline." *Mosaic* 30.1 (1997): 81–96.

Holler, Linda. *Erotic Morality: The Role of Touch in Moral Agency*. New Brunswick, NJ: Rutgers University Press, 2002.

Holmes, Caroline. *The Not so Little! Book of Dung*. Thrupp: Sutton, 2006.

Horner, Shari. "Voices from the Margins: Women and Textual Enclosure in *Beowulf*." In *The Postmodern Beowulf: A Critical Casebook*, edited by Eileen A. Joy and Mary K. Ramsay, 467–500. Morgantown: West Virginia University Press, 2006.

Hsy, Jonathan. *Trading Tongues: Merchant, Multilingualism, and Medieval Literature*. Columbus, OH: Ohio State University Press, 2013.

Inglis, David. *A Sociological History of Excretory Experience: Defecatory Manners and Toiletry Technologies*. Lewiston, ME: Edwin Mellen Press, 2001.

Iovino, Serenella. "Bodies of Naples: Stories, Matter, and the Landscapes of Porosity." In *Material Ecocriticism*, edited by Serenella Iovino and Serpil Oppermann, 97–113. Bloomington: Indiana University Press, 2014.

———. "Naples 2008, or, the Waste Land: Trash, Citizenship, and an Ethic of Narration." *Neohelicon* 36 (2009): 335–346.

Iovino, Serenella, and Serpil Oppermann. *Material Ecocriticism*. Bloomington: Indiana University Press, 2014.

———. "Introduction." In *Material Ecocriticism*, edited by Serenella Iovino and Serpil Oppermann, 1–17. Bloomington: Indiana University Press, 2014.

———. "Material Ecocriticism: Materiality, Agency, and Models of Narrativity." *Ecozon@* 3.1 (2012): 75–91.

———. "Theorizing Material Ecocriticism: A Diptych." *Interdisciplinary Studies in Literature and Environment* 19.3 (Summer 2012): 448–475.

James, Mervyn. "Ritual, Drama and the Social Body in the Late Medieval English Town." *Past & Present* 98 (1983): 3–29.

Johnson, Alison. *Scarista Style*. London: Futura, 1987.

Johnson, Barbara. *The Feminist Difference: Literature, Psychoanalysis, Race, and Gender*. Cambridge, MA: Harvard University Press, 1998.

Johnson, Eleanor. "The Poetics of Waste: Medieval English Ecocriticism." *PMLA* 127.3 (2012): 460–476.

Johnson, Ken. "Andres Serrano." *The New York Times*, September 19, 2008. Accessed October 18, 2014. http://www.nytimes.com/2008/09/19/arts/design/19gall.html?_r=2&pagewanted=2&ref=arts&oref.

Johnson, Steven. "Tool for Thought." *The New York Times Book Review*, January 30, 2005. Accessed October 18, 2014. http://www.nytimes.com/2005/01/30/books/review/30JOHNSON.html.

Jørgensen, Dolly. "'All Good Rule of the Citee': Sanitation and Civic Government in England, 1400–1600." *Journal of Urban History* 36 (2010): 300–315.

———. "Cooperative Sanitation: Managing Streets and Gutters in Late Medieval England and Scandinavia." *Technology and Culture* 49 (2008): 547–567.

———. "The Metamorphosis of Ajax." *Early English Studies* 3 (2010): 1–31.

Joughin, John J. "Shakespeare's Genius: *Hamlet*, Adaptation and the Work of the Following." In *The New Aestheticism*, edited by John J. Joughin and Simon Malpas, 131–150. Manchester: Manchester University Press, 2003.

Joy, Eileen A. "'In His Eyes Stood a Light, Not Beautiful': Levinas, Hospitality, *Beowulf*." In *Levinas and Medieval Literature: The "Difficult Reading" of English and Rabbinic Texts*, edited by Ann Astell and J. A. Jackson, 57–84. Pittsburgh, PA: Duquesne University Press, 2009.

———. "You Are Here: A Manifesto." In *Animal, Vegetable, Mineral: Ethics and Objects*, edited by Jeffrey Jerome Cohen, 152–172. Washington, DC: Oliphaunt Books, 2012.

Joy, Eileen A., and Anna Kłosowska, eds. *On Style: An Atelier*. Brooklyn, NY: punctum books, 2013.

Joy, Eileen A., and Christine Neufeld, "A Confession of Faith: Notes toward a New Humanism." *Journal of Narrative Theory* 37 (2007): 161–190.

Joy, Eileen A., and Mary K. Ramsay. "Introduction." In *The Postmodern Beowulf: A Critical Casebook*, edited by Eileen A. Joy and Mary K. Ramsay, xxix–lxvii. Morgantown, WV: West Virginia University Press, 2006.

———. eds. *The Postmodern Beowulf: A Critical Casebook*. Morgantown, WV. West Virginia University Press, 2006.

Kadir, Djelal. "Comparative Literature Hinternational." In *Comparative Literature: States of the Art*, edited by Djelal Kadir, Special issue of *World Literature Today*, 69.2 (1995): 245–303.

Karolides, Nicholas J., Margaret Bald, and Dawn B. Sova. *100 Banned Books: Censorship Histories of World Literature*. New York: Checkmark Books, 1999.

Karras, Ruth Mazo. *From Boys to Men: Formations of Masculinity in Late Medieval Europe*. Philadelphia: University of Pennsylvania Press, 2003.

Kaufman, Frederick. "Wasteland: A Journey through the American Cloaca." *Harper's Magazine*, February 2008: 46–59.

Kearney, Richard. *States of Mind: Dialogues with Contemporary Thinkers*. New York: New York University Press, 1995.

Kelly, Robert L., and David Hurst Thomas. *Archeology*, 6th ed. Belmont, CA: Wadsworth, 2012.

Kelso, John. "These Residents Just Aren't Going with the Flow." *Austin American-Statesman*, March 28, 2008.

Kendall, Tina, and Kristin Koster. "Critical Approaches to Cultural Recycling: Introduction." *Other Voices* 3.1 (2007). Accessed April 18, 2012. http://www.othervoices.org/3.1/guesteditors/index.php.

Kilroy, Gerard. "Requiem for a Prince: Rites of Memory in *Hamlet*." In *Theatre and Religion: Lancastrian Shakespeare*, edited by Richard Dutton, Alison Findlay, and Richard Wilson, 143–160. Manchester: Manchester University Press, 2003.

Kiser, Lisa J. "The Garden of St. Francis: Plants, Landscape, and Economy in Thirteenth Century Italy." *Environmental History* 8 (2003): 229–245. Accessed October 19, 2014. http://rbedrosian.com/Gardens/Garden_Italy_Saint_Francis.pdf.

Kłosowska, Anna. "Style as Third Element." In *On Style: An Atelier*, edited by Eileen A. Joy and Anna Kłosowska, 25–36. Brooklyn, NY: punctum books, 2013.

Kristeva, Julia. *Powers of Horror: An Essay on Abjection*. Translated by Leon S. Roudiez. New York: Columbia University Press, 1987.

Kristof, Nicholas D. "Where Sweatshops Are a Dream." *The New York Times*, January 14, 2009. Accessed April 8, 2010. http://www.nytimes.com/2009/01/15/opinion/15kristof.html.

Kuhn, Thomas S. *The Structure of Scientific Revolutions*. 2nd ed. Chicago, IL: University of Chicago Press, 1970.

Kutzbach, Konstanze, and Monika Mueller, eds. *The Abject of Desire: The Aestheticization of the Unaesthetic in Contemporary Literature and Culture*. Amsterdam and New York: Rodopi, 2007.

Lacan, Jacques. *Television*, edited by Joan Copjec, translated by Denis Hollier, Rosalind Krauss, and Annette Michelson. New York: W. W. Norton, 1990.

Lacoeque, André. *The Feminine Unconventional: Four Subversive Figures in Israel's Tradition*. Minneapolis, MN: Fortress Press, 1990.

LaFraniere, Sharon. "For Chinese Women, a Basic Need and Few Places to Attend to It." *The New York Times*, March 1, 2012. Accessed March 5, 2012. http://www.nytimes.com/2012/03/01/world/asia/chinese-women-demand-more-public-toilets.html.

Lakoff, George, and Mark Johnson. *Metaphors We Live By*. 2nd ed. Chicago, IL: University of Chicago Press, 2003.

Lakoff, George, and Mark Turner. *More than Cool Reason: A Field Fuide to Poetic Metaphor*. Chicago, IL: University of Chicago Press, 1989.

Lamb, Gregory M. "The Need to Feed Hungry Families Cultivates New Interest in Gleaning." *Christian Science Monitor*, November 2, 2009. Accessed November 20, 2014. http://www.csmonitor.com/World/Making-a-difference/2009/1102/p07s01-lign.html.

Landry, Donna. "Mud, Blood, and Muck: Country Filth." *Genre* 27 (1994): 315–332.

Lang, Daniel. "'Give Us the Dumpsters—Or—Give Us Life': *Res Derilictae* and the Trash of Free Trade." *Other Voices* 3.1 (2007). Accessed April 18, 2012. http://www.othervoices.org/3.1/dlang/index.php.

Laporte, Dominique. *History of Shit*. Translated by N. Benabid and R. El-Khoury. Cambridge, MA: MIT Press, 2000.

Lathers, Marie. "Toward an Excremental Posthumanism: Primatology, Women, and Waste." *Society & Animals* 14.4 (2006): 417–436.

Latour, Bruno. *Pandora's Hope: Essays on the Reality of Science Studies.* Cambridge, MA: Harvard University Press, 1999.

———. "*Realpolitik* to *Dingpolitik*—Or How to Make Things Public." 1–31. Web. Accessed January 29, 2013. http://www.bruno-latour.fr/sites/default/files/96-DINGPOLITIK-GB.pdf.

———. *Reassembling the Social: An Introduction to Actor-Network-Theory.* Oxford: Oxford University Press, 2005.

———. *We Have Never Been Modern.* Cambridge, MA: Harvard University Press, 1993.

———. "Why Has Critique Run Out of Steam? From Matters of Fact to Matters of Concern." *Critical Inquiry* 30 (2004): 225–248.

Leake, Eric. "Humanizing the Inhumane: The Value of Difficult Empathy." In *Rethinking Empathy Through Literature*, edited by Meghan Marie Hammond and Sue J. Kim, 175–185. New York: Routledge, 2014.

Leguay, Jean-Pierre. *La Pollution au Moyen Âge.* 4th ed. Paris: Editions Jen-Paul Gisserot, 2005.

Leitch, Vincent B. *Literary Criticism in the 21st Century: Theory Renaissance.* London: Bloomsbury, 2014.

Lerer, Seth. "Grendel's Glove." *English Literary History* 61 (1994): 721–751.

———. "'On fagne flor': The Postcolonial *Beowulf*." In *Postcolonial Approaches to the European Middle Ages: Translating Cultures*, edited by Ananya Jahanara Kabir and Deanne Williams, 77–102. Cambridge: Cambridge University Press, 2005.

Levinas, Emmanuel. *Autrement qu'être ou au-delà l'essence. Phaenomenologica*, Vol. 54. The Hague: Martinus Nijhoff, 1974. In *Otherwise than Being or Beyond Essence*, translated by Alphonso Lingis. The Hague: Martinus Nijhoff, 1981.

———. *Ethics and Infinity: Conversations with Philippe Nemo.* Translated by R. A. Cohen. Pittsburgh: Duquesne University Press, 1985.

———. "Ethics of the Infinite." In *States of Mind: Dialogues with Contemporary Thinkers*, edited by Richard Kearney. New York: New York University Press, 1995: 177–199.

———. *Humanism of the Other*, Translated by Nidra Poller. Introduction by Richard A. Cohen. Urbana: University of Illinois Press, 2006.

———. *On Escape De l'évasion.* Translated by Bettina Bergo. Stanford, CA: Stanford University Press, 2003.

———. *Totality and Infinity: An Essay on Exteriority.* Translated by Alphonso Lingis. Dordrecht: Kluwer, 1991.

Levinas, Emmanuel, and Richard Kearney. "Dialogue with Emmanuel Levinas." In *Face to Face with Levinas*, edited by Richard A. Cohen, 13–34. Albany, NY: State University of New York Press, 1986.

Lévi-Strauss, Claude. *The Savage Mind.* Chicago, IL: University of Chicago Press, 1966.

Lewin, Ralph A. *Merde: Excursions in Scientific, Cultural, and Sociohistorical Coprology.* New York: Random House, 1999.

Lewis, Felice Flanery. *Literature, Obscenity, & the Law.* Carbondale, IL: Southern Illinois University Press, 1976.

Little, William G. *The Waste Fix: Seizures of the Sacred from Upton Sinclair to the Sopranos.* New York: Routledge, 2002.

Lochrie, Karma. *Covert Operations: The Medieval Uses of Secrecy*. Philadelphia: University of Pennsylvania Press, 1999.

Long, Thomas G. "Chronicle of a Death We Can't Accept." *The New York Times*, October 31, 2009. Accessed October 18, 2014. http://www.nytimes.com/2009/11/01/opinion/01long.html.

Lyall, Sarah. "Whimsical Works of Art, Found Sticking to the Sidewalk." *The New York Times*, June 14, 2011. Accessed November 22, 2014. http://www.nytimes.com/2011/06/14/world/europe/14muswell.html?_r=1&ref=global-home.

———. "Why Can't the English Just Give Up that Class Folderol?" *The New York Times*, April 26, 2007. Accessed October 18, 2014. http://www.nytimes.com/2007/04/26/world/europe/26britain.html?pagewanted=print.

Mack, Teresa. "Toward an Understanding of the 'Fiction of the Space Between' An(O)ther Kind of Agency." April 2010. Unpublished paper.

Malkin, Elisabeth. "For Some in Mexico, Trash Is a Treasure Worth Defending." *The New York Times*, February 16, 2012. Accessed October 18, 2014. http://www.nytimes.com/2012/02/17/world/americas/for-some-in-mexico-trash-is-a-treasure-worth-defending.html?pagewanted=all.

Margulis, Lynn, and Dorion Sagan. *What Is Life?* Berkeley: University of California Press, 2000.

Marland, Pippa, and John Parham. "Remaindering: The Material Ecology of Junk and Composting." *Green Letters: Studies in Ecocriticism* 18.1 (2014): 1–8.

Marotti, Arthur F. "Shakespeare and Catholicism." In *Theatre and Religion: Lancastrian Shakespeare*, edited by Richard Dutton, Alison Findlay, and Richard Wilson, 218–241. Manchester: Manchester University Press, 2003.

Martin, Andrew. "What's Eating Me." *The Guardian Weekend*, January 10, 2004: 36–39.

Martin, Emily. "The Egg and the Sperm: How Science Has Constructed a Romance Based on Stereotypical Male-Female Roles." In *Feminist Theory and the Body: A Reader*, edited by Janet Price and Margrit Shildrick, 179–189. New York: Routledge, 1999.

Masten, Jeffrey. "Is the Fundament a Grave?" In *The Body in Parts: Fantasies of Corporeality in Early Modern Europe*, edited by David Hillman and Carla Mazzio, 129–145. New York: Routledge, 1997.

Mayhew, Henry. *London Labour and London Poor*, edited by Victor Neuburg. London: Penguin, 1985.

Mazzola, Elizabeth. *The Pathology of the English Renaissance: Sacred Remains and Holy Ghosts*. Leiden and Boston: Brill, 1998.

McAvoy, Liz Herbert. *Authority and the Female Body in the Writings of Julian of Norwich and Margery Kempe*. Cambridge: D. S. Brewer, 2004.

McCarthy, Tom. "'Ulysses' and Its Wake." *London Review of Books*, June 19, 2014: 39–41.

McGowen, Todd. "The Obsolescence of Mystery and the Accumulation of Waste in Don DeLillo's *Underworld*." *Critique: Studies in Contemporary Fiction* 46.2 (Winter 2005): 123–145.

Melillo, Edward D. "Nutrient Rifts." Accessed October 18, 2014. http://discardstudies.com/discard-studies-compendium/.

Mellor, Anne K. "Possessing Nature: The Female in *Frankenstein*." Reprinted in Mary Shelley. *Frankenstein: A Norton Critical Edition*, edited by J. Paul Hunter, 355–368. 2nd ed. New York: W. W. Norton, 2012. From *Romanticism and Feminism*, edited by Anne K. Mellor. Bloomington: Indiana University Press, 1988.

Mentz, Steve. "Brown." In *Prismatic Ecology: Ecotheory Beyond Green*, edited by Jeffrey Jerome Cohen, 193–212. Minneapolis, MN: University of Minnesota Press, 2013.

Miller, William Ian. *The Anatomy of Disgust*. Cambridge, MA: Harvard University Press, 1997.

Mitchell, W. J. T. "Medium Theory: Preface to the 2003 *Critical Inquiry* Symposium." *Critical Inquiry* 30 (2004): 324–335.

———. *What Do Pictures Want? The Lives and Loves of Images*. Chicago, IL: University of Chicago Press, 2005.

Moi, Toril. *What Is a Woman? And Other Essays*. Oxford: Oxford University Press, 1999.

Montroso, Alan. "Human." In *Inhuman Nature*, edited by Jeffrey Jerome Cohen, 39–59. Washington, DC: Oliphaunt Books, 2014.

Morley, Simon, ed. *The Sublime*. London and Cambridge, MA: Whitechapel Gallery/MIT Press, 2010.

Morrison, Susan Signe. "Disabling the Medieval: The Excremental Middle Ages." In *Rubbish, Waste and Litter: Culture and Its Refuse/als*, edited by Tadeusz Rachwał, 113–121. Warsaw: Wydawnictwo SWPS Academica, 2008.

———. *Excrement in the Late Middle Ages: Sacred Filth and Chaucer's Fecopoetics*. New York: Palgrave Macmillan, 2008.

———. "Exkremente im Spannungsfeld von Privatsphäre und öffentlichem Raum." In *TABU—Über den gesellschaftlichen Umgang mit Ekel und Scham*. Translated by Uta Romer, 57–73. Berlin: Kulturverlag Kadmos, 2009.

———. "The Pleasures of Fecopoet[h]ics." http://www.siue.edu/babel/Kzoo09RemarksMorrison.htm. August 24, 2009.

———. "Postmedieval Fecopoet[h]ics." *postmedieval: a journal of medieval cultural studies* 1 (2010): 150–156.

———. "Waste Aesthetics: Form as Restitution." *Interdisciplinary Studies in Literature and Environment* 20.3 (2013): 464–478.

———. "Waste Studies and Chaucer's Fecopoetics: A New Paradigm for Literary Analysis." http://www.inthemedievalmiddle.com/. October 22, 2008.

———. "Waste Studies—A New Paradigm for Literary Analysis: Something Is Rotten in the Denmark of *Beowulf* and *Hamlet*." The 2009 Compass Interdisciplinary Virtual Conference. Accessed June 4, 2010. http://compassconference.files.wordpress.com/2009/10/civc-paper-waste-studies-a-new-paradigm-for-literary-analysis-something-is-rotten-in-the-denmark-of-beowulf-and-hamlet-susan-signe-morrison.pdf.

Morton, Timothy. *The Ecological Thought*. Cambridge, MA: Harvard University Press, 2010.

———. *Ecology without Nature: Rethinking Environmental Aesthetics*. Cambridge, MA and London: Harvard University Press, 2007.

———. "The Liminal Space between Things: Epiphany and the Physical." In *Material Ecocriticism*, edited by Serenella Iovino and Serpil Oppermann, 269–279. Bloomington: Indiana University Press, 2014.

———. "An Object-Oriented Defense of Poetry." *New Literary History* 43 (2012): 205–224.

———. "Queer Ecology." *PMLA* 125.2 (2010): 273–282.

———. *Realist Magic: Objects, Ontology, Causality.* University of Michigan Library, Ann Arbor: Open Humanities Press, 2013. http://openhumanitiespress.org/realist-magic.html.

———. "Treating Objects Like Women: Feminist Ontology and the Question of Essence." In *International Perspectives in Feminist Ecocriticism*, edited by Greta Gaard, Simon C. Estok, and Serpil Oppermann, 56–69. New York: Routledge, 2013.

Moser, Walter. "Garbage and Recycling: From Literary Theme to Mode of Production." *Other Voices* 3.1 (2007). Accessed April 18, 2012. http://www.othervoices.org/3.1/wmoser/index.php.

Mumford, Lewis. *The City in History: Its Origins, Its Transformations, and Its Prospects.* New York: Harcourt, Brace & World, 1961.

Murphy, Ann. "The Political Significance of Shame." *Borderlands e-journal* 3 (2004). Accessed October 21, 2014. http://www.borderlands.net.au/vol3no1_2004/murphy_shame.htm.

Murray, Jacqueline. "Men's Bodies, Men's Minds: Seminal Emissions and Sexual Anxiety in the Middle Ages." *Annual Review of Sex Research* 8 (1997): 1–26.

Nardizzi, Vin. "Medieval Ecocriticism." *postmedieval: a journal of medieval cultural studies* 4.1 (2013): 112–123.

Neudecker, Richard. *Die Pracht der Latrine. Zum Wandel öffentlicher Bedürfnisanstalten in der kaiserzeitlichen Stadt.* Studien zur antiken Stadt, 1. Munich: Verlag Dr. Friedrich Pfeil, 1994.

Neufeld, Christine. "Always Accessorize: In Defense of Scholarly *Cointise*." In *On Style: An Atelier*, edited by Eileen A. Joy and Anna Kłosowska, 87–109. Brooklyn, NY: punctum books, 2013.

Newcomb, W. *The Indians of Texas: From Prehistoric to Modern Times.* Austin, TX: University of Texas Press, 1961/2002.

Ngai, Sianne. "Raw Matter: A Poetics of Disgust." In *Telling It Slant*, edited by Mark Wallace and Steven Marks, 161–190. Tuscaloosa, AL: University of Alabama Press, 2002.

Niles, John D. "Locating *Beowulf* in Literary History." In *The Postmodern Beowulf: A Critical Casebook*, edited by Eileen A. Joy and Mary K. Ramsay, 131–161. Morgantown, WV: West Virginia University Press, 2006.

Nolan, Rachel. "Refuse Heap Is Archive for Night of Hatred." *The New York Times*, October 28, 2008. Accessed October 18, 2014. http://www.nytimes.com/2008/10/28/world/europe/28germany.html?pagewanted=print.

"Non-Hazardous Waste Management Hierarchy." Accessed November 22, 2014. http://www.epa.gov/waste/nonhaz/municipal/hierarchy.htm.

Nussbaum, Martha C. *From Disgust to Humanity: Sexual Orientation and Constitutional Law.* Oxford/New York: Oxford University Press, 2010.

———. *Poetic Justice: The Literary Imagination and Public Life*. Boston, MA: Beacon Press, 1995.

———. *Upheavals of Thought: The Intelligence of Emotions*. Cambridge: Cambridge University Press, 2001.

O'Brien O'Keefe, Katherine. "*Beowulf*, Lines 702b–836: Transformations and the Limits of the Human." *Texas Studies in Literature and Language* 23.4 (1981): 484–494.

Olin, M. "Touching Photographs: Roland Barthes's 'Mistaken' Identification." *Representations* 80 (2002): 99–118.

Olsen, Bjønar. *In Defense of Things: Archaeology and the Ontology of Objects*. Lanham, ML: AltaMira Press, 2010.

Oppermann, Serpil. "From Ecological Postmodernism to Material Ecocriticism: Creative Materiality and Narrative Agency." In *Material Ecocriticism*, edited by Serenella Iovino and Serpil Oppermann, 21–36. Bloomington: Indiana University Press, 2014.

O'Reilly, Édouard Magessa. "*Molloy*, Part II, Where the Shit Hits the Fan." *Genetic Joyce Studies* 6 (2006). Accessed April 6, 2010. http://www.geneticjoycestudies.org/GJS6/GJS6OReilly.htm.

O'Reilly, Kathleen. "'Traditional' Women, 'Modern' Water: Linking Gender and Commodification in Rajasthan, India." *Geoforum* 37 (2006): 958–972.

Orlando, Francesco. *Obsolete Objects in the Literary Imagination: Ruins, Relics, Rarities, Rubbish, Uninhabited Places, and Hidden Treasures*. Translated by Gabriel Pihas and Daniel Seidel, with Alessandra Grego. New Haven, CT: Yale University Press, 2006.

Ortega y Gasset, José. "An Essay in Esthetics by Way of a Preface." In *Phenomenology and Art*, translated by P. Silver, 127–160. New York: Norton, 1975.

Ortner, Sherry B. "Is Female to Male as Nature is to Culture?" *Feminist Studies* 1.2 (Fall 1972): 5–21.

Oswald, Dana M. "'Wigge under Wætere': Beowulf's Revision of the Fight with Grendel's Mother." *Exemplaria* 21 (2009): 63–82.

Owens, Margaret E. *Stages of Dismemberment: The Fragmented Body in Late Medieval and Early Modern Drama*. Newark: University of Delaware Press, 2005.

Panek, Richard. "Out There." *The New York Times Magazine*, March 11, 2007. Accessed October 18, 2014. http://www.nytimes.com/2007/03/11/magazine/11dark.t.html?pagewanted=all.

Paster, Gail Kern. *The Body Embarrassed: Drama and the Disciplines of Shame in Early Modern England*. Ithaca, NY: Cornell University Press, 1993.

Patterson, Lee. "Perpetual Motion: Alchemy and the Technology of the Self." *Studies in the Age of Chaucer* 15 (1993): 25–57.

Paul, C. N. James, and Murray L. Schwartz. *Federal Censorship: Obscenity in the Mail*. New York: Free Press of Glencoe, 1961.

Petroff, Elizabeth Alvilda. *Medieval Women's Visionary Literature*. New York: Oxford University Press, 1986.

Phillips, Dana. "Excremental Ecocriticism and the Global Sanitation Crisis." In *Material Ecocriticism*, edited by Serenella Iovino and Serpil Oppermann, 172–185. Bloomington: Indiana University Press, 2014.

Phillips, Dana, and Heather I. Sullivan. "Material Ecocriticism: Dirt, Waste, Bodies, Food, and Other Matter." *Interdisciplinary Studies in Literature and Environment* 19.3 (Summer 2012): 445–447.

Pinker, Steven. *The Stuff of Thought: Language as a Window into Human Nature.* New York: Viking, 2007.

Ploss, Emil Ernst, et al. *Alchimia: Ideologie und Technologie.* Munich: Heinz Moos Verlag, 1970.

Pollan, Michael. "The Vegetable-Industrial Complex." *The New York Times Magazine,* October 15, 2006. Accessed October 18, 2014. http://www.nytimes.com/2006/10/15/magazine/15wwln_lede.html.

Poore, Carol. "Who Belongs? Disability and the German Nation in Postwar Literature and Film." *German Studies Review* 26 (2003): 21–42.

Poovey, Mary. *Making a Social Body.* Chicago, IL: University of Chicago Press, 1995.

Pope, Rob. *The English Studies Book: An Introduction to Language, Literature and Culture.* 2nd ed. London: Routledge, 2002.

Pops, Martin. "The Metamorphosis of Shit." *Salmagundi* 56 (1982): 26–61.

Povinelli, Elizabeth A. "The Child in the Broom Closet: States of Killing and Letting Die." *South Atlantic Quarterly* 107.3 (2008): 509–530.

Puff, Helmut. "The Sodomite's Clothes: Gift-Giving and Sexual Excess in Early Modern Germany and Switzerland." In *The Material Culture of Sex, Procreation, and Marriage in Premodern Europe,* edited by Anne L. McClanan and Karen Rosoff Encarnación, 251–272. New York: Palgrave, 2002.

Quinn, Naomi. "The Cultural Basis of Metaphor." In *Beyond Metaphor: The Theory of Tropes in Anthropology,* edited by James W. Fernandez, 56–93. Stanford, CA: Stanford University Press, 1991.

Rachwał, Tadeusz, ed. *Rubbish, Waste and Litter: Culture and Its Refuse/als.* Warsaw: Wydawnictwo SWPS Academica, 2008.

Raitt, Suzanne. "Psychic Waste: Freud, Fechner, and the Principle of Constancy." In *Culture and Waste: The Creation and Destruction of Value,* edited by Gay Hawkins and Stephen Muecke, 73–83. Lanham, MD: Rowman & Littlefield, 2003.

Ramazani, Jahan. "Modernist Bricolage, Postcolonial Hybridity." *Modernism* 13.3 (2006): 445–463.

Rambo, Shelly. "Haunted (by the) Gospel: Theology, Trauma, and Literary Theory in the Twenty-First Century." *PMLA* 125.4 (2010): 936–941.

Ramsay, Raylene. *Robbe-Grillet & Modernity: Science, Sexuality, & Subversion.* Gainesville, FL: University Press of Florida, 1992.

Randall, Marilyn. "Recycling Recycling, or *plus ça change...*" *Other Voices* 3.1 (2007). Accessed April 18, 2012. http://www.othervoices.org/3.1/mrandall/index.php.

Ratcliffe, Barrie M. "Perceptions and Realities of the Urban Magic: The Rag Pickers of Paris in the First Half of the Nineteenth Century." *Canadian Journal of History* 27 (1992): 197–233.

Ratcliffe, Sophie. *On Sympathy.* Oxford: Oxford University Press, 2008.

Rathje, William, and Cullen Murphy. *Rubbish! The Archaeology of Garbage.* New York: HarperCollins, 1992.

Reinhard, Kenneth. "Kant with Sade, Lacan with Levinas." *MLN* 110.4 (1995): 785–808.

Reklaityte, Ieva. "The Hygienic Situation in Medieval Muslim Towns: A Case of al-Andalus." The Medieval Institute, Western Michigan University, May 2008.

Reno, Joshua Ozias. "Toward a New Theory of Waste: From 'Matter out of Place' to Signs of Life." *Theory Culture Society* 31.6 (2014): 3–27.

Reyes, Paul. "Bleak Houses: Digging through the Ruins of the Mortgage Crisis." *Harper's Magazine*, October 2008: 31–45.

Reynolds, Reginald. *Cleanliness and Godliness*. Garden City, NY: Doubleday, 1946.

Rhode, David. "Coprolites from Hidden Cave, Revisted: Evidence for Site Occupation History, Diet and Sex of Occupants." *Journal of Archaeological Science* 30 (2003): 909–922.

Ricoeur, Paul. *The Rule of Metaphor: The Creation of Meaning in Language*. Translated by Robert Czerny, Kathleen McLaughlin, and John Costello, S. J. London: Routledge, 2003.

Robbe-Grillet, Alain. "Nature, Humanism and Tragedy" (1958). In *Snapshots and towards a New Novel*, translated by Barbara Wright. London: Calder & Boyars, 1965.

Robertson, Eric. "Volcanoes, Guts and Cosmic Collisions: The Queer Sublime in *Frankenstein* and *Melancholia*." *Green Letters: Studies in Ecocriticism* 18.1 (2014): 63–77.

Robertson, Kellie. "Exemplary Rocks." In *Animal, Vegetable, Mineral: Ethics and Objects*, edited by Jeffrey Jerome Cohen, 91–121. Washington, DC: Oliphaunt Books, 2012.

Rogers, Heather. *Gone Tomorrow: The Hidden Life of Garbage*. New York: New Press, 2005.

Rollo, David. *Kiss My Relics: Hermaphroditic Fictions of the Middle Ages*. Chicago, IL: University of Chicago Press, 2011.

Ross, Thomas W. "Taboo-Words in Fifteenth-Century English." In *Fifteenth-Century Studies: Recent Essays*, edited by Robert F. Yeager, 137–160. Hamden, CT: Archon, 1984.

Rothstein, Edward. "In a Collection of Memorabilia, Politics at Its Most Boisterous." *The New York Times*, June 28, 2008. Accessed October 18, 2014. http://www.nytimes.com/2008/06/28/arts/design/28muse.html?pagewanted=print.

Royte, Elizabeth. *Garbage Land: On the Secret Trail of Trash*. New York: Little, Brown, 2005.

Ruebush, Mary. *Why Dirt is Good: 5 Ways to Make Germs Your Friends*. New York: Kaplan, 2009.

Sabar, Ariel. "One Man's Trash is Another Man's Two-Bit 'Trashball.'" *The New York Times*, April 29, 2007. Accessed October 18, 2014. http://www.nytimes.com/2007/04/29/us/29trash.html.

Sabine, Ernest L. "City Cleaning in Mediaeval London." *Speculum* 12 (1937): 19–43.

———. "Latrines and Cesspools of Mediaeval London." *Speculum* 9 (1934): 303–321.

Sandilands, Catriona. "Eco Homo: Queering the Ecological Body Politic." *Social Philosophy Today* 19 (2004): 17–39.

Sandner, David. "Tracking Grendel: The Uncanny in *Beowulf*." *Extrapolation* 40 (1999): 162–176.

Sarason, Rabbi Richard. *Asher Yatsar [Our Bodily Needs]*. Mishkan T'filah, 194–195. June 26, 2008, Week 241, Day 4 23 Sivan 5768. Accessed October 23, 2014. http://tmt.urj.net/archives/4jewishethics/062608.htm.

Scanlan, John. "Commentary on Susan Signe Morrison's Waste Studies—A New Paradigm for Literary Analysis: Something Is Rotten in the Denmark of *Beowulf* and *Hamlet*." The 2009 Compass Interdisciplinary Virtual Conference. Accessed June 4, 2010. http://compassconference.files.wordpress.com/2009/10/civc-commentary-john-scanlan-manchester-metropolitan-university-on-waste-studies-a-new-paradigm-for-literary-analysis-susan-signe-morrison.pdf.

———. *On Garbage*. London: Reaktion, 2005.

Scanlan, John, and John F. M. Clark, eds. *Aesthetic Fatigue: Modernity and the Language of Waste*. Newcastle upon Tyne: Cambridge Scholars, 2013.

Schaffer, Guy. "Camp." http://discardstudies.com/discard-studies-compendium/.

Schiffer, Michael B. "Is There a 'Pompeii Premise' in Archaeology?" *Journal of Anthropological Research* 41.1 (1985): 18–41.

———. "Toward the Identification of Formation Processes." *American Antiquity* 48.4 (1983): 675–706.

Schmidt, Christopher. "'Baby, I *Am* the Garbage': James Schuyler's Taste for Waste." Accessed March 9, 2012. http://www.uiowa.edu/~ijcs/waste/schmidt.html.

———. *The Poetics of Waste: Queer Excess in Stein, Ashbery, Schuyler, and Goldsmith*. New York: Palgrave Macmillan, 2014.

Schoenfeldt, Michael. "Fables of the Belly in Early Modern England." In *The Body in Parts: Fantasies of Corporeality in Early Modern Europe*, edited by David Hillman and Carla Mazzio, 242–261. New York: Routledge, 1997.

Scribner III, Charles. "Celestial Eyes—From Metamorphosis to Masterpiece." *Princeton Library Chronicle* 53.2 (1992): 141–155.

Seaman, Myra J. "Becoming More (than) Human: Affective Posthumanisms, Past and Future." *Journal of Narrative Theory* 37.2 (2007): 246–257.

Sebald, W. G. "An Attempt at Restitution." *The New Yorker*, December 20 and 27, 2004. Accessed October 19, 2014. http://www.newyorker.com/magazine/2004/12/20/an-attempt-at-restitution.

———. *Campo Santo*, edited by Sven Meyer, translated by Anthea Bell. London: Hamish Hamilton, 2005.

Sebold, Alice. "Living with the Dead." New York Times, September 11, 2005. Accessed October 16, 2014. http://www.nytimes.com/2005/09/11/opinion/11sebold.html?_r=0.

Sedinger, Tracey. "Theory Terminable and Interminable: On Presentism, Historicism, and the Problem of *Hamlet*." *Exemplaria* 19 (2007): 455–473.

Seng, Peter. "Ophelia's Songs in *Hamlet*." *Durham University Journal* 25 (1964): 77–85.

Serres, Michel. *Malfeasance: Appropriation Through Pollution?* Translated by Anne-Marie Feenberg-Dibon. Stanford, CA: Stanford University Press, 2011.

Shaviro, Steven. *Passion & Excess: Blanchot, Bataille, and Literary Theory*. Tallahassee: Florida State University Press, 1990.

Shoaf, R. Allen. *Chaucer's Body: The Anxiety of Circulation in the Canterbury Tales*. Gainesville, FL: University Press of Florida, 2001.

Shove, Elizabeth, Matthew Watson, Martin Hand, and Jack Ingram. *The Design of Everyday Life*. Oxford: Berg, 2007.

Sławek, Tadeusz. "The Vase and Broken Pieces: Productivity and the Margin of Waste." In *Rubbish, Waste and Litter: Culture and Its Refuse/als*, edited by Tadeusz Rachwał, 9–29. Warsaw: Wydawnictwo SWPS Academica, 2008.

Slovic, Scott. "Editor's Note." *Interdisciplinary Studies in Literature and Environment* 20.3 (2013): 453–457.

Smith, Daniel B. "Is There an Ecological Unconscious?" *The New York Times*, January 31, 2010. Accessed October 18, 2014. http://www.nytimes.com/2010/01/31/magazine/31ecopsych-t.html?pagewanted=all.

Smith, Peter J., "Review of *Excrement in the Late Middle Ages: Sacred Filth and Chaucer's Fecopoetics*:" *Times Higher Education*, January 15, 2009.

Smith, Rob. Interview. Wellcome Collection, London, England. May 14, 2011. Facilitator: Rosie Cox.

Smith, Virginia, et al. *Dirt: The Filthy Reality of Everyday Life*. London: Profile Books, 2011.

Snediker, Michael. "To Peach or Not to Peach: Style and the Interpersonal." In *On Style: An Atelier*, edited by Eileen A. Joy and Anna Kłosowska, 47–53. Brooklyn, NY: punctum books, 2013.

Snyder, Gary. *A Place in Space*. Washington, DC: Counterpoint, 1995.

———. *The Real Work: Interviews and Talks: 1964–1979*. New York: New Directions, 1980.

Soper, Kate. "Response Essay: The Postmedieval Project: Promise and Paradox." *postmedieval: a journal of medieval cultural studies* 1 (2010): 256–261.

———. "Waste Matters." *Capitalism, Nature, Socialism* 14.2 (2003): 129–134.

Sova, Dana B. *Banned Books: Literature Suppressed on Social Grounds*, revised edition. New York: Facts on File, 2006.

Spelman, Elizabeth V. "Combing Through Trash: Philosophy Goes Rummaging." *The Massachusetts Review* 52.2 (2011): 313–325.

———. "Woman as Body: Ancient and Contemporary Views." *Feminist Studies* 8.1 (1982): 109–131.

Stallabrass, Julian, "Trash." In *The Object Reader*, edited by Fiona Candlin and Raiford Guins, 406–424. London: Routledge, 2009.

Stam, Robert. "Hybridity and the Aesthetics of Garbage: The Case of Brazilian Cinema." Accessed April 20, 2010. http://www.tau.ac.il/eial/IX_1/stam.html.

Stanley, Alessandra. "Undercover Boss: He's Good at Pushing Paper, but Can He Pick Up Trash?" *The New York Times*, February 6, 2010. Accessed February 8, 2010. http://tv.nytimes.com/2010/02/06/arts/television/06under.html?_r=1.

Steel, Karl. "With the World, or Bound to Face the Sky: The Postures of the Wolf-Child of Hesse." In *Animal, Vegetable, Mineral: Ethics and Objects*, edited by Jeffrey Jerome Cohen, 9–34. Washington, DC: Oliphaunt Books, 2012.

Stell, P. M., and Louise Hampson. *Probate Inventories of the York Diocese 1350–1500*. YML Occasional Series, Dean and Chapter of York, 1998.

Stockton, Will. *Playing Dirty: Sexuality and Waste in Early Modern Comedy*. Minneapolis, MN: University of Minnesota Press, 2011.

Stoekl, Allan. *Bataille's Peak: Energy, Religion, and Postsustainability*. Minneapolis, MN: University of Minnesota Press, 2007.

Strasser, Susan. *Waste and Want: A Social History of Trash*. New York: Henry Holt, 2000.

Strohm, Paul. "Sovereignty and Sewage." In *Lydgate Matters: Poetry and Material Culture in the Fifteenth Century*, edited by Lisa H. Cooper and Andrea Denny-Brown, 57–70. New York: Palgrave Macmillan, 2007.

Sullivan, Heather I. "Dirt Theory and Material Ecocriticism." *Interdisciplinary Studies in Literature and Environment* 19.3 (Summer 2012): 515–531.

———. "The Ecology of Colors: Goethe's Materialist Optics and Ecological Posthumanism." In *Material Ecocriticism*, edited by Serenella Iovino and Serpil Oppermann, 80–94. Bloomington: Indiana University Press, 2014.

Sullivan, Robert. *The Meadowlands: Wilderness Adventures on the Edge of a City*. New York: Doubleday, 1998.

Summit, Jennifer. "Leland's *Itinerary* and the Remains of the Medieval Past." In *Reading the Medieval in Early Modern England*, edited by Gordon McMullan and David Matthews, 159–176. Cambridge: Cambridge University Press, 2007.

Tabuchi, Hiroko. "Japan Finds Story of Hope in Undertaker Who Offered Calm Amid Disaster." *The New York Times*, March 11, 2012. Accessed March 19, 2012. http://www.nytimes.com/2012/03/11/world/asia/a-year-later-undertakers-story-offers-japan-hope.html.

Tally, Robert T., Jr. "World Literature and Its Discontents." *English Language and Literature* 60.3 (2014): 401–419.

Tambling, Jeremy. "Letters and Litter." In *Rubbish, Waste and Litter: Culture and Its Refuse/als*, edited by Tadeusz Rachwał, 30–49. Warsaw: Wydawnictwo SWPS Academica, 2008.

Tanoukhi, Nirvana. "The Movement of Specificity." *PMLA* 128.3 (2013): 668–674.

Temko, Christine. "Regulation and Refuse Matter in Don DeLillo's *Underworld* and Eugene Marten's *Waste*." *Interdisciplinary Studies in Literature and Environment* 20.3 2013: 494–512.

Thompson, Edward P. "Commitment in Politics." *Universities and Left Review* 6 (1959): 50–55.

Thompson, Michael. *Rubbish Theory: The Creation and Destruction of Value*. Oxford: Oxford University Press, 1979.

Tiffany, Daniel. "Lyric Substance: On Riddles, Materialism, and Poetic Obscurity." *Critical Inquiry* 28.1 (2001): 72–98.

Tolan, John Victor. *Petrus Alfonsi and His Medieval Readers*. Gainesville, FL: University Press of Florida, 1993.

Turner, Bryan S. "The Body in Western Society: Social Theory and Its Perspectives." In *Religion and the Body*, edited by Sarah Coakley, 15–41. Cambridge: Cambridge University Press, 1997.

Tyler, T. *Ciferae: A Bestiary in Five Fingers*. Minneapolis, MN: University of Minnesota Press, 2012.

Uebel, Michael. *Ecstatic Transformation: On the Uses of Alterity in the Middle Ages*. New York: Palgrave Macmillan, 2005.

———. Visit to "Waste Studies: Cultural Refuse/als in Medieval Literature." February 11, 2010, Texas State University, San Marcos.

Van Dyke, Carolynn. "Touched by an Owl? An Essay in Vernacular Ethology." *postmedieval: a journal of medieval cultural studies* 7 (2014): 1–24.

Van Esterik, Penny. "No Free Lunch." *Agriculture and Human Values* 22 (2005): 207–208.

Vanhoutte, Jacqueline. "Denmark's Rotting Reconsidered." *Philological Quarterly* 91.3 (2012): 393–418.

Veblen, Thorstein. *Conspicuous Consumption* (1902). Accessed October 18, 2014. http://www.fordham.edu/halsall/mod/1902veblen00.asp.

Viney, Will. "The Future of Ruins." In *Aesthetic Fatigue: Modernity and the Language of Waste*, edited by John Scanlan and John F. M. Clark, 138–158. Newcastle upon Tyne: Cambridge Scholars, 2013.

———. "'Unproductive and Uninhabited': Wastes of Place and Time." The Rubbish Symposium. July 30, 2011 at Birkbeck College, University of London. Accessed September 13, 2011. http://narratingwaste.wordpress.com/.

———. *Waste: A Philosophy of Things*. London: Bloomsbury, 2014.

Vergine, Lea, "From Junk to Art." In *Trash: From Junk to Art*, edited by Lea Vergine, 19–27. Corte Madera, CA: Gingko Press, 1997.

Von Staden, Heinrich. "Women and Dirt." *Helios* 19 (1992): 7–30.

Waller, Gary. *The Virgin Mary in Late Medieval and Early Modern English Literature and Popular Culture.* Cambridge: Cambridge University Press, 2011.

Wanner, Kevin J. "Purity and Danger in Earliest Iceland: Excrement, Blood, Sacred Space, and Society in *Eyrbyggja saga*." *Viking and Medieval Scandinavia* 5.5 (2009): 213–250.

Watkins, Calvert. *The American Heritage Dictionary of Indo-European Roots*. 3rd ed. Boston: Houghton Mifflin Harcourt, 2011.

———. *How to Kill a Dragon: Aspects of Indo-European Poetics.* Oxford: Oxford University Press, 1995.

Weiner, Eric. "Slum Visits: Tourism or Voyeurism?" *The New York Times*, March 9, 2008. Accessed October 18, 2014. http://www.nytimes.com/2008/03/09/travel/09heads.html.

Wheeler, Wendy. "Natural Play, Natural Metaphor, and Natural Stories: Biosemiotic Realism." In *Material Ecocriticism*, edited by Serenella Iovino and Serpil Oppermann, 67–79. Bloomington: Indiana University Press, 2014.

White, Hayden. "The Forms of Wildness: Archaeology of an Idea." In *Tropics of Discourse: Essays in Cultural Criticism*, 157–162. Baltimore, MD: Johns Hopkins University Press, 1978.

Whiteley, Gillian. *Junk: Art and the Politics of Trash*. London: I. B. Tauris, 2010.

Williams, Raymond. *Marxism and Literature*. Oxford: Oxford University Press, 1977.

Wines, Michael."China's Impolitic Artist, Still Waiting to be Silenced." *The New York Times*, November 28, 2009. Accessed October 18, 2014. http://www.nytimes.com/2009/11/28/world/asia/28weiwei.html.

Wohlfarth, Irving. "Et Cetera? The Historian as Chiffonnier." *New German Critique* 39 (1986): 143–168.

Wollan, Malia. "Vandals Create a Pungent Problem in San Francisco." *The New York Times*, January 21, 2009. Accessed October 18, 2014. http://www.nytimes.com/2009/01/22/us/22potties.html.

Worth, Robert F. "Despite Caste-Less Society in Yemen, Generations Languish at Bottom of Ladder." *The New York Times*, February 27, 2008. Accessed October 18, 2014. http://www.nytimes.com/2008/02/27/world/africa/27iht-27yemen.10456434.html?pagewanted=all.

Yaeger, Patricia. "Editor's Column: The Death of Nature and the Apotheosis of Trash; or, Rubbish Ecology." *PMLA* 123 (2008): 321–339.

Yates, Julian. "Sheep-Tracks—A Multi-Species Impression." In *Animal, Vegetable, Mineral: Ethics and Objects*, edited by Jeffrey Jerome Cohen, 173–209. Washington, DC: Oliphaunt Books, 2012.

Zamora, Lois Parkinson. "Eccentric Periodization: Comparative Perspectives on the Enlightenment and the Baroque." *PMLA* 128.3 (2013): 690–697.

Zantingh, Matthew. "When Things Act Up: Thing Theory, Actor-Network Theory, and Toxic Discourse in Rita Wong's Poetry." *Interdisciplinary Studies in Literature and Environment* 20.3 (2013): 622–646.

Zapf, Hubert. "Creative Matter and Creative Mind: Cultural Ecology and Literary Creativity." In *Material Ecocriticism*, edited by Serenella Iovino and Serpil Oppermann, 51–66. Bloomington: Indiana University Press, 2014.

Žižek, Slavoj. *Living in the End Times*. London: Verso, 2010.

———. "In Astra Taylor." *The Examined Life*. 2008. https://www.youtube.com/watch?v=iGCfiv1xtoU.

———. "Knee-Deep." *London Review of Books*. September 2, 2004: 12.

Index

CPSIA information can be obtained
at www.ICGtesting.com
Printed in the USA
LVOW10*1015111116
512603LV00005B/24/P

9 781137 405661